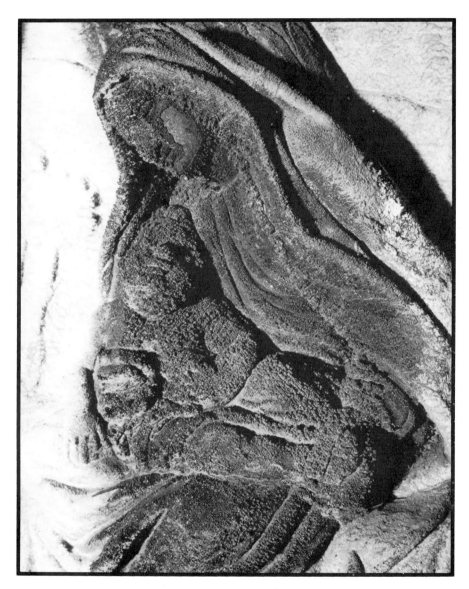

The marble Madonna in this photograph was carved about 1650 beneath a buttress of the Cathedral of Milan, Italy. The combustion of high-sulfur oil and coal in this industrial area of Italy, as in many cities, results in the generation of sulfur dioxide. This combines with rain water to produce sulfuric acid, which, in turn, reacts with the marble, eventually disintegrating it after prolonged exposure. The lighter areas show less severe damage.

Chemical Contamination in the Human Environment

Chemical Contamination
IN THE
Human Environment

MORTON LIPPMANN

RICHARD B. SCHLESINGER

Department of Environmental Medicine
New York University Medical Center

New York
OXFORD UNIVERSITY PRESS
1979

Copyright © 1979 by Oxford University Press, Inc.

Printed in the United States of America

Library of Congress Cataloging in Publication Data

Lippmann, Morton.
    Chemical contamination in the human environment.

    Bibliography: p.
    Includes index.
    1. Pollution.   2.   Environmental health.
I. Schlesinger, Richard B., joint author.   II.   Title.
TD174.L56   363.6   78-7394
ISBN 0-19-502441-9
ISBN 0-19-502442-7 pbk.

Preface

The focus of concern in this book is the extent and significance of chemical contamination of our environment. The discussion is presented in terms of the underlying physical, chemical, and biological processes which determine the behavior, fate, and ultimate effects on human health and welfare of the multitude of chemicals discharged into the environment.

The book provides a comprehensive, current overview of chemical contaminants in the total environment in a way which has seldom been attempted. Many works in this area either have concentrated on one medium (i.e., air, water, food, occupational environment) or concentrated on a single element or group of compounds (e.g., mercury, lead, pesticides). Of those books which have attempted to provide comprehensive coverage, most have been multiple-author compilations which usually are deficient with respect to uniformity of style and/or depth of coverage.

Our concentration on chemical contamination is not meant to belittle additional aspects of the overall environmental problem which are also of great concern. These include the increase in actual or potential contamination by physical agents (e.g., noise, ultrasound, γ and X radiations, ultraviolet radiation, microwaves) and disease transmission by infectious agents through the air and water, and via animate vectors. The human environment is also clearly shaped by social and economic factors, and the health effects of chemical contaminants may be greatly influenced or even overwhelmed by the effects of tobacco smoke, alcohol, and drugs. Finally, there is currently great concern about the possible development and inadvertent release of potent mutants produced in research with recombinant DNA. Each of these other areas is sufficiently different and complex to warrant separate treatment and discussion. They will be brought into this book only where they interact with the effects of chemical contaminants.

This book has been prepared as a text for an introductory graduate level course given in the Interdepartmental Program in Environmental Health Sciences of the Graduate School of Arts and Science of New York University. However, it should also be suitable for undergraduate courses in environmental science. Scientific notation and nomenclature familiar to those who have completed introductory courses in the basic sciences are used throughout. Where a greater degree of sophistication is needed in order to adequately discuss a specific issue, it is provided within the discussion.

The major focus is on the potential for effects on human health and welfare. Relatively little discussion will be devoted to disturbances within the environment that have little direct impact on man. This is not to say that such disturbances may not eventually be shown to have such effects, or that they are not important in themselves. It is simply not possible to spread the focus of this book to include them while still providing the depth of coverage desired on each topic. An extensive supplementary bibliography is provided in the Appendix for those readers who wish to delve into specific topics in greater depth than we were able to do in a book of this nature.

In fields as broad and controversial as environmental contamination and environmental health, the perspective offered here will undoubtedly disappoint many. We have attempted to avoid presenting attractive hypotheses and plausible assumptions as proven facts. This may lead to disagreement with some current consensus views on particular issues, and may upset those who approach specific issues with an advocacy perspective.

In the current state of knowledge on chemical contaminants in the environment and their potential effects on human health and welfare, the data base is almost always inadequate in scope and depth, and those data which are available are frequently unreliable. The uncritical and/or selective use of these data by individuals with different perspectives accounts for the frequent incidences of scientists coming to diametrically opposed conclusions at public hearings on specific issues.

Thus, our major objective in this book is to provide the reader with the technical background needed to intelligently evaluate the scientific issues, and thereby be able to develop informed judgments when there is sufficient reliable information available.

In the preparation of this book, we benefited from the constructive criticisms and comments of numerous people. Of special value were the careful reviews of some of our colleagues at the Institute of Environmental Medicine; Beverly Cohen, Joan Daisey, Joshua Gurman, Marc Halpern, and Michael McCawley. We also gratefully acknowledge the assistance of Mary Bader and Donna Halpern, who typed the several drafts of the manuscript.

M. L.

R. B. S.

Tuxedo, New York
November, 1978

Contents

1   Introduction and Historical Perspective   3

2   Characterization of Contaminants, Environments, and Health   16

3   Sources of Contaminants   58

4   Dispersion of Contaminants   117

5   Fate of Contaminants: Transformations and Sinks   152

6   Effects of Contaminants on Human Health   190

7   Effects of Contaminants on Environmental Quality   232

8   Contamination Criteria and Exposure Limits   270

9   Sampling and Measurement of Contaminants   311

10   Control of Contamination   381

   Appendix   442

   Index   449

1
Introduction and Historical Perspective

INTRODUCTION

Chemical contamination of the environment did not originate at any particular time. Natural processes have been generating contaminants, which may be defined as toxic or otherwise harmful materials present in measurable concentrations, throughout most of the history of the earth. Some natural waters remote from human activities have levels of dissolved chemicals which make them unsafe by today's standards, and the air above swamps and near geothermal springs is contaminated by sulfur gases. There have been some poisonings of humans and grazing animals by natural chemical toxins in the fruits and foliage they consumed.

Of course, most current problems of chemical contamination arise from anthropogenic sources, those attributable either directly or indirectly to human activity. In shaping the environment to our needs and convenience, we have used the earth's resources to feed, clothe, and heat us, to power our machines, and to produce material goods. Our ability to mold our environment has enabled us to greatly increase our numbers and improve our standard of living. However, in our often careless use of natural resources we have created contamination. Although contamination is often used interchangeably with pollution, the latter is better defined as contamination to a degree which renders something unfit for its desired use.

When early man discovered how to control and use fire, he or she must have found that one of its undesirable side effects was the exposure to its smoke. When fire was brought indoors to heat caves or shelters, the problem of smoke exposure became more severe, and at least some additional ventilation usually had to be provided in order to utilize the benefits of the fireplace. However, success in this regard was only partial; mummified human lungs from the pre-industrial age show considerable carbonaceous pigmentation. Such people were undoubtedly exposed to significant levels of carbon monoxide as well.

In using the words "clean," "dirty," "contaminated," and "polluted," we are using terms which are subjective or, in specific contexts, have arbitrary meanings. Thus, a human environment can never be absolutely clean because, by definition, it includes at least one contaminant source, a person. An environment passes from clean to contaminated when the source of contamination, relative to its rate of elimination, is sufficiently large, or where there are enough sources whose aggregate output is sufficiently large, to exceed some sensory or physical concentration limit. Thus, practices that are environmentally acceptable in rural areas, such as wood fires, discharge of sanitary waste liquids into septic tanks or underground drainage fields, and unleashed pet dogs, become unacceptable in densely populated urban regions.

The degree of environmental contamination which a society finds acceptable is highly variable in both space and time, and depends more upon the rate of change in contaminant levels than their absolute amounts. It also depends upon whether there appears to be a realistic alternative. Very high levels of soot from inefficient coal combustion were tolerated for centuries, as long as cleaner alternatives were not available at prices deemed reasonable.

With the large scale conversion of heating and electric utility boilers to oil and gas combustion in the past few decades, there was a major reduction in sulfur dioxide, dustfall, and suspended airborne particulates. As we partially convert back to coal in the next few decades, it is not likely that our more affluent and environmentally aware society will permit significantly increased levels of contaminants. It will therefore have to pay for the relatively expensive control technologies needed to prevent air pollution.

Public acceptance of environmental contamination also depends upon whether it is perceived to be "natural." Fire and its effluents have been part of the human environment throughout recorded history. Thus, the potential health effects of inhaled combustion products create relatively little concern, even though they contain many readily detectible toxicants.

On the other hand, the nuclear power industry has kept most discharges of radioactive waste products to a very low level, and produces electric power without generating any of the chemical toxicants cited above. However, a large proportion of the public prefers greater reliance on coal-fired power plants than on nuclear-power plants. This is because many people associate nuclear power with bombs, and exposure to radioactive wastes with cancer. The fear of bombs and cancer, in effect, leads to public acceptance of increased dependence on fossil fuels, even when their currently documented adverse effects on human health and environmental quality can be shown to be much greater.

Similar considerations apply to food safety. Under the Delaney clause of the Food and Drug Act, a variety of food additives, colors, and packaging materials have been banned because they were demonstrated to have the potential of causing cancer in laboratory animals when administered in high doses. On the other hand, many natural foods contain carcinogens and other toxicants of equal or greater potency, yet neither the Food and Drug Administration nor the general public shows much interest in banning their distribution.

If the general reader is confused about the effects of chemical contaminants on environmental quality and human health, it is not surprising, since most reports in the popular media are selective with respect to content and emphasis, and directly or indirectly reach the media for the purpose of advancing a particular point of view. In the current inadequate and incomplete state of knowledge about most environmental issues, it is relatively easy to provide plausible documentation for either the pro or con side of almost every issue.

Some of the current confusion also arises from the differing backgrounds and perspectives of the component disciplines and people involved, and to the absence of common terminology and reference conditions. For example, the chemist generally uses a reference temperature of 0° C (32° F), the meteorologist uses 15° C (59° F), while the industrial hygienist uses "room temperature," 70° F (about 21° C).

HISTORICAL BACKGROUND

One of the earliest written discussions of the relation between environment and health was the essay "On Airs, Waters and Places" written by Hippocrates, the Father of Medicine, in about 460 B.C. He advised physicians to consider the winds, seasons, and sources of water when evaluating the health of their patients. Hippocrates also described lead colic in miners, as well as diseases of other occupational groups.

Other authors of the Classic Period also recognized that some of the materials used in metallurgy were toxic. Pliny the Elder discussed the dangers in handling sulfur and zinc, and Galen recognized the dangers of acid mists among copper miners. Although the ancients were not aware of the possibility, chronic lead poisoning may have been widespread. Leaded glazes were widely used on kitchen pottery, and acidic wines and foods extracted some of the lead. The Romans also used lead pipes for the delivery of drinking water, and some of the lead was dissolved into the water. The exposures would have been greatest among the more prosperous Romans, who more often had running water and glazed vessels. The decline of the Roman Empire has been attributed by some to chronic lead intoxication, on the basis that the recorded decline in fertility among the upper classes was consistent with the effects of ingested lead.

Occupational Diseases

The association between exposure to chemical contaminants and human health effects has most frequently been made for people with occupational exposures, where the levels of exposure are generally higher than for the general population. Treatises on occupational diseases began to appear in the Middle Ages in Europe. In 1472, Ulrich Ellenbog of Augsburg wrote an eight-page booklet which discussed the toxic actions of carbon monoxide, nitric acid vapors, lead, mercury, and other metals.

A classic description of mining technology and its hazards, *De Re Metallica*, was published in 1556 by the heirs of Georg Bauer, a native of Saxony who was more commonly known by his Latin name of Georgius Agricola. From 1526 to his death in 1555, Agricola had been the official physician of the Bohemian mining town of Joachimstal, a major source of European silver, and more recently of radium and uranium. The silver coins of Joachimstal were known as Thalers, which in English later became dollars. *De Re Metallica* is a scholarly work of 12 books. It was translated from the Latin into English in 1912 by an American mining engineer and his wife. (The engineer, Herbert C. Hoover, eventually gave up engineering for politics.) In the last part of the sixth book, Agricola described the diseases of the lungs, joints, and eyes which were common among the miners. It appears from descriptions of the lung diseases that the men had silicosis, tuberculosis, lung cancer, and combinations thereof.

In 1567, another posthumous work appeared which was entitled *Von der Bergsucht und anderen Bergkrankheiten* (On the Miners' Sickness and Other Diseases of Miners). It was written by Theophrastus Bombastus von Hohenheim, better known as Paracelsus, an itinerant physician and

A—FURNACE. B—STICKS OF WOOD. C—LITHARGE. D—PLATE. E—THE FOREMAN
WHEN HUNGRY EATS BUTTER, THAT THE POISON WHICH THE CRUCIBLE EXHALES MAY NOT
HARM HIM, FOR THIS IS A SPECIAL REMEDY AGAINST THAT POISON.

Fig. 1–1 Woodcut from Agricola, Book X. The technology of lead smelting was considerably more advanced than the recommended prophylaxis for lead poisoning. Note the barrier plate, D, which protects against splatter burns.

From: Agricola, G. *De Re Metallica*, Basel, 1556. Translated by H. C. Hoover and L. H. Hoover for the *Mining Magazine*, London, 1912. Reprinted by Dover Press.

alchemist of Swiss descent. This monograph was specifically devoted to the occupational diseases of mine and smelter workers. Paracelsus did not consider dust exposure to be the causative factor in the lung diseases he observed in miners, but rather explained them in terms of alchemy and the stars. He was considerably more astute in his description of the diseases among smelter workers, however, and differentiated between acute and chronic poisonings. His detailed descriptions of mercurialism covered

A—Hearth. B—Heap. C—Slag-vent. D—Iron mass. E—Wooden mallets.
F—Hammer. G—Anvil.

Fig. 1-2 Woodcut from Agricola, Book IX. Note respiratory protection of the furnace worker.

From: Agricola, G. *De Re Metallica*, Basel, 1556. Translated by H. C. Hoover and L. H. Hoover for the *Mining Magazine*, London, 1912. Reprinted by Dover Press.

most of the currently recognized symptoms. Paracelsus is also well-known for his enunciation of a basic tenet of toxicology: "All substances are poisons; there is none which is not a poison. The right dose differentiates a poison and a remedy."

The most comprehensive description of occupational diseases of its time and for well over a century thereafter was a book of 40 chapters entitled *De Morbis Artificum* (Diseases of Workers), published in 1700 by the Italian, Bernardino Ramazzini, a professor of medicine at the University of Modena and, after 1700, at the University of Padua. Ramazzini is the generally acknowledged "Father of Occupational Medicine." His descriptions of diseases covered most of the trades practiced in his time, including those of dirty and humble trades, such as corpse carriers, porters, and laundresses. He stated: "When a doctor visits a working-class home he should be content to sit on a three-legged stool, if there isn't a gilded chair, and he should take time for his examination; and to the questions recommended by Hippocrates, he should add one more — What is your occupation?" Unfortunately, there is still a great deal of unrecognized occupational disease today because too many physicians still neglect to ask that single important question.

Hazardous working conditions and occupational diseases were also common in the more technologically advanced countries in Asia. Extensive descriptions of the operations involved in the mining and refining of metals were described in the *Atlas of Important Products in Mountains and Sea of Japan, 1754*, and *Atlas of Mining and Refining of Copper, 1801*, and are illustrated by woodcuts, such as the one reproduced in Figure 1–3.

In 1775, an English physician, Sir Percival Pott, provided the first description of occupationally induced cancer, that of scrotal cancer in chimney sweeps.

In 1831, Charles Turner Thackrah made a special contribution to this era by the publication of his 200-page book, *The Effects of Arts, Trades and Professions and All Civic States and Habits of Living on Life and Longevity*, based mainly on his experience in the manufacturing district of Leeds, England.

Scattered reports on occupational diseases appeared in the British, French, German, and American literature through the balance of the nineteenth century. Before the end of the century it was clear that it was desirable to anticipate problems associated with industrial exposures to toxic chemicals before they happened, rather than after their effects were apparent in workers, and that this could usually be accomplished through

Fig. 1–3   The refining of copper in the Besshi Copper Mine in Japan, circa 1800.
From: "Atlas of Mining and Refining of Copper, 1801," Reprinted in: Miura, T. A. Short History of Occupational Health in Japan (Part I). *The Jour. of Science of Labour* 53:509–525, 1977.

the systematic exposure of laboratory animals. Pioneering work along these lines began in the 1880's under K. L. Lehmann in Würzburg, and by 1884 he had published data on the results of studies with 35 gases and vapors.

With the rapid growth in industrialization in the nineteenth century, more and more workers were being exposed to a broadening spectrum of toxic materials at increasing concentration levels. The obvious effect was a great increase in occupational disease and disability. This was first apparent in England, and by 1833 the first of the English Factory Acts was passed by Parliament. It established the principle that people injured at work are entitled to compensation. While there was no requirement to prevent the conditions which led to the need for compensation, it became more profitable for many businesses to reduce the compensation costs through preventive measures rather than through paying claims. The need for positive preventive measures was recognized later, and the English Factory Act of 1878 created a centralized Factory Inspectorate. Most of the major European countries followed the British lead, but it wasn't until 1911 that Wisconsin became the first U.S. state to establish workmen's compensation, and not until 1948 that the last one did so. The first state programs to inspect industry for occupational exposures began in 1913 in New York and Ohio, but nationwide coverage was not achieved until the passage of the federal Occupational Safety and Health Act of 1970.

The Federal Government's involvement in occupational health began with the creation of the Bureau of Mines in 1910, and the establishment of an Office of Industrial Hygiene and Sanitation within the Public Health Service in 1914. In that year, they jointly conducted the first of a series of comprehensive studies of certain lung disorders in the dusty trades, under the direction of Dr. Anthony J. Lanza. Pioneering texts began to appear in 1914 with W. Gilman Thompson's *The Occupational Diseases*. The first text on industrial toxicology was *Industrial Poisons* by Alice Hamilton, which appeared in 1925. The first American scientific journal devoted specifically to occupational health, the *Journal of Industrial Hygiene*, first appeared in 1919. In 1918, Harvard became the first university to establish degree programs in occupational health, programs later broadened to environmental health.

Air Contamination and Health

Community air contamination arising from the combustion of fossil fuels first received official recognition at the end of the thirteenth century, when Edward I of England issued an edict to the effect that, during sessions of Parliament, there should be no burning of sea coal or channel coal, so-called because it was brought from Newcastle to London by sea transport via ports on the English Channel. Despite a succession of further royal edicts, taxes, and even occasional prison confinements and torture, the use of coal for producing heat continued in London, especially as the increase in population led to a depletion in the availability of wood for fuel. The first scholarly report on the problem: *Fumifugium or the Inconvenience of Aer and Smoak of London Dissipated, together with some Remedies Humbly Proposed* by John Evelyn, was published by the royal command of Charles II in 1661. Evelyn, one of the founding members of the Royal Society, recognized and discussed the problem in terms of the sources, effects, and feasibility of controls.

Unfortunately, no effective controls were instituted until after the report of the Royal Commission on the effects of the "killer fog" of December, 1952, which attributed approximately 4,000 excess deaths in London to the episode. Retrospective examinations of vital statistics demonstrated that there had been numerous prior episodes involving excess deaths during periods of air stagnation in London and other British cities. The reductions in smoke levels achieved in Britain since 1952 have eliminated observable excess deaths attributable to fossil fuel combusion.

Excess deaths attributable to smoke had occurred elsewhere prior to 1952, but fortunately involved smaller populations. As discussed in Chapter 6, the most notable of these were in the Meuse Valley in Belgium in 1930, and at Donora in Pennsylvania in 1948.

In the 1940's it became apparent that Southern California had an air pollution problem, and that it was a very different kind of pollution from that long known in London and the eastern United States. The California variety was characterized by oxidant gases, such as ozone, rather than by reducing gases, such as sulfur dioxide. Furthermore, the oxidants were formed in the atmosphere by photochemical processes.

Water Contamination and Health

Most contamination problems in water have, until quite recently, centered on infectious water-borne diseases rather than on chemical contamination. Historically, the growth of cities was limited by their ability to prevent contamination of their water supplies by fecal wastes. The more successful ancient cities had their drinking waters supplied from distant sources by aqueducts and enclosed pipes and channels.

In more modern times, the first association between water contamination and human disease was made by John Snow in his classic epidemiologic investigations of the cholera epidemic in London in 1853. It was at about this same time that water pollution was becoming a matter of serious concern in England for esthetic reasons. The rapid growth of London in the first half of the nineteenth century and the adoption of the practice of discharging the effluent from newly installed water closets into sewers constructed for carrying away storm drainage led to such an overpowering stench from the Thames at Westminster that Parliament found it difficult to meet in 1858. The solution for that immediate problem was to extend the sewer system, so as to transfer the wastes downstream far enough from Parliament to alleviate the nuisance.

The development of bacteriology in the latter half of the nineteenth century provided a scientific basis for understanding the role of water in the transmission of typhoid and other enteric bacterial diseases. Water-filtration processes capable of reducing bacterial concentrations by one or two orders of magnitude were developed. By the turn of the century, these processes were starting to come into widespread use. But at cities such as Pittsburgh and Cincinnati, which still used unfiltered river water, annual death rates from typhoid fever remained around 100 per 100,000 in 1900, and in the nation as a whole, the typhoid death rate was about 35 per 100,000.

In the first 30 years of this century, most of the cities and towns using rivers as sources of water built water-treatment plants. In 1908, the use of chlorine as a water disinfectant was introduced, and it became a standard treatment operation. This made possible the production of bacteriologically safe water at very little cost, even when raw water of very poor quality was being treated.

Outbreaks of water-borne disease are relatively rare in the United States today. In the 15 years from 1946 to 1960, there were 228 known outbreaks of disease or poisoning, affecting some 26,000 persons, while in the preceding eight-year period, 327 outbreaks, involving 110,000 cases, were reported. The improvement took place in spite of increased population and opportunity for contamination.

Although much has been done to reduce disease risks from viable agents, very little is known about the possible health effects of the variety of largely unidentified chemical compounds that enter water-supply sources with sewage and industrial wastes. Many of these compounds are not effectively removed by today's water-treatment plants. Recent reports indicating the presence of trace amounts of known carcinogens in most drinking water supplies may lead to a change in the design and performance requirements of our water treatment and waste-disposal facilities.

Food Contamination and Health

The nutritional values and safety of foods have always been important, and the intentional use of chemical preservation techniques, such as salting and smoking, can be traced back to earliest recorded history. Recognition of natural toxicants was often incorporated into dietary laws, which were generally enforced as religious taboos.

Within the last century, modern preservation techniques have made it possible for foods to be processed and distributed for mass markets. It became economically advantageous to make the processed foods attractive in appearance and to keep them that way for extended periods of time. The separation in time and space between production and consumption also created temptations to adulterate the products with fillers and less than wholesome raw materials.

While the Federal Government did not begin to enforce food safety regulations until 1906, some individual states recognized their responsibility to provide a safe food supply. As early as 1764, the Massachusetts Bay Colony established a sanitary code for slaughterhouses. California became a state in 1850, and in the same year passed a pure food and drug law. In 1856, Massachusetts prohibited the adulteration of milk. Following the enactment of the British Pure Food and Drug Law in 1875, other states passed laws regulating the handling and production of food. By 1900, most states had regulations designed to protect the consumer from unsanitary and adulterated foods.

National recognition of the need for federal regulation in the United States essentially began with the appointment of Dr. Harvey W. Wiley as the chemist for the United States Department of Agriculture in 1883. He campaigned vigorously against misbranded and adulterated foods. Finally, in 1906, the Wiley or Sherman Act, regulating interstate transportation of food, was passed. But Dr. Wiley resigned from the Department of Agriculture in 1912, embittered by the struggle necessary to pass the legislation.

Most manufacturers observed the law of 1906, and adulteration with known harmful substances was rare. However, economic cheating was widespread, and adulterants, sometimes toxic, were common in some foods. Unfortunately, the Wiley Act did not provide for legal standards or identification of foods. In 1938, the U.S. Congress passed the Food, Drug, and Cosmetic Act. This law, with its various amendments, is administered by the Food and Drug Administration (FDA) which was also organized in 1938. The principal amendments relating to foods are the Pesticide Amendment Act of 1954 (revised 1972), the Food Additive Amendment of 1958, and the Color Additive Amendments of 1960 and 1972. These regulations affect about 60% of the food produced in the United States. The remaining 40% is under state regulation, which in many cases parallels federal legislation.

In the 1958 Food Additive Amendment exceptions were made for all additives that, because of years of widespread use in foods, were "Generally Recognized As Safe (GRAS) by experts qualified by scientific training and experience." These exceptions, numbering more than 600 items, comprise the so-called GRAS list, which has been and continues to be the subject of controversy.

The Food Additive Amendment of 1958 also included the cancer or Delaney clause which states: " . . . No additive shall be deemed to be safe if it is found to induce cancer when ingested by man or animal, or if it is found, after tests which are appropriate for the evaluation of the safety of food additives, to induce cancer in man or animal. . . ." Although the aim of the Delaney clause is excellent, serious problems exist in the evaluation of data, protocols for testing for carcinogenicity, and the extrapolation of the data to man. These problems have polarized scientific and legislative authorities, as well as the consuming public, into proponents and opponents of the measure. The controversy has reerupted with each application of the Delaney clause, such as when cyclamates were banned in 1969, red dye #2 in 1976, and when a proposed ban on saccharin was announced in 1977.

2
Characterization of Contaminants, Environments, and Health

CHARACTERIZATION OF CHEMICAL CONTAMINANTS

Concentration Units

In environmental science, confusion often arises from the use of the same or similar sounding terms having different meanings in different contexts. This is especially true when considering the concentrations of air and water contaminants. Both are frequently expressed in parts per million (ppm) or parts per billion (ppb). However, when used for air contaminants, the units are molar or volume fractions, while when used for water contaminants, they are weight fractions. In order to avoid unnecessary confusion, the concentration units used in this book have appropriate subscripts: ppm_v to indicate parts per million parts of air, a volume ratio, and ppm_w to indicate parts per million parts of water, a weight ratio. This problem is often avoided altogether by expressing all fluid contaminant concentrations as the weight of contaminant per unit volume (e.g., cubic meter, m^3, or liter, l,) of fluid. In air, the units generally used are mg/m^3 or $\mu g/m^3$, while in water they are most often mg/l. or $\mu g/l$.

Air Contaminants

Chemical contaminants can be dispersed in air at normal temperatures and pressures in gaseous, liquid, and solid forms. The latter two represent suspensions of particles in air, and were given the generic term "aerosols" by Gibbs (1) on the basis of analogy to the term hydrosol, used to describe disperse systems in water. On the other hand, gases and vapors, which are present as discrete molecules, form true solutions in air. Particles consisting of moderate to high vapor-pressure materials tend to evaporate rapidly, since those small enough to remain suspended in air for more than a few minutes (smaller than about 10 μm) have large surface-to-volume ratios. Some materials with relatively low vapor pressures can have appreciable fractions in both the vapor and aerosol forms simultaneously.

GASES AND VAPORS

Once dispersed in air, contaminant gases and vapors generally form mixtures so dilute that their physical properties, such as density, viscosity, enthalpy, etc., are indistinguishable from those of clean air. Such mixtures may be considered to follow ideal gas-law relationships. There is no practical difference between a gas and a vapor, except that the latter is generally considered to be the gaseous phase of a substance which is normally a solid or liquid at room temperature. While dispersed in the air, all molecules of a given compound are essentially equivalent in their size and capture probabilities by ambient surfaces, respiratory tract surfaces, and contaminant collectors or samplers.

AEROSOLS

Aerosols, being dispersions of particles in air, have the very significant additional variable of particle size. Size affects particle motion and, hence, the probabilities for physical phenomena such as coagulation, dispersion, sedimentation, impaction onto surfaces, interfacial phenomena, and light-scattering properties. It is not possible to characterize a given particle by a single size parameter. For example, a particle's aerodynamic properties depend on density and shape as well as linear dimensions, while the effective size for light-scattering is dependent on refractive index and shape.

In some special cases, all of the particles are essentially the same in size. Such aerosols are considered to be monodisperse. Examples are natural pollens and some laboratory-generated aerosols. More typically, aerosols are composed of particles of many different sizes, and hence are called heterodisperse or polydisperse. Different aerosols have different degreees of size dispersion. It is, therefore, necessary to specify at least two parameters in characterizing aerosol size: a measure of central tendency, such as a mean or median, and a measure of dispersion, such as an arithmetic or geometric standard deviation.

Particles generated by a single source or process generally have diameters following a log-normal distribution, i.e., the logarithms of their individual diameters have a Gaussian distribution. In this case, the measure of dispersion is the geometric standard deviation, which is the ratio of the

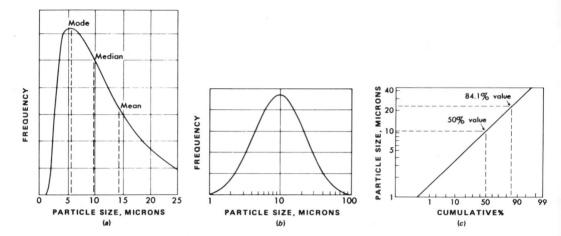

Fig. 2–1   Particle size distribution data. (a) Plotted on linear coordinates (b) Plotted on a logarithmic-size scale (c) In practice, logarithmic probability coordinates are used to display the percent of particles less than a specific size versus that size. The geometric standard deviation (σ_g) of the distribution is equal to the 84.1% size/50% size.

From: Perkins, H. C. *Air Pollution.* New York: McGraw-Hill, © 1974. Used with permission of McGraw-Hill Book Co.

84.1 percentile size to the 50 percentile size (Figure 2–1). When more than one source of particles is significant, the resulting mixed aerosol will usually not follow a single log-normal distribution, and it may be necessary to describe it by the sum of several distributions.

PARTICLE CHARACTERISTICS

There are many properties of particles, other than their linear size, which can greatly influence their airborne behavior and their effects on the environment and health. These include:

Surface: For spherical particles, the surface varies as the square of the diameter. However, for an aerosol of given mass concentration, the total aerosol surface increases with decreasing particle size. Airborne particles have much greater ratios of external surface to volume than do bulk materials and, therefore, the particles can dissolve or participate in surface reactions to a much greater extent than would massive samples of the same materials. Furthermore, for nonspherical solid particles or aggregate particles, the ratio of surface to volume is increased, and for particles with internal cracks or pores, the internal surface area can even be greater than the external area.

Volume: Particle volume varies as the cube of diameter; therefore, the few largest particles in an aerosol tend to dominate its volume concentration.

Shape: A particle's shape affects its aerodynamic drag as well as its surface area, and therefore its motion and deposition probabilities.

Density: A particle's velocity due to gravitational or inertial forces increases as the square root of its density.

Aerodynamic Diameter: The diameter of a unit-density sphere having the same terminal settling velocity as the particle under consideration is equal to its aerodynamic diameter. Terminal settling velocity is the equilibrium velocity of a particle which is falling under the influence of gravity and fluid resistance. Aerodynamic diameter is determined by the actual particle size, the particle density, and an aerodynamic shape factor.

TYPES OF AEROSOLS

Aerosols are generally classified in terms of their processes of formation. While the following classification is neither precise, nor comprehensive, it is commonly used and accepted in the industrial hygiene and air pollution fields:

Dust: An aerosol formed by mechanical subdivision of bulk material into airborne fines having the same chemical composition. A general term for the process of mechanical subdivision is comminution, and it occurs in operations such as crushing, grinding, drilling, and blasting. Dust particles are generally solid and irregular in shape and have diameters greater than 1 μm.

Fume: An aerosol of solid particles formed by condensation of vapors formed at elevated temperatures by combustion or sublimation. The primary particles are generally very small (less than 0.1 μm), and have spherical or characteristic crystalline shapes. They may be chemically identical to the parent material, or they may be composed of an oxidation product, such as a metal oxide. Since they may be formed in high number concentration, they often rapidly coagulate, forming aggregate clusters of low overall density.

Smoke: An aerosol formed by combustion of organic materials. The particles are generally liquid droplets with diameters of less than 0.5 μm.

Mist: A droplet aerosol formed by mechanical shearing of a bulk liquid, for example, atomization, nebulization, bubbling, spraying, or by condensation. The droplet size can cover a very large size range, usually from about 2 μm to greater than 50 μm.

Fog: A water aerosol formed by condensation on atmospheric nuclei at high relative humidities. The droplet sizes are generally greater than 1 μm.

Smog: A popular term for a pollution aerosol derived from a combination of smoke and fog.

Haze: A submicrometer-sized aerosol of hygroscopic particles which takes up water vapor at relatively low relative humidities.

Aitken or Condensation Nuclei (CN): Very small atmospheric particles (mostly smaller than 0.1 μm) formed by combustion processes and by chemical conversion from gaseous precursors.

Accumulation Mode: A term given to the particles in the ambient atmosphere which range from 0.1 to about 2.5 μm. These particles generally are spherical, have liquid surfaces, and form by coagulation and condensation of smaller particles which derive from gaseous precursors.

Coarse Particle Mode: Ambient air particles larger than about 2.5 μm, and generally formed by mechanical processes and surface dust resuspension.

AEROSOL CHARACTERISTICS

Aerosols have integral properties that depend upon the concentration and size distribution of the particles. In mathematical terms, these properties can be expressed in terms of certain constants or "moments" of the size distribution (2). Some integral properties such as light-scattering ability or electrical charge depend on other particle parameters as well. Some of the important integral properties are:

Number Concentration: The total number of airborne particles per unit volume of air, without distinction as to their sizes, is the zeroth moment of the size distribution. In current parctice, instruments are available which count the numbers of particles of all sizes from about 0.005 to 50 μm. In many specific applications, such as fiber counting for airborne asbestos, a more restricted size range is specified (3).

Surface Concentration: The total external surface area of all the particles in the aerosol, which is the second moment of the size distribution, may be of interest when surface catalysis or gas adsorption processes are of concern. Aerosol surface is one factor affecting light-scatter and atmospheric-visibility reductions.

Volume Concentration: The total volume of all the particles, which is the third moment of the size distribution, is of little intrinsic interest in itself. However, it is closely related to the mass concentration, which for many environmental effects is the primary parameter of interest.

Mass Concentration: The total mass of all the particles in the aerosol is frequently of interest. The mass of a particle is the product of its volume and density. If all of the particles have the same density, the total mass concentration is simply the volume concentration times the density. In some cases, such as "respirable" dust sampling (4), the parameter of interest is the mass concentration over a restricted range of particle size. In this application, particles too large to be considered "respirable" by man are excluded from the integral.

Dustfall: The mass of particles depositing from an aerosol onto a unit surface per unit time is proportional to the fifth moment of the size distribution. Dustfall has long been of interest in air pollution control because it provides an indication of the soiling properties of the aerosol.

Light-Scatter: The ability of airborne particles to scatter light and cause a visibility reduction is well known. Total light-scatter can be determined by integrating the aerosol surface distribution with the appropriate scattering coefficients.

Water Contaminants

Chemical contaminants can be found in water in solution or as hydrosols; the latter are immiscible solid or liquid particles in suspension. An aqueous suspension of liquid particles is generally called an emulsion. Many materials with relatively low aqueous solubility will be found in both dissolved and suspended forms.

DISSOLVED CONTAMINANTS

Water is known as the universal solvent. While there are many compounds that are not completely soluble in water, there are few which do not have some measurable solubility. In fact, the number of chemical contaminants in natural waters is primarily a function of the sensitivity of the analyses. For organic compounds in rivers and lakes, it has been observed that as the limits of detection decrease by an order of magnitude, the numbers of compounds detected increase by an order of magnitude, so that one might expect to find at least 10^{-12} gm/l. (approximately 10^{10} molecules per liter) of each of the million organic compounds reported in the literature (5). Similar considerations undoubtedly apply to inorganic chemicals as well.

DISSOLVED SOLIDS

Water-quality criteria generally include a nonspecific parameter called "dissolved solids." However, it is customary to exclude natural mineral salts, such as sodium chloride, from this classification. Also, water criteria

Table 2.1   Dissolved Oxygen and Oxygen Demand Parameters

BOD	Biochemical or biological oxygen demand. The oxygen consumed by a waste through bacterial action
COD	Chemical oxygen demand. The oxygen consumed by a waste through chemical oxidation
TOD	The theoretical oxygen demand required to completely oxidize a compound to CO_2, H_2O, PO_4^{-3}, SO_4^{-2}, and NO_3^-
5-day BOD	The BOD consumed by a waste in 5 days
Ult. BOD	The ultimate BOD consumed by a waste in an infinite time
IOD	The "Immediate" oxygen demand consumed in 15 minutes (without using chemical oxidizers or bacteria)
DO	Dissolved oxygen. (A "negative" DO is a positive IOD)

for specific toxic chemicals dissolved in water are frequently exceeded without an excessive total dissolved-solids content.

DISSOLVED GASES

Compounds dissolved in water may also exist in the gaseous phase at normal temperatures and pressures. Some of these, such as hydrogen sulfide and ammonia, which are generated by decay processes, are toxicants.

Oxygen is, of course, the most critical of the dissolved gases with respect to water quality. It is essential to most higher aquatic life forms and is needed for the oxidation of most of the organic chemical contaminants to more innocuous forms. Thus, a critical parameter of water quality is the concentration of dissolved oxygen (DO). Another important parameter is the extent of the oxygen "demand" associated with contaminants in the water. The most commonly used index of oxygen demand is the 5-day BOD (Biochemical Oxygen Demand after 5 days of incubation). Another is the COD (Chemical Oxygen Demand). The basic oxygen demand parameters are defined in Table 2–1.

SUSPENDED PARTICLES

A nonspecific water-quality parameter which is widely used is "suspended solids." The stability of aqueous suspensions depends on particle size, density, and charge distributions. The fate of suspended particles depends on a number of factors, and particles can dissolve, grow, coagulate, or be ingested by various life forms in the water. They can become "floating solids" or part of an oil film, or they can fall to the bottom to become part of the sediments.

There are many kinds of suspended particles in natural waters, and not all of them are contaminants. Any moving water will have currents which cause bottom sediments to become resuspended. Also, natural runoff will carry soil and organic debris into lakes and streams. In any industrialized area such sediment and surface debris will always contain some chemicals considered to be contaminants. However, a large proportion of the mass of such suspended solids would usually be "natural," and would not be considered as contaminants.

The suspended particles can have densities which are less than, equal to, or greater than that of the water, so that the particles can rise as well as fall. Furthermore, the effective density of particles can be reduced by the attachment of gas bubbles.

Gas bubbles form in water when the water becomes saturated and can not hold any more of the gas in solution. The solubility of gases in water varies inversely with temperature. For example, oxygen saturation of fresh water is 14.2 mg/l. at 0° C and 7.5 mg/l. at 30° C, while in sea water the corresponding values are 11.2 and 6.1 mg/l.

A special class of suspended solids are the colloids.

COLLOIDS

Colloids are true sols in that the suspensions are extremely stable. They have a very small, uniform particle size and possess unipolar charges which cause the individual particles to repel one another. The particles are so small, on the order of 0.01 μm, that they pass through most filters. For practical purposes, colloids behave like true solutions and are usually not measured as suspended solids in most assays.

OIL AND FLOATING SOLIDS

Oil, grease, and other organic immiscible liquids can exist as discrete droplets or globules, or can coalesce on the surface as a film. In extreme cases they can produce a fire hazard. In lesser amounts, such films can block the absorption of atmospheric oxygen, can retard photosynthesis in aquatic plants and thereby reduce oxygen production, can coat and destroy algae and other plankton, and make the waters unfit for fish life, swimming, and other recreational uses. Floating solids can simply be an aesthetic blight, or, if they contain oxygen-consuming or toxic materials, can contribute to the degradation of the water quality.

SEDIMENTS

Particles which fall to the bottom of a body of water remain accessible to the water for partial dissolution and periodic resuspension during storms and flow surges. If there is enough sedimentary fallout relative to the depth of the overlying water, the sediments can gradually reduce the depth of the water and thereby affect its velocity and flow patterns. Finally, the sediment layer can become anaerobic and, therefore, a source of toxic compounds which can diffuse into the overlying water.

Food Contaminants

Chemical contaminants of almost every conceivable kind can be found in most types of human food. Food can acquire these contaminants at any of several stages in its production, harvesting, processing, packaging, transportation, storage, cooking, and serving. In addition, there are many naturally occurring toxicants in foods, as well as compounds which can become toxicants upon conversion by chemical reactions with other constituents or additives, or by thermal or microbiological conversion reactions during processing, storage, or handling.

Each food product has its own natural history. Most foods are formed by selective metabolic processes of plants and animals. In forming tissue, these processes can act to either enrich or discriminate against specific toxicants in the environment. For animal products, where the flesh of interest in foods was derived from the consumption of other life forms, there are likely to be several stages of biological discrimination and, therefore, large differences between contaminant concentrations in the ambient air and/or water, and the concentrations within the animals.

Since both natural and anthropogenic toxicants are present in almost all foods, it is important to distinguish between toxicity and hazard. Toxicity is an intrinsic property of the chemical, while hazard is the capacity to produce injury under the circumstances of exposure. A hazard occurs when the inherent toxicity of the chemical is expressed, and is dependent on many factors, especially on the amount ingested.

NATURAL TOXICANTS

No segment of the environment to which humans are exposed is as chemically complex as food. Food products contain both nutrient and nonnutrient components and, until quite recently, relatively little attention was paid to the latter. The humble potato, a food staple for millions of people, contains more than 100 nonnutrient chemical substances, including solanine alkaloids, oxalic acid, arsenic, tannins, and nitrates. Table 2-2 lists some of the toxic compounds present naturally in common plant foods, and their toxic effects.

Despite the presence of a multitude of toxic chemicals in the diets of

Table 2.2   Some Toxic Compounds Present Naturally in Common Plant Foods

Compound	Major source	Main toxic effect
Safrole	Oil of sassafras	Carcinogen
Myristicin	Nutmeg, parsley	Hallucinogen
Carotatoxin	Carrots	Nerve toxin
Synephrine	Lemons	Vasoconstriction
Hemagglutinins (protein)	Soybeans, other legumes	Agglutination of red blood cells
Norepinephrine	Bananas	Vasoconstriction
Solanine	Irish potatoes	Neural transmitter inhibition
Goitrin*	Cabbage, turnips	Antithyroid activity
Hydrogen cyanide*	Lima beans	Asphyxiant

*This is not present in the original plant but is formed by enzymatic reactions from nontoxic precursors following harvest, during processing, or following digestion.

Compiled from: Campbell, A. D., Horwitz, W., Burke, J. A., Jelinek, C. F., Rodricks, J. V., and Shibko, S. I. Food Additives and Contaminants, pp. 167–79 in *Handbook of Physiology, Section 9, Reactions to Environmental Agents.* D. H. K. Lee, H. L. Falk, and S. D. Murphy (eds.). Bethesda, Md.: American Physiological Society, 1977.

normal humans, there is little hazard because the consumption of each is very low. Problems generally arise from an excessive dependence on a limited number of foods during an extended period of time. For example, people consuming large amounts of cabbage have developed goiter, and people with extended daily consumption of a half gallon of tomato juice developed lycopenemia, a skin discoloration (6).

A balanced diet not only helps to keep the level of each natural toxicant below hazardous levels, but also provides greater opportunities for interactions between toxicants. For example, the toxic effects of cadmium are reduced by an accompanying elevated level of zinc, and copper reduces the toxic effects of molybdenum.

NATURAL CONTAMINANTS

Foods can be contaminated by a variety of natural processes not involved either direct or indirect human intervention. Such processes can result in contamination by: (a) products of decay and decomposition which are generated between the growth of the food and its consumption; (b) microbiological and animal pests and/or their residues, wastes and metabolites; (c) chemical congeners of normal nutrient materials which become incorporated into the food during growth, such as excessive levels of nitrates, mercury, selenium, etc., taken up from soils having high concentrations

Table 2.3 Some Mycotoxin Contaminants of Food

Mycotoxin	Genus of producing mold	Typical substrate	Toxic effect
Aflatoxin B$_1$	*Aspergillus*	Peanuts, oil seeds, corn	Carcinogen
Ochratoxin A	*Aspergillus*	Grains	Kidney toxin
Sterigmatocystin	*Aspergillus*	Grains	Carcinogen
Patulin	*Penicillium*	Apples	Carcinogen
Cyclopiazonic acid	*Penicillium*	Grains	Tremors, paralysis
Luteoskyrin	*Penicillium*	Rice	Liver toxin
Islandotoxin	*Penicillium*	Rice	Carcinogen

Compiled from: Campbell, A. D., Horwitz, W., Burke, J. A., Jelinek, C. F., Rodricks, J. V., and Shibko, S. I. Food Additives and Contaminants, pp. 167–79 in *Handbook of Physiology, Section 9, Reactions to Environmental Agents*. D. H. K. Lee, H. L. Falk, and S. D. Murphy (eds.). Bethesda, Md.: American Physiological Society, 1977.

due to geochemical anomalies. As an example of natural contaminants, Table 2–3 lists some mycotoxins which are produced by molds growing on common plant foods.

ANTHROPOGENIC CONTAMINANTS

Human activities can greatly increase the concentrations of the contaminants already present naturally in food, and can also introduce entirely new ones. Some of these latter materials, such as pesticides, are applied intentionally during the growth of the food, and become contaminants to the extent that they persist as residues (within or on the surfaces of foods) long after their intended function has been completed. Others, such as the polychlorinated biphenyls (PCB's), are completely inadvertent food contaminants, since they were never intentionally applied to foods. However, they have been widely used in consumer products, have become ubiquitous environmental contaminants, and have reached excessive levels in some fish and birds through biological concentration processes. Table 2–4 presents PCB levels in selected food products.

AIR CONTAMINANTS

Contaminants in the ambient air may produce pervasive contamination of foods, but they are not likely to result in acute intoxications. The atmosphere is a continuous envelope reaching essentially all of the human environment; but it is also immense, and capable of rapidly diluting the contaminants discharged into it.

Table 2.4   Polychlorinated Biphenyls in Selected Foods, 1975

Food	Maximum residue (ppm_w in fat)	Percentage containing any measurable residue
Milk	1.9	0.7
Fish	9.0	17.8
Meats and Poultry	0.06% had residues > 5 ppm_w	0.3

Compiled from: Proceedings of National Conference on Polychlorinated Biphenyls, 1975. Washington, D. C.: U.S. Environmental Protection Agency, EPA-560/6-75-004.

WATER CONTAMINANTS

Contamination of food via water takes place in several distinct pathways. One is by the consumption of fish and shellfish which have taken up chemicals from the water or from lower aquatic life forms. Another is via the consumption of fruits and grains which took up contaminants from irrigation waters. A more indirect path is the consumption of animal products from livestock consuming contaminated irrigation waters, and the crops grown with these waters. Finally, residual contaminants in drinking waters can reach us either directly or through transfers from water used in cooking.

PESTICIDE RESIDUES

Chemical pesticides are applied to crops and soil to increase the quantity and quality of the harvest. They may also be applied to the harvested food during transportation and/or storage to minimize spoilage and losses. Some of the specific residue limits (tolerances) established by the Food and Drug Administration (FDA) for the major pesticides are listed in Table 2-5. When these limits are exceeded, the foods involved are subject to seizure and withdrawal from the market.

Pesticide residues in and on foods have resulted in readily measurable human body burdens (Table 2-6). Since most of the chlorinated hydrocarbon pesticides are known or suspected carcinogens, there is a great concern and effort devoted to the monitoring and control of their residues in the food supply, and there are continuing efforts to ban more of them from agricultural usage.

RESIDUES OF DRUGS AND GROWTH STIMULANTS

Veterinary drugs and growth stimulants are given to domestic fowl and animals to increase the quantity and/or quality of the meat, and to control epidemic disease. They may be applied by injection, implantation, or

Table 2.5 Selected Pesticides Whose Residues May Be Present in Foods

Pesticide	Major food source	Typical tolerance, ppm_w	Primary toxic effect	FAO/WHO acceptable daily intake, mg/kg body weight
Organochlorine	All agricultural products; concentrate in fat: fish, animal tissue, milk, eggs		Central nervous system depressant; stores in body fat	
Dieldrin		0.1		0–0.0001
BHC (Lindane)		1		0–0.125
Toxaphene		7		
Heptachlor		0		0–0.0005
DDT		7		0–0.005
Organophosphorus	Fruits, vegetables, grains		Cholinesterase inhibition	
Malathion		8		0–0.02
Parathion		1		0–0.05
Diazinon		0.75		0–0.002
Carbamate	Fruits and vegetables		Cholinesterase inhibition (reversible)	
Carbaryl		10		0–0.1 (temporary)
Carbofuran		0.1		
Chlorophenoxy acid	Grains (infrequent)		Not clearly defined	
2,4-D		0.5		0–0.3
Triazine	Meat products (infrequent)		Not clearly defined	
Atrazine		0.25		

*Defined as the daily dose of the chemical that with practical certainty will not result in injury after a lifetime of exposure, on the basis of the facts known at the time of this establishment.

From: Campbell, A. D., Horwitz, W., Burke, J. A., Jelinek, C. F., Rodricks, J. V., and Shibko, S. I. Food Additives and Contaminants, pp. 167–79 in *Handbook of Physiology, Section 9, Reaction to Environmental Agents*. D. H. K. Lee, H. L. Falk, and S. D. Murphy (eds.). Bethesda, Md.: American Physiological Society, 1977.

Table 2.6   Selected Pesticide Residues in Human Tissues

Nation	Residue (ppm$_w$)*			
	DDT (in fat)	Lindane	Aldrin/dieldrin	Haptachlor/ heptachlor epoxide
Australia	2	—	0.05	—
Canada	4	0.06	0.2	0.07
France	5	—	—	—
Great Britain	3 – 6	0 – 0.3	0.2 – 0.3	Trace
Holland	2 – 7	0 – 0.1	0.2	0.009
India	12 – 28	1 – 2	0.03 – 0.06	—
Israel	5 – 8	—	—	—
Italy	8	0.06	0.5	0.2
United States	5 – 10	0.2 – 0.5	0.1 – 0.3	0.1 – 0.2

*The residues were measured in the period 1963 – 1969.

Compiled from: Edwards, C. *Persistent Pesticides in the Environment. Critical Reviews in Environmental Control*, vol. 1. Cleveland: Chemical Rubber Publishing Co., 1970.

Table 2.7   Some Drugs Used in Animal Feeds for Prophylactic, Therapeutic, and Growth-Promotion Purposes

I. Antimicrobial
    Antibiotics
      Bacitracin
      Erythromycin, oleandomycin, tylosin
      Novobiocin
      Nystatin
      Penicillins
      Streptomycin, dihydrostreptomycin, neomycin
      Tetracyclines: oxytetracycline, chlortetracycline
    Sulfonamides
      Sulfachloropyridazine
      Sulfadimethoxine
      Sulfamethazine
      Sulfathiazole
    Nitrofurans
      Furaltadone
      Furazolidone
      Nihydrazone
      Nitrofurazone
II. Antiprotozoal (Coccidiostats)
    Aklomide
    Buquinolate and related compounds
    Clopidol
    Sulfanitran
    Zoalene

III. Anthelmintics
    Coumaphos
    Dichlorvos
    Haloxon
    Levamisole
    Piperazine
    Thiabendazole
IV. Growth promoters
    Hormones
      Diethylstilbestrol
      Estradiol
      Melangesterol
      Progesterone
      Testosterone
      Zeranol
    Arsenicals
      Arsanilic acid
      Carbarsone
      Nitarsone
      Roxarsone

From: Campbell, A. D., Horwitz, W., Burke, J. A., Jelinek, C. F., Rodricks, J. V., and Shibko, S. I. Food Additives and Contaminants, pp. 167 – 79 in *Handbook of Physiology, Section 9, Reactions to Environmental Agents*. D. H. K. Lee, H. L. Falk, and S. D. Murphy (eds.). Bethesda, Md.: American Physiological Society, 1977.

ingestion, and some fraction may remain as a detectable residue in the flesh, or in the eggs or milk products produced by the animals. Some of the drugs used in animal feeds are listed in Table 2 – 7.

A chemical of particular concern has been the artificial hormone, diethylstilbestrol (DES), which is a widely used growth stimulant. Its use as a human drug in high doses has been implicated in a higher-than-normal cervical cancer incidence among the daughters of the women treated. Its presence at low levels in meat, therefore, raises the potential of an increase in cancer among the general population.

Concern has also been expressed about residues of antibiotics in meat, resulting from their use to control disease. The residues may be capable of causing adverse reactions among allergic individuals. Also, continued ingestion of low levels in foods may lead to a reduced potency if the antibiotics are used therapeutically in the people at a later time.

PACKAGING MATERIAL RESIDUES

Packaged foods can take up chemicals from the materials used to package or contain them. Problems may arise in canned foods; for example, condensed milk has been contaminated by lead in the solder used to seal the cans. More recently, most of the focus of concern has centered on plastic packaging materials, especially the polymeric materials such as polyvinyl chloride (PVC) and acrylonitrile. The polymers themselves have long interlocking-chain structures and are not considered toxic. However, the polymers are always contaminated, to some extent, with unpolymerized monomer, and some of the monomers are known carcinogens.

ACCIDENTS AND MISUSE

Acute chemical intoxications via food are generally attributable to accidental cross-contamination of foods and industrial chemicals or pesticides in transportation or storage, mislabeling by manufacturers, or by a failure to understand or follow label instructions.

There have been many cases of cross-contamination between bread flour and pesticides which have led to acute poisonings and some deaths. A notable example of mislabelling occurred in Michigan in 1975, when a fire retardant containing polybrominated biphenyls (PBB's) was mistakenly bagged in containers intended for a cattle feed supplement. The insidious poisoning of the cattle which resulted was not immediately recognized, and a large number of farmers, their families, and people in the general population were severely contaminated and suffered health effects from consuming meat and milk from these animals.

There have been numerous cases of mercury intoxication resulting from the direct consumption of bread made from seed grains treated with mercurial pesticides. Some occurred because the grain was not properly labelled. Others have occurred when the labels were not read, sometimes because they were in the wrong language, or because the people involved were illiterate.

Finally, contamination sometimes has resulted from a major release due to an explosion or other industrial accident. A recent example is the factory explosion in Seveso, Italy, in July, 1976, which spread 2–3 kg of the highly toxic chemical dioxin downwind over a considerable distance. The extent of the contamination was not disclosed to the farmers affected and, therefore, domestic animals grazed on contaminated ground, and contaminated milk and eggs reached many people.

FOOD ADDITIVES

Food additives are not contaminants in the sense that they are intentionally applied to food products for a particular purpose. At the time their usages were approved, there was presumably no evidence that they would produce significant adverse effects. However, in recent years, many food additives have been banned by the FDA, or are being reevaluated because of evidence that they were capable of producing cancer in laboratory animals. Notable examples which attracted considerable public attention in the mid-1970's are the food coloring dye known as red #2, the artificial sweetener saccharin, and the nitrites used to preserve red meat. The latter can combine with secondary amines in the disgestive tract to form carcinogenic nitrosamines.

The controversy about food additive safety, and the occasional banning of food additives on the basis of evidence for carcinogenic potential, has led to understandable confusion and concern among the general public about the safety of food additives which are still approved, and has resulted, in part, in the proliferation of "natural food" retail stores, cooperatives, and restaurants. It has also led to an increased effort by the FDA to more thoroughly evaluate the safety of commonly used food additives.

Excessive use of food additives may cause increases in diseases other than cancer. Sodium chloride has been associated with human hypertension, while an excessive daily intake of phosphates can lead to premature cessation of bone growth in children (6).

Many processed foods are fortified with vitamins. When these are added to the vitamins naturally occurring in foods, and to the vitamins ingested in tablets or as liquids, large daily intakes may occur. Excessive intakes of vitamins A and D in particular have caused a variety of acute and chronic health effects (6). However, it is important to remember that vitamin deficiencies cause many more health problems than do excessive ingestions.

SPECIAL PROBLEMS

Genetic manipulations of plants to introduce more desirable properties and to increase yields may result in alterations of their levels of essential nutrients and natural toxicants. Common plants such as tomatoes and potatoes, which have toxic foliage, need special attention with respect to altered toxicant levels. On the other hand, selective breeding can be employed to reduce the concentrations of toxic substances in natural foods. Lima beans having low cyanogenetic glycoside content have been developed in order to minimize their cyanide-generating capacity (6).

While toxicants in food represent public health problems, the more important dangers associated with food consumption arise from other considerations. The major problem is undoubtedly overeating, especially of foods rich in fats and sugars, resulting in obesity, cardiovascular disease, hypertension, and dental caries. Problems also arise frequently in special populations, such as teenagers, who may consume large amounts of "junk" food resulting in unbalanced diets which permit the development of deficiency diseases.

CHARACTERIZATION OF ENVIRONMENTS

The abiotic, or nonliving, environment can be divided into three components: (a) the atmosphere, or air; (b) the hydrosphere, or waters; and (c) the lithosphere, or earth's crust. The aggregate of all of the life forms within all these three components is termed the biosphere.

Although these components are often separated in terms of many discussions in this book, it is important to realize that transport of contaminants takes place within and between all media. Thus, release of a contaminant into one component does not insure that it will remain there; contamination must be considered in terms of the environment as a whole. Eventually, the contaminants are chemically transformed or are physically trapped into a permanent storage site. The ultimate disposition site or removal mechanism for a contaminant is known as a sink.

Characteristics of the Atmosphere

The earth's atmosphere is simple in some respects and complex in others. It is relatively uniform in composition with respect to its major mass components (oxygen and nitrogen), yet extremely variable in some minor components, such as water vapor and ozone (O_3), which play major roles in its heat and radiation fluxes. It has a complex structure based on temperature gradients. This structure governs its mixing characteristics and the buildup of contaminants, yet is usually invisible to us, except when light-scattering particles make it visible.

STRUCTURE OF THE ATMOSPHERE

The structure of the atmosphere, while invisible, is real and of major importance to the dilution of contaminants. It is governed by the lapse rate, which is the rate of change of air temperature with height above the ground. The changes in temperature and pressure with height are shown in Figure 2–2.

The lowest of the atmospheric layers is the troposphere. It contains about 75% of the mass of the atmosphere and almost all of its moisture. It extends to a height which varies from about 9 km at the poles to about 15 km at the equator, and it has an average lapse of about −6.5° C/km. The boundary between the troposphere and the next layer, the stratosphere, is known as the tropopause. The stratosphere contains essentially all of the remainder of the mass of the atmosphere; it is nearly isothermal (the temperature does not change with altitude) in lower regions and shows a temperature increase with height in the upper regions.

CHEMICAL COMPOSITION

The major constituents of dry air at ground level are nitrogen (N_2) at 78.1% by volume, oxygen (O_2) at 21.0%, and argon (Ar) 0.9%. Carbon dioxide is present at about 330 ppm_v (0.033%), neon (Ne) at 18 ppm_v, helium (He) at about 5 ppm_v and methane (CH_4) at about 1.5 ppm_v. All other gases are present at less than 1 ppm_v.

About 3% of the total mass of the lower atmosphere is water vapor (H_2O), but the concentration is extremely variable in both space and time. In general, the warmer portions of the atmosphere contain more water vapor. The water vapor content becomes lower with increasing altitude and with increasing latitude. Water vapor plays a critical role in governing the earth's heat exchange and the motion of the atmosphere, due to its

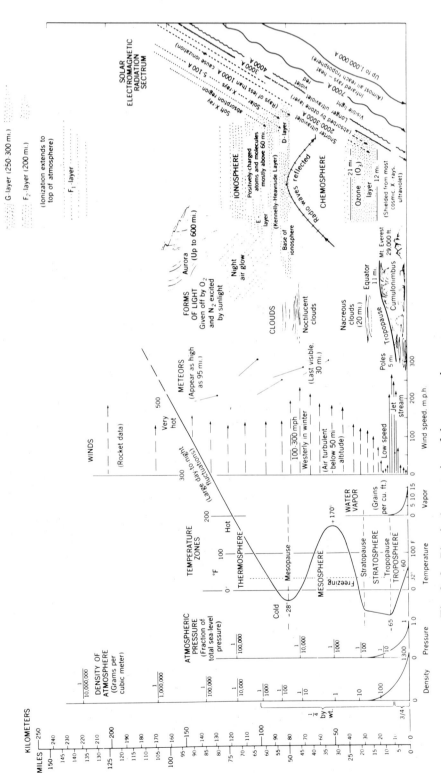

Fig. 2–2 Physical properties of the atmosphere.

From: Strahler, A. N. *The Earth Sciences*, 2nd ed. New York: Harper and Row, 1971. © 1963, 1971 by Arthur N. Strahler. Reprinted by permission of Harper & Row, Publishers, Inc.

high heat capacity, absorption of infrared radiation, and heat of vaporization. Further effects attributable to atmospheric water result when the air motions create clouds — aerosols of water droplets in which the energy received as sunshine in one place is liberated as the latent heat of vaporization in another.

EARTH-ATMOSPHERE ENERGY BALANCE

The sun is the source of essentially all of the energy which reaches the earth. Radiant energy from the sun covers the entire electromagnetic spectrum. However, most of it occurs around the visible portion, i.e., wavelengths from 0.4 μm to 0.7 μm (Figure 2 – 3).

Of the incoming radiant energy, about 30 – 50% is scattered back towards space, reflected by the atmosphere, due primarily to clouds and to some extent, by solid particles, or by the earth's surface. On a global basis, the average reflectivity, or albedo, of the earth's surface and atmosphere is about 35%. The actual albedo of any specific surface is highly dependent

Fig. 2–3   Emission spectra to and from the earth. The assumption is that the sun and earth radiate as "black bodies" having temperatures of 6,000° K and 250° K respectively. Note the change in the character of the radiation from the short-wave to the long-wave portion of the spectrum.

From: Perkins, H. C. *Air Pollution.* New York: McGraw-Hill, © 1974. Used with permission of McGraw-Hill Book Co.

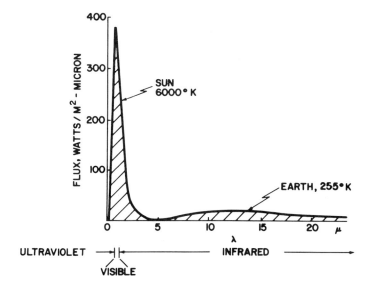

upon the particular area and its characteristics; ice- and snow-covered polar regions have high reflectivity, while the reflectivity of oceans is relatively low, with most incident energy being absorbed.

About 20% of the incident radiant energy is absorbed as it passes through the atmosphere. Stratospheric ozone absorbs about 1–3%, primarily in the short-wave ultraviolet (UV) portion of the spectrum; this effectively limits further penetration to those wavelengths greater than 0.3 μm.

In the troposphere, 17–19% of the incoming radiation is absorbed, due primarily to water vapor and secondarily to CO_2. The total atmospheric absorption for radiant energy with wavelengths of 0.3–0.7 μm is not very large, and these effectively penetrate an essentially "transparent" atmospheric window.

In total, about 50% of the incoming solar radiation reaches the surface

Fig. 2–4 Earth-atmosphere energy balance.

From: Perkins, H. C. *Air Pollution.* New York: McGraw-Hill, © 1974. Used with permission of McGraw-Hill Book Co.

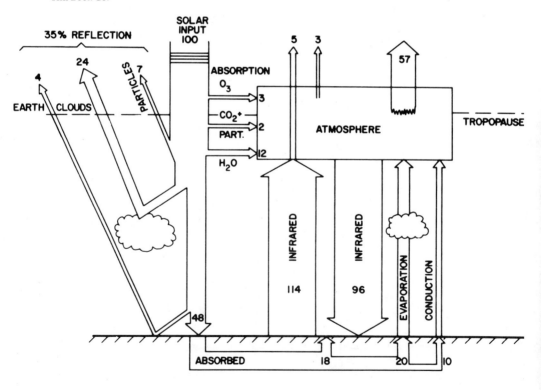

of the earth, and is absorbed. The surface reradiates energy through a broad range of wavelengths, but with a flat maximum in the long-wave, infrared portion of the spectrum at about $10-12$ μm (Figure $2-3$). The atmosphere is nearly opaque to this radiation; most is absorbed by water vapor and droplets, and by CO_2. Some is then reradiated back to earth, or out into space. The earth-atmosphere energy balance is summarized in Figure $2-4$.

By allowing effective penetration of the short-wave solar radiation, yet retaining a large fraction of the reradiated long-wave radiation, the atmosphere acts as an insulator, keeping heat near the surface of the earth. This phenomenon is known as the "greenhouse effect."

The solar flux, which is the intensity of radiation as measured by the amount of energy transfered per unit area per unit time, decreases with increasing latitude. For example, the average mid-winter solar flux is over 800 cal/cm²/day at the equator, and less than 200 cal/cm²/day at 50° N. latitude. On the other hand, the flux of infrared radiation from the atmosphere to space only drops from about 470 cal/cm²/day at the equator to about 400 at 50° N.

The average radiation into space must equal that absorbed from the sun. Thus, it follows that a substantial amount of energy must flow from the tropics towards the poles within the troposphere. This flow of energy is accomplished primarily by systems of warm air currents toward the poles and cool currents toward the tropics, and partially by the corresponding ocean currents.

CIRCULATION OF THE ATMOSPHERE

At altitudes above about 500 m, the atmosphere exhibits a general pattern of circulation characterized by several dominant wind systems. These are the trade winds, the jet stream or midlatitudinal westerlies, and the polar easterlies. Below 500 m, the surface of the earth exerts effects of varying degrees upon air circulation. A diagram of the general circulation of the atmosphere is shown in Figure $2-5$.

Throughout the troposphere, there is a rapid decrease in temperature towards the poles. One result of this is a high-speed wind known as the subtropical jet stream. As seen in Figure $2-6$, the tropopause is discontinuous in the region of the jet stream; it is through these "gaps" in the midlatitudes that much of the circulation occurs between the stratosphere and the troposphere.

In addition to this general circulation, the atmosphere has secondary and small-scale circulation patterns. Secondary circulations are exempli-

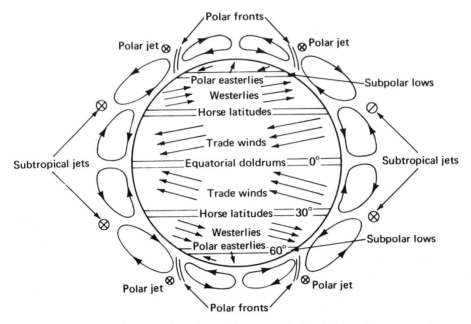

Fig. 2–5 Schematic representation of the general circulation of the atmosphere.
From: Seinfeld, J. H. *Air Pollution: Physical and Chemical Fundamentals.* New York: McGraw-Hill, ©
1975. Used with permission of McGraw-Hill Book Co.

fied by the migratory high- and low-pressure areas seen in daily weather
charts. Examples of small-scale circulations are land-sea breezes, moun-
tain-valley winds, and thunderstorms.

Characteristics of the Hydrosphere

HYDROLOGIC CYCLE

The hydrosphere includes a variety of distinctly different aquatic environments which are essential to human life and economic productivity. It is a
dynamic system due to the hydrologic cycle, in which liquid water evapo-
rates and is transported through the atmosphere. The cycle is illustrated
schematically in Figure 2–7. While the atmospheric water vapor mass
constitutes only 10^{-5} of the hydrosphere, it is critical both to the heat balance of the globe as previously discussed, and to the replenishment of
fresh water needed for drinking and for agriculture purposes.

As indicated in Table 2–8, most of the hydrosphere is contained in the
oceans, which are saline. Furthermore, most of the fresh water is relatively

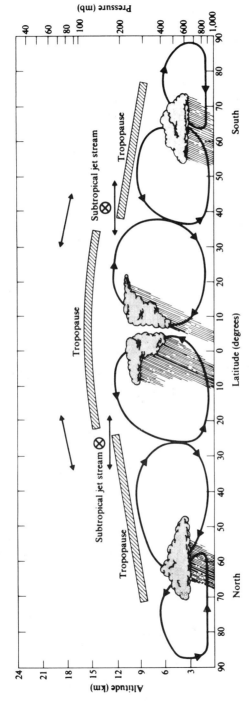

Fig. 2–6 Schematic representation of the circulatory system in the troposphere during summer in the Northern Hemisphere, showing the discontinuity in the troposphere at the subtropical jet stream.

From: Williamson, S. J. *Fundamentals of Air Pollution*. Reading, Mass.: Addison-Wesley, © 1973.

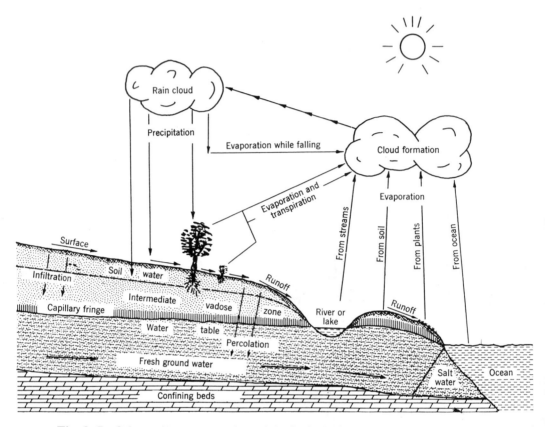

Fig. 2-7   Schematic representation of the hydrologic cycle.
From: Briggs, G. F. Terminology. *Air Conditioning, Heating and Ventilating*, August, 1967.

inaccessible, being in groundwater or in polar ice. Its inaccessibility is indicated by its rate of turnover, i.e., by the time required for each part to be exchanged in the hydrologic cycle.

FRESH SURFACE WATERS

Chemically pure waters, i.e., containing only H_2O, are not found in nature. Water's great power as a solvent insures that natural waters contain many other substances which are largely derived from the lithosphere. For example, Table 2-9a presents a listing of some major chemicals found in natural, fresh waterways in the United States. The table applies to those waters which support a varied and profuse fish fauna, and are thus considered nonpolluted. The concentrations of minerals present at lower con-

Table 2.8 The Hydrosphere

Component	Approximate volume (10^3 km³)	Turnover rate (years)
Oceans	1,370,000	3,000
Groundwater	60,000	5,000★
(groundwater zones with active turnover)	(4,000)	(330†)
Polar ice caps	24,000	8,000
Surface waters of land	280	7
Rivers	1.2	0.031
Soil moisture	80	1
Water vapor in atmosphere	14	0.027
Total hydrosphere:	1,454,000	2,800

★Inclusive of groundwater runoff to oceans, thus bypassing rivers – 4,200 years.
†Inclusive of groundwater runoff to oceans, thus bypassing rivers – 280 years.
Compiled from: Lvovitch, M. I. World Water Resources, Present and Future. *Ambio* 6:13–21, 1977.

Table 2.9a Some Average Properties of Natural Waterways in the United States★

	5% of the waters contain less than	95% of the waters contain less than
Total dissolved solids	72	400
Bicarbonate (HCO_3^-)	40	180
Sulfate (SO_4^{-2})	11	90
Nitrate (NO_3^-)	0.2	4.2
Calcium (Ca^+) Magnesium (Mg^+)	18.5	66.0
Sodium and potassium ($Na^+ + K^+$)	6	85
Free carbon dioxide (CO_2)	0.1	5.0
Ammonia (NH_3)	0.5	2.5
Chloride (Cl^-)	3	170
Iron (Fe)	0.1	0.7

★Numbers are in ppm$_w$.

Table 2.9b   Trace Metals in River Water

Metal	Concentration (global average) (μg/l.)
Cd	<1
Co	0.1
Cr (VI)	1
Cu	7
Hg	0.002
Mo	0.6
Pb	3
V	0.9
Zn	20

Compiled from: Miettinen, J. K. Inorganic Trace Elements as Water Pollutants, pp. 113–36 in *Water Quality*. F. Coulston and E. Mrak (eds.). New York: Academic Press, 1977.

centrations, including many of the toxic chemicals of interest, are more variable and cannot readily be summarized. Table 2–9b presents average heavy-metal concentrations in nonpolluted river waters.

The surface runoff into lakes and rivers are used extensively for drinking water supplies. Sometimes they are used with little if any treatment, as in New York City and Boston which collect their water in distant watersheds isolated from contaminated runoff. In other cities located on rivers or lakes with extensive commercial traffic and upstream sources of contamination, the water requires physical and chemical treatment and disinfection before it is safe to use in water-supply systems.

Lakes and other enclosed bodies of water, especially those in temperate zones, show a thermal stratification during different seasons of the year. This pattern is due to the interplay of temperature, water density, and wind, and is shown in Figure 2–8.

The maximum density of fresh water occurs at a temperature of about 4° C; density decreases as the temperature is further reduced to the freezing point, 0° C. Thus, in winter the lake surface will be covered by a layer of ice, while waters slightly warmer than 0° C will remain at various depths below the surface. This ice layer prevents wind-induced circulation, suppressing vertical and horizontal movements and further loss of heat to the atmosphere. This situation is known as winter stagnation.

As the weather gets warmer in the spring and the ice begins to break up, the surface water starts to warm up. As its temperature increases towards 4° C and it becomes denser, this water sinks below the colder, and less dense, layer beneath it. This overturning of the water, which may last sev-

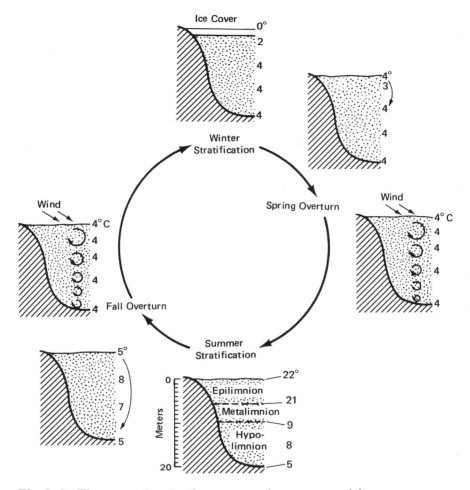

Fig. 2-8 The seasonal cycle of temperature in a temperate lake.
From: Kormondy, E. J. *Concepts of Ecology.* p. 149 Englewood Cliffs, NJ: Prentice-Hall, © 1969. Reprinted by permission of Prentice-Hall.

eral weeks, results in a temperature profile that is essentially uniform at all depths of the lake; as the ice melts, vertical circulation is further aided by wind action. As the weather gets progressively warmer towards summer, the surface water becomes warmer still, and thus less dense than underlying colder waters. The water is directly stratified, in terms of temperature, in the condition known as summer stratification or stagnation. This condition generally persists from April to November in northern latitudes of the Northern Hemisphere. The upper water layer is termed the epilimnion, and has a relatively uniform warm temperature. The bottom layer, the hypolimnion, is characterized by a relatively uniform cold temperature. Between these two layers is the metalimnion, a zone of maximum temperature change. The rapid temperature change itself is termed a thermocline.

As autumn approaches, the surface water layer once again cools and sinks. The water becomes mixed and stirred to increasing depths, and equality of surface and bottom temperatures occur once again, as it did in the spring, in a process termed fall overturn. When the surface freezes, winter stagnation is reestablished.

This temperature cycle has important implications in terms of essential nutrient circulation and contaminant levels in temperate aquatic environments. Thermal gradients are also gradients for concentrations of dissolved gases. Oxygen absorbed at the surface layers is distributed throughout the lake by water circulating within the epilimnion. Waste gases produced during the decomposition of bottom deposits are released by contact with air over surface waters. Within the thermocline, there is a sharp drop in DO (and a rise in concentration of gases of decomposition) until DO reaches a minimum below the level of the thermocline. Because the stability of the waters during summer stratification reduces mixing, the hypolimnion tends to become depleted of oxygen. This effect may be enhanced by the discharge of certain contaminants.

As part of the semiannual overturn and mixing of the entire lake, nutrients are mixed and redistributed. Those in lower waters are brought to surface waters, and made available for use in photosynthesis. Thus, overturns may be periods of decreased water quality, as sudden blooms of aquatic plants, such as algae, occur. Lakes, therefore, often exhibit seasonal differences in water quality. In some inland lakes, water in the lower depths remains unmixed with the main water mass during turnovers, contributing to the stagnation of nutrient cycling.

The biological productivity of surface waters depends heavily on temperature, dissolved and suspended nutrients and contaminants, and oxygen concentration. There is continual interchange of nutrients and contaminants between the water, bottom sediments, suspended particles, and various life forms.

GROUNDWATER

While most of the fresh water is underground (Table 2-8), surface waters supply over 80% of current usage. Thus, as surface water supplies become

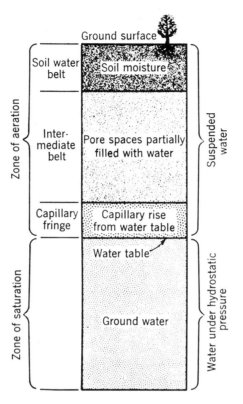

Fig. 2–9   Groundwater is the part of subsurface water within the zone of saturation.

From: Briggs, G. F. Terminology. *Air Conditioning, Heating and Ventilating*, August, 1967.

inadequate, groundwater usage is bound to increase. Groundwater is generally free of significant contamination by bacteria and suspended solids but will usually contain significant amounts of dissolved solids. These characteristics result from the intimate and extended contact with the minerals that constitute the lithosphere. The soil particles filter out the bacteria and suspended particles, and increase the mineral content by dissolution to the point of chemical equilibrium.

Not all subsurface water is groundwater. As shown in Figure 2–9, groundwater lies in the zone of saturation, where it fills essentially all the voids in the rock stratum. Its upper limit is known as the water table. When usable volumes of water can be extracted from a saturated zone, it is called an aquifer. When the groundwater is under considerable pressure,

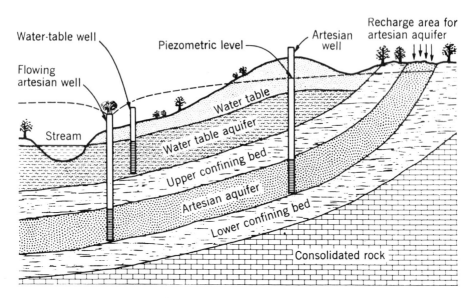

Fig. 2–10 Subsurface and groundwater phase of the hydrologic cycle.
From: Briggs, G. F. Terminology. *Air Conditioning, Heating and Ventilating,* August, 1967.

the water will flow upward without mechanical pumping through a well pipe drilled into it. This is known as an artesian well (Figure 2 – 10).

The depth of the water table can be quite variable in time, depending on the rates of water extraction and recharge. Natural recharge through precipitation and infiltration into the ground may be inadequate in areas with heavy use of groundwater, and the resulting fall in the water table may limit the supply and/or cause significant subsidence of the land surface. These problems can be partially or completely overcome in some areas by installing recharge basins or wells. Recharge basins are designed to catch storm water, which percolates through their base to the water table. Recharge wells are used for industrial waste waters and cooling waters, which generally are pretreated to an acceptable quality before being pumped into the groundwater.

COASTAL WATERS AND ESTUARIES

The relatively shallow waters of the continental shelves and the estuaries, where tidal action results in mixing of fresh water runoff with saline ocean water, are of particular importance because of their very high biological productivity. They are the major harvest grounds for many commercial shellfish and finned fish. In addition, many ocean fish migrate into estuaries to spawn. Their life cycles can be interrupted by changes in oxygen content, contaminant concentrations, or the benthic environment.

Estuaries are heavily used for marine commerce, and are located within our major population centers. New York, Philadelphia, and Washington, D.C. are situated on the estuaries of the Hudson, Delaware, and Potomac Rivers, respectively.

The flow in an estuary is cyclic, rising and falling for 6.2 hours from ebb slack (sea-water efflux) to flood slack (sea-water influx) and back to ebb slack. A full tidal cycle is completed in 12.4 hours.

THE OCEANS

The surface waters of the oceans, those up to 200 m deep, are relatively uniform in temperature, density, and salinity because of the mixing resulting from surface winds. There are characteristic surface currents of the oceans (Figure 2–11) which largely coincide with the surface wind patterns. The intermediate zone, down to a depth of about 1,000 m is characterized by decreasing temperature and increasing density and salinity with depth. This zone has a very low vertical diffusivity and, therefore, acts in effect as a barrier between the surface and deep waters.

The deep ocean waters, with temperatures of 1–4° C, occupy about 75% of the total volume. Thus, most of the ocean waters contribute little to the hydrologic cycle and, in fact, have been relatively inaccessible to human activity. Dredging of the ocean bottom in deep water is just beginning on a semicommercial basis. These operations will stir up large volumes of bottom sediment and bring some of this sediment into the surface layers. Their impacts on the environment could conceivably be significant, but are not known at this time.

Characteristics of the Lithosphere

The lithosphere is quite important to our understanding of chemical contamination of the human environment. In terms of contamination, the soil is the most significant part of the lithosphere.

The characteristics of the soil determine the rate of spread of chemical contaminants placed on or in it and, therefore, their access to the groundwater. Soil, which is derived from the weathering of rocks, is a very complex system, having solid, liquid, and gaseous phases. Its solids consist of particles of different chemical and mineralogical composition which also vary in particle size, shape, and packing characteristics. The particle con-

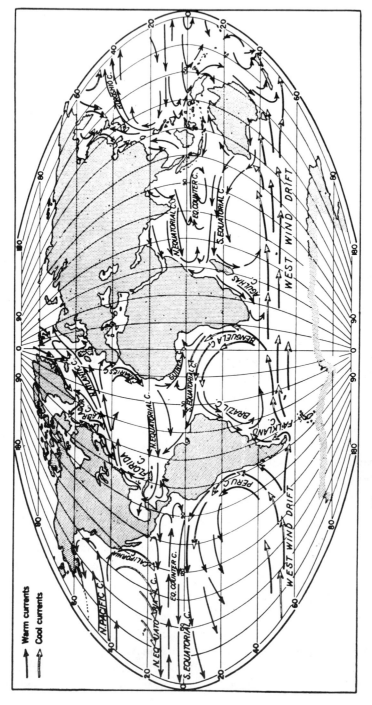

Fig. 2–11 The principal ocean currents of the world.
From: Pettersen, S. *Introduction to Meteorology.* New York: McGraw-Hill, © 1958. Used with permission of McGraw-Hill Book Co.

figuration determines the characteristics of the channels (pores) which contain air and/or water. The composition of both fluid phases is also variable, both temporally and spatially.

Soils consist of horizons, or layers, which are roughly parallel to the ground surface, with each differing in properties, such as color and texture, from its adjoining layers. A succession of horizons is termed the soil profile. Transfer of minerals and organic material in solution or suspension occurs between horizons via water moving through the pores.

The solid phase of soil may be separated into inorganic and organic components. The inorganic particles are generally classified into fractions, depending on their size. In the U.S. Department of Agriculture classification, particles smaller than 2 μm in diameter are clay, those between 2 and 50 μm are silt, those between 50 μm and 2 mm are sand, and those larger than 2 mm in diameter are gravel. The overall textural designation of a soil depends on the ratio of the masses of the less than 2 mm fractions, and these are generally expressed in terms of a textural triangle, as illustrated in Figure 2–12.

The fraction which largely governs the physical properties and chemical retentions of the soil is the colloidal clay. While sand and silt are composed mainly of quartz and other primary mineral particles, clay is composed of a large group of minerals. Some are amorphous, but many are highly structured microcrystals. The most prevalent of these are the layered aluminosilicates. The clay particles have the highest specific surface. They adsorb water, causing the soil to swell upon wetting and shrink upon drying. Most clays have negative electrical charges, and form electrostatic double layers with exchangeable cations in the water phase.

The quantity of cations adsorbed per unit mass of soil is known as the cation exchange capacity, which may range as high as 0.60 mEq per gram. The attraction of a cation to a negatively charged clay particle increases with increasing chemical valence, while more highly hydrated cations are more easily displaced than less hydrated ones. Thus, the order of preference of the common soil cations would be: $Al^{+3} > Ca^{+2} > Mg^{+2} > K^+ > Na^+ > Li^+$. When lime is added to an acid soil, Ca^{+2} ions displace many of the hydrogen ions on the surface of the clay particles. Soils with little colloidal materials, such as those consisting largely of sand, have low natural exchange capacities.

Aside from living creatures, the organic fraction of soil consists of plant and animal tissues undergoing decay, plus a relatively chemically stable colloidal fraction called humus. The humic component also plays a part in ion exchange, with the relative role of inorganic vs organic fractions in

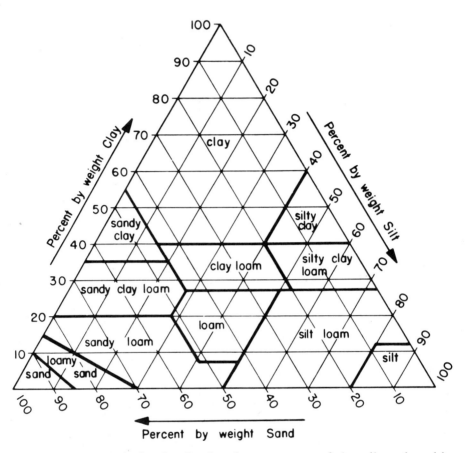

Fig. 2–12 Textural triangle, showing the percentages of clay, silt, and sand in the basic soil textural classes.

From: Hillel, D. *Soil and Water: Physical Principles and Processes.* New York: Academic Press, 1971.

total exchange capacity depending upon the soil type; about 20–70% of the capacity may be due to the humus.

Although ion exchange capacities of soils are usually measured in terms of cations, positively charged soil particles also occur, and soils thus have some degree of anion exchange capacity.

The very large surface area of soil and its great ion exchange capacity permit it to retain nutrient elements to a far greater extent than that retained in the soil moisture.

Characteristics of the Biosphere

All life forms interact with their abiotic and biotic environments. Thus, the biosphere frequently plays a critical role in the spread of chemical contaminants, especially via those forms which undergo transport across or within abiotic components during daily or life cycles. For example, phytoplankton can rise and fall within the aquatic environment in diurnal cycles; many animals can consume their prey in one place, migrate and be consumed by larger predators at another.

The biosphere tends to conserve those chemical elements which are essential to life. This occurs by a process of cycling in characteristic fashion in pathways termed biogeochemical cycles. Very often contaminant residues also cycle, and may thus be retained in the biosphere for long periods of time.

The various biota may also influence contaminant dispersion by their ability to greatly concentrate certain chemical elements or compounds by active metabolic processes. Organic chemicals may accumulate in fatty tissues, while cationic radionuclides which are essential elements, or are close chemical congeners of these elements (Sr for Ca is an example of the latter case), will accumulate in tissues which store those elements. Metabolic discrimination can work both ways; some chemicals will concentrate in tissues relative to their environmental levels, while others will be found at much lower levels in the biological tissues.

Characteristics of Occupational Environments

Occupational environments have been major sources of excessive exposures of people to toxic chemicals. While the populations exposed are much smaller than those exposed to contaminants in community air, drinking water, and foods, the levels of exposure are potentially much larger. In fact, most cases of chemical intoxication have resulted from occupational exposures, the largest portion from inhalation, but a substantial number have also been due to exposure of the skin. The effects can be localized in the lungs or on the skin, or they can be systemic, after translocation throughout the body via the blood or lymph.

In some ways, the occupational environment is less variable than the community environment outside the factory walls. The excursions of air temperature, humidity, and air velocity are generally less extreme, and patterns of activity tend to be róutine. On the other hand, the changes in air contaminant concentrations with location may be very large because of the proximity of contaminant sources, and the workers may be more susceptible to intoxication because of the stresses of the work environment. These include physical exertion with its alterations in the pattern and depth of respiration, and the psychological stresses which may result from keeping up with fixed work schedules. Additional stresses may result from shift work which conflicts with normal circadian rhythms.

CHARACTERIZATION OF HEALTH, DISEASE, AND DISABILITY

Definitions of Health

There is no universally accepted definition of health. Perhaps the most widely accepted one today is that of the World Health Organization, which describes health as a state of complete physical, mental, and social well-being, and not merely the absence of disease or infirmity. Unfortunately, by a strict interpretation of this rather idealistic definition, very few people could be considered healthy.

The discussion to follow will be limited largely to physical well-being. The health effects to be discussed are those that can be recognized by clinical signs, symptoms, or decrements in functional performance. Thus, for all practical purposes, we will be considering health as the absence of measurable disease, disability, or dysfunction.

Health Effects

Recognizable health effects in populations are generally divided into two categories: mortality and morbidity. The former refers to the number of deaths per unit of population per unit time, and to the ages at death. Morbidity refers to nonfatal cases of reportable disease.

Accidents, infectious diseases, and massive overexposures to toxic chemicals can cause excess deaths to occur within a short time after the exposure to the hazard. They can also result in residual disease and/or dysfunction. In many cases, the causal relationships are well-defined, and it may be possible to develop quantitative relationships between dose and subsequent response.

The number of people exposed to chemical contaminants at low levels is, of course, much greater than the number exposed at levels high enough to produce overt responses. Furthermore, low-level exposures are often continuous or repetitive over periods of many years. The responses, if any, are likely to be nonspecific, i.e., an increase in the frequency of chronic diseases which are also present in nonexposed populations. For example, any small increase in the incidence of heart disease or lung cancer attributable to a specific chemical exposure would be difficult to detect, since these diseases are present at high levels in nonexposed populations. In smokers they are likely to be influenced more by cigarette exposure than by the chemical in question.

The increases in the incidence of diseases from low-level long-term exposure to environmental chemicals invariably occur among a very small percentage of the population, and can only be determined by large-scale epidemiological studies (epidemiology is the study of the distribution and frequency of a disease in a specific population) involving thousands of person-years of exposure. The only exceptions are chemicals which produce very rare disease conditions, where the clustering of a relatively few cases may be sufficient to identify the causative agent. Notable examples of such special conditions are the industrial cases of chronic berylliosis caused by the inhalation of beryllium-containing dusts, liver cancers which resulted from the inhalation of vinyl chloride vapors, and pleural cancers which resulted from the inhalation of asbestos fibers. If these exposures had produced more commonly seen diseases, the specific materials might never have been implicated as causative agents.

Low-level chemical exposures may play contributory, rather than primary, roles in the causation of an increased disease incidence, or they may not express their effects without the coaction of other factors. For example, the excess incidence of lung cancer is very high in uranium miners and asbestos workers who smoke cigarettes, but is only marginally elevated among nonsmoking workers with similar occupational exposures. For epidemiological studies to provide useful data, they must take appropriate account of smoking histories, age, and sex distributions, socioeconomic levels, and other factors which affect mortality rates and disease incidence.

MORTALITY

In industrialized societies, there is generally good reporting of mortality and age at death, but with few exceptions, quite poor reporting of cause-of-death. In studies which are designed to determine associations between exposures and mortality rates, it is usually necessary to devote a major part of the effort to follow-up investigations of cause-of-death. The productivity of these follow-ups is often marginal, limiting the reliability of the overall study.

MORBIDITY

Difficult as it may be to conduct good mortality studies, it is far more difficult, in most cases, to conduct studies involving other health effects criteria. While there is generally little significant variability in the definition of death, there is a great deal of variation in the diagnosis of many chronic diseases. There are variations between and within countries and states, and these are exacerbated by the differences in background and outlooks of the physicians making the individual diagnoses. Furthermore, there are some important chronic diseases which cannot be definitively diagnosed in vivo.

Many epidemiological studies rely on standardized health status questionnaires, and the success of these studies depends heavily on the design of the questionnaires. Of equal importance in many studies is the training and motivation of the persons administering the questionnaires.

Similar considerations apply to the measurement of functional impairment. The selection of the measurements to be used is very important; those functions measured should be capable of providing an index of the severity of the disease. Equally important here are the skills of the technicians administering these tests and their maintenance and periodic recalibration of the equipment.

Some studies try to avoid bias from the administrators of the questionnaires and functional tests by having the selected population enter the desired information themselves. They may be asked to make appropriate notations in notebook diaries, or to call a central station whenever they develop the symptoms of interest. Other investigations use nonsubjective indices such as hospital admissions, clinic visits, and industrial absenteeism as their indicators of the health effects to be associated with the environmental variables.

Dose-Response Relationships

Studies of the specific responses of biological systems to varying levels of exposure can provide a great deal of information on the nature of the responses, their underlying causes, and the possible consequences of various levels of exposure. However, it must be remembered that the data are

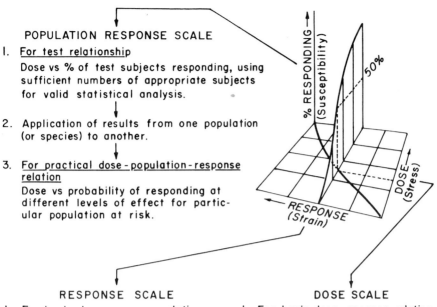

POPULATION RESPONSE SCALE

1. For test relationship
   Dose vs % of test subjects responding, using
   sufficient numbers of appropriate subjects
   for valid statistical analysis.

2. Application of results from one population
   (or species) to another.

3. For practical dose-population-response
   relation
   Dose vs probability of responding at
   different levels of effect for particular population at risk.

RESPONSE SCALE

1. For basic dose-response relation
   Quantitative scale of response, expressed in terms of most basic
   event at critical site; response
   increments of equal value over
   entire scale.

2. Consequent events, leading to ill-
   health.

3. For practical dose-response relation
   Magnitude of health risk (immediate
   or potential) to exposed individual.

DOSE SCALE

1. For basic dose-response relation
   Quantitative scale of dose at critical site in body, expressed in
   terms of active form of stress
   agent at site.

2. Intervening events between portal
   of entry into body and critical
   site, which mediate between external level of exposure and effective dose at critical site.

3. For practical dose-response relation
   Magnitude of exposure to external conditions that give rise to
   stress, measured in appropriate
   terms for translation into effective dose.

Fig. 2–13 Dose (population) response relationship with suggested distinction
between basic (toxicological) and practical (health) scales on the three axes. The
illustrative curve on the horizontal plane portrays the dose-response relationship
for the middle (50%) of the exposed population; the curve on the vertical plane
shows the percentages of population response of the indicated degree over the
whole range of doses. The vertical line from the dose scale indicates the magnitude of dose needed to produce the indicated degree of response at the 50% population level.

From: Hatch, T. F. Significant Dimensions of the Dose-Response Relationship. *Arch. Env. Health 16:*
571–578, 1968. © 1968, American Medical Association.

most reliable only for the conditions of the test, and for the levels of exposure which produced clear-cut responses.

Generally, in applying experimental data to low-level environmental exposure conditions, it is necessary to extrapolate back to doses which are orders of magnitude smaller than those which produced the effects in the test system. Since the slope of the curve becomes increasingly uncertain the further one extends it beyond the range of experimental data, the extrapolated effects estimate may be in error by a very large factor.

The basic dimensions of the dose-response relationship for populations are shown in Figure 2–13. There are many important factors which affect each of the basic dimensions.

FACTORS AFFECTING DOSE

The effective dose is the amount of toxicant reaching a critical site in the body. It is proportional to the concentrations available in the environment: in the air breathed, the water and food ingested, etc. However, the uptake also depends on the route of entry into the body and the physical and chemical forms of the contaminant. For airborne contaminants, for example, the dose to the respiratory tract depends on whether they are present in a gaseous form or as an aerosol. For contaminants which are ingested, uptake depends on transport through the membranes lining the gastrointestinal tract, which, in turn, depends on both aqueous and lipid solubilities.

For contaminants which penetrate membranes, reach the blood, and are transported systemically, subsequent retention in the body depends on their metabolism and toxicity in the various tissues in which they are deposited.

In all of these factors, there are great variations within and between species, and therefore great variations in effective dose for a given environmental level of contamination.

FACTORS AFFECTING RESPONSE

The response of an organism to a given environmental exposure can also be quite variable. It can be influenced by age, sex, the level of activity at the time of exposure, and the competence of the various defense mechanisms of the body. The competence of the body's defenses may, in turn, be influenced by the prior history of exposures to chemicals having similar effects, since those exposures may have reduced the reserve capacity of some important functions. The response may also depend on other environmental factors, such as heat stress and nutritional deficiencies.

REFERENCES

1. Gibbs, W. E. *Clouds and Smokes*. New York: Blakiston Co., 1924.
2. Friedlander, S. K. *Smoke, Dust and Haze*. New York: Wiley, 1977.
3. Joint AIHA-ACGIH Aerosol Hazards Evaluation Committee. Recommended Procedures for Sampling and Counting Asbestos Fibers. *Amer. Ind. Hyg. Assoc. J. 36*:83–90, 1975.
4. Lippmann, M. "Respirable" Dust Sampling. *Amer. Ind. Hyg. Assoc. J. 31*: 138–159, 1970.
5. Second Task Force for Research Planning in Environmental Health Science. *Human Health and the Environment—Some Research Needs*. DHEW Publ. #NIH 77–1277, 1977.
6. Committee on Food Protection, NRC. *Toxicants Occurring Naturally in Foods*, 2nd ed. Washington, D.C.: National Academy of Sciences, 1973.

3

Sources of Contaminants

INTRODUCTION

There are a variety of ways to classify the chemical contaminants found in the environment. They can be divided (a) into natural and anthropogenic; (b) by the medium in which they are dispersed — air, water, soil, food; (c) by specific elements and/or compound classes; (d) by their state of matter — solid, liquid, gas, or dispersion (colloid, suspension, etc.). In this chapter the contaminants are classified according to their sources, and it, therefore, becomes necessary to consider all physical and chemical forms and all media simultaneously. This approach provides the best opportunity to associate specific contaminants with the activity which generates them. Furthermore, it permits an appreciation of the total impact of a given sphere of activity and of the possible implications of alternative control strategies in those cases where current environmental contamination levels are excessive.

Some chemical contaminants have no known primary sources which could account for their observed concentrations. Ozone in urban air and chlorinated hydrocarbons in drinking water are examples of secondary contaminants formed by chemical reactions within the environment.

PRIMARY CONTAMINANTS

Natural Contaminants

Chemically pure air and water are not found in nature, since there is always an abundance of sources of both chemical and biologically active

Table 3.1 Sources and Concentrations of Atmospheric Trace Gases

Contaminant	Main anthropogenic source	Natural source	Estimated global emissions (tons/yr)		Atmospheric background concentration	Remarks
			Anthropogenic	Natural		
Sulfur dioxide (SO_2)	combustion of coal and oil	volcanoes	146×16^6	no estimate	0.2 ppb$_v$	—
Hydrogen sulfide (H_2S)	chemical processes, sewage treatment	volcanoes, biological action in swamps	3×10^6	100×10^6	0.2 ppb$_v$	Background conc. based on only one set of data
Carbon monoxide (CO)	auto exhaust, other combustion	forest fires, oceans, terpene reactions	304×10^6	33×10^6	0.1 ppm$_v$	Ocean contribution to natural source very small
Nitric oxide (NO); Nitrogen dioxide (NO_2)	combustion	bacterial action in soil (?)	53×10^6	NO: 430×10^6 NO_2: 658×10^6	NO: 0.2–2 ppb$_v$ NO_2: 0.5–4 ppb$_v$	Little data on natural sources
Ammonia (NH_3)	waste treatment	biological decay	4×10^6	1160×10^6	6–20 ppb$_v$	—
Nitrous oxide (N_2O)	none	bacterial action in soil	—	590×10^6	0.25 ppm$_v$	—
Hydrocarbons	combustion, chemical processes	biological processes	88×10^6	CH_4: 1.6×10^9 terpenes: 200×10^6	CH_4: 1.5 ppm$_v$ non-CH_4: < 1 ppb$_v$	"Reactive" hydrocarbon emissions from anthropogenic sources = 27×10^6 tons/year
Carbon dioxide (CO_2)	combustion	biological decay, release from ocean	1.4×10^{10}	10^{12}	330 ppm$_v$	Atmospheric conc. is increasing yearly

Adapted from: Robinson, E. R. and Robbins, R. C. Emissions, Concentrations and Fate of Gaseous Atmospheric Pollutants, pp. 1–93 in *Air Pollution Control*, Part II, W. Strauss (ed.). New York: Wiley-Interscience, 1972.

agents. The life cycles of plants generate pollens, the decomposition of vegetable and animal matter releases various organic and inorganic compounds, and the hydrologic cycle leaches metals from the earth's crust. Thus, unless the compound of interest is unknown in nature, it is generally difficult to apportion measured levels at any particular location to their natural and anthropogenic sources. Tables 3–1 and 3–2 show that, on a global basis, the emissions from natural sources for atmospheric particulates and most trace gases greatly exceed those from anthropogenic sources; the notable exceptions are the gases SO_2 and CO. However, in any given region, the local sources often have a much greater influence on concentrations, so the increment from natural sources of, for example, gases such as NH_3, NO/NO_2, and H_2S may be negligible. In order to appreciate when and to what extent natural sources are important to the observed environmental levels, it is necessary to discuss contaminants released by natural processes.

Table 3.2   Sources of Particles < 40 μm Diameter in the Atmosphere

Sources	Global emissions (10^6 metric tons/year)
Natural	
Direct Emission of Particles (primary contaminant)	
Soil and rock debris*	100–500
Forest fires and slashburning debris*	3–150
Sea salt	300
Volcanoes	25–150
Particles Formed from Gaseous Emissions (secondary contaminant)	
Sulfate from H_2S	130–200
Ammonium salts from NH_3	80–270
Nitrate from $NO + NO_2$	60–430
Hydrocarbons from plant exudations	75–200
Subtotal	773–2200
Anthropogenic	
Direct Emission of Particles	10–90
Particles Formed from Gaseous Emissions	
Sulfate from SO_2	130–200
Nitrate from $NO + NO_2$	30–35
Hydrocarbons	15–90
Subtotal	185–415
Total	958–2615

*Includes unknown amounts of indirect anthropogenic contributions.

From: *Study of Man's Impact on Climate (SMIC): Inadvertent Climate Modification.* Cambridge, Mass.: MIT © 1971. Reprinted by permission of the M.I.T. Press.

While many "natural" contaminants can be found in areas not significantly influenced by human activity, it does not necessarily follow that the observed levels are independent of human activity. A notable example is air contamination by ragweed pollen. Ragweed grows well in recently disturbed soil, but is at a competitive disadvantage in soils with established covers of natural vegetation. In the sense that it grows best on construction sites, abandoned fields, and urban lots, and coexists with cereal grains on some agricultural fields, it is more of an anthropogenic contaminant than a natural one.

VIABLE PARTICLES

A lengthy discussion of viable particles is beyond the scope of this book. However, since the prevalence of some, such as ragweed pollen, is often largely a function of human activity, and since their impact on human health is greater than all of the other air contaminants combined, they will be discussed briefly in order to place them in perspective with the nonviable chemical contaminants. The size ranges of viable particulates are shown in Table 3-3.

Hay fever (pollinosis) severely affects about 10% of the U.S. population, and has been estimated to cause more than 10 million days of lost time from work each year in the United States. Additional costs are attributed to the need for medical consultations and therapeutic drugs. While the dispersion of pollen is seasonal and most of the health effects disappear at

Table 3.3   Size Range of Viable Particulates

Particulate	Stokes' diameter* (μm)
Viruses	0.015-0.45
Bacteria	0.3-15
Fungi	3-100
Algae	0.5
Protozoa	2-10,000
Moss spores	6-30
Fern spores	20-60
Pollen grains (wind-borne)	10-100
Plant fragments, seeds, insects, other microfauna	100+

*The Stokes' diameter for a particle is the diameter of a sphere having the same bulk density and same terminal settling velocity as the particle.

Compiled from: Edmonds, R. L. pp. 6-11 in *Ecological Systems Approaches to Aerobiology,* W. S. Benninghoff and R. L. Edmonds (eds.). Ann Arbor, Michigan: University of Michigan, 1972.

the end of the season, some cases of untreated hay fever lead to the development of bronchial asthma.

Airborne microorganisms cause a substantial share of the infectious disease incidence, especially in indoor environments. Pathogenic bacteria, viruses, fungi, spores, and molds are responsible for a wide variety of diseases.

Spores of fungi can become airborne by active processes or by the action of wind, rain, or insects. Bacteria and viruses become airborne in aqueous aerosols formed by a variety of processes, including sneezing and coughing, spraying of sewage in treatment and disposal, and wind action on standing water in ponds and lakes.

The viable contaminants discussed here can, of course, also be ingested in drinking water or with contaminated food products and may produce similar effects following either inhalation or ingestion. The differences will be in the degree of respiratory tract and gastrointestinal tract involvement, the overall pathogenic potential, and the geographic pattern of disease incidence.

PRODUCTS OF PLANT AND ANIMAL METABOLISM

All living organisms create metabolic waste products which are discharged into their environments and normally recycled or neutralized in the immediate vicinity without significant impact on their own or the human environment. Difficulties usually arise when human activities alter the natural patterns of life cycles or the density of the populations. Examples are poultry farms and animal feed lots where the population density is so great that the volume of waste products cannot be effectively disposed of or recycled by natural processes. These problems will be discussed further under the topic of agricultural wastes.

The respiratory activity of animals and plants and the photosynthetic activity of green plants results in the exchange of oxygen and carbon dioxide. These processes are in approximate balance on a global basis.

On a local scale, there is a considerable diurnal variation in CO_2 concentration. It is taken in by plants during daylight at a rate dependent upon the type and density of vegetation, the amount of solar radiation, and the CO_2 concentration. It is released to the atmosphere from decomposing organic material, largely via bacterial activity in the soil. The release rate is dependent on the type of soil, its moisture content, and its temperature. Above a wheat field in sunny weather, for example, the concentration of CO_2 near the ground can reach 500 ppm_v. Much of the CO_2 released is taken up by plants just above the ground and is thus short-circuited. The

CO_2 in soil air can be two orders of magnitude greater than in the surface air.

Vegetation is also responsible for the release of substantial quantities of hydrocarbons. Perhaps the most significant of these are a class known as terpenes. Although released as vapors, they readily condense in the atmosphere to form fine particles. A fairly concentrated light-scattering aerosol may be formed over areas where terpenes are released in significant quantities, such as coniferous forests. These aerosols can be the most characteristic feature of the region, as in the Great Smoky Mountains of North Carolina. Estimated terpene emissions are presented in Table 3 – 1.

PRODUCTS OF ORGANIC DECOMPOSITION

All living things eventually die, and their tissue constituents are recycled to the environment. Some decomposition takes place by physical processes such as evaporation or incineration. However, in the natural state, most organic matter is utilized as an energy and materials source by a succession of microorganisms, and is eventually broken down into simple and stable inorganic chemical entities, such as water and carbon dioxide.

When proteins are broken down in nature by aerobic microorganisms, the process is known as decay. Oxidation continues until stable end products are produced, and there are no odorous gases. The nitrogen in the protein undergoes conversion to ammonium, nitrite, and, finally, nitrate forms. The sulfur compounds are converted to sulfates.

In anaerobic decomposition, or putrefaction, the process of oxidation is incomplete, and the breakdown products include many unstable and odorous gases. The initial products are organic acids, acid carbonates, carbon dioxide, and hydrogen sulfide. Intermediate products include ammonia, acid carbonates, carbon dioxide, and sulfides, including mercaptans. The final products include carbon dioxide, humus ammonia, methane, and hydrogen sulfide. Ammonia, hydrogen sulfide, and the mercaptans have strong and unpleasant odors. Hydrogen sulfide is toxic to many animals, including man. Methane is odorless and nontoxic, but is combustible and can build up to explosive concentrations in underground air spaces.

EROSION

Wind action on land can resuspend settled dust and can erode materials from surfaces. Similarly, wave action and flowing waters can erode the surfaces over which they pass, increasing the dissolved and suspended particulate burdens of the water. While these are natural processes, they can be greatly assisted by human activities. It is much easier to suspend tilled or trampled soil than soil with a natural cover of vegetation, and tilled soil may also be much more susceptible to erosion by storm water. Erosion can also take place with man-made wind or water motion, such as that created by high-speed vehicles, like cars, trucks, and motorboats. Mining and construction activities can also result in heavy local concentrations of dust.

FIRE

Lightning strikes and spontaneous combustion of organic matter have always caused and continue to cause fires in forest and brush lands. Such fires generate carbon monoxide, carbon dioxide, and a wide variety of organic materials which are products of incomplete combustion and/or pyrosynthesis. While most forest fires are attributable to humans in recent years, human intervention has also limited the spread of fires started by natural causes.

SEA SPRAY

The breaking of waves in the oceans creates aqueous aerosols. Most of the droplets fall back into the water, but many of the smaller ones remain airborne long enough for the water to evaporate. A large proportion of the resulting salt particles are sufficiently small to remain airborne for considerable times and distances.

NATURAL RADIOACTIVITY

Radioactive isotopes, or radionuclides, are continually being released to the atmosphere from two sources. One is the interaction of cosmic rays with atmospheric gases to produce tritium (^3H), carbon-14 (^{14}C), and a variety of other nuclides; these are listed in Table 3–4. Carbon-14, with a half life ($T_{1/2}$) of 5,730 years, is produced from nitrogen-14, and is continually incorporated into the biota. The proportion of ^{14}C to ^{12}C, therefore, indicates the elapsed time since the carbon was incorporated into living tissue, and has been used as an index of the age of archeological specimens.

In addition to the nuclides produced continually by cosmic radiation, there are a variety of radionuclides which have been in the lithosphere since it was formed. The ones of greatest interest to man are uranium-235 (^{235}U, $T_{1/2} = 7.1 \times 10^8$ years), uranium-238 (^{238}U, $T_{1/2} = 4.5 \times 10^9$ years), thorium-232 (^{232}Th, $T_{1/2} = 1.4 \times 10^{10}$ years), and potassium-40 (^{40}K, $T_{1/2} = 1.26 \times 10^9$ years). Some radionuclides, e.g., potassium-40, decay directly to stable daughter isotopes. Others, e.g., ^{235}U, ^{238}U, and ^{232}Th, decay through complicated chains and have numerous daughters with varying half-lives

Table 3.4   Natural Radionuclides Produced by Cosmic Rays

Radionuclide	Half-Life	Concentration in troposphere (pCi/kg air)*
^{3}H	12.3 years	3.2×10^{-2}
^{7}Be	56.3 days	0.28
^{10}Be	2.5×10^6 years	3.2×10^{-8}
^{14}C	5730 years	3.4
^{22}Na	2.6 years	3.0×10^{-5}
^{24}Na	15.0 hours	—
^{32}Si	~650 years	5.4×10^{-7}
^{32}P	14.3 days	6.3×10^{-3}
^{33}P	24.4 days	3.4×10^{-3}
^{35}S	88 days	3.5×10^{-3}
^{36}Cl	3.1×10^5 years	6.8×10^{-9}
^{38}S	2.9 hr	—
^{38}Cl	37.3 min	—
^{39}Cl	55.5 min	—

*The basic unit of activity of a radionuclide, i.e., the number of atoms which disintegrate per unit time, is the curie, Ci. One curie equals the amount of a radioactive material which undergoes 3.7×10^{10} disintegrations per second; 1 pCi or picocurie equals 0.037 disintegrations/sec.

Compiled from: Perkins, R. W. and Nielson, J. M. Cosmic-Ray Produced Radionuclides in the Environment. *Health Physics, 11*:1297–1304, 1965.

and emissions. Each chain includes several α-emitters, which are of special concern in terms of human carcinogenesis.

Certain radionuclides are large contributors to natural atmospheric radioactivity via formation of radioactive noble gases which diffuse from the earth's crust.

When radium-226 (^{226}Ra), a member of the ^{238}U chain, decays, it becomes radon gas (^{222}Rn). Similarly, radium-224 (^{224}Ra), a member of the ^{232}Th chain, decays to thoron gas (^{220}Rn), another isotope of the element radon. These gases can diffuse into the atmosphere out of the solid matrix holding the parent isotopes. The ^{222}Rn, with a half-life of 3.8 days, is released to a greater extent than the ^{220}Rn, with its much shorter half-life of 54 seconds. The daughter products of radon and thoron decay, which are also radioactive, are molecular-sized particles and rapidly diffuse onto available surfaces, primarily other airborne particles. Since most of these airborne particles are very small and remain in the atmosphere for a relative long time, these daughter products can stay airborne for significant time intervals.

The concentration of radon daughters in the atmosphere is quite variable. It depends on the strength of the local sources, i.e., the concentration of radium in the soil and/or building materials, the porosity of the matrix containing the radium, and meteorological factors. Very high concentrations can be found in enclosed spaces such as mines, caves, and the basements of homes, schools, and commercial buildings where the underlying soil and/or building materials have relatively high radium contents. On a population basis, radon and daughters account for 50 to 90% of the total background radiation exposure to the lungs.

VULCANISM

Volcanic eruptions occur relatively infrequently and at irregular intervals; those with significant releases to the troposphere average a few each year. Even less frequently, perhaps once or twice per century, there is a major eruption which occurs with sufficient energy to inject ash and sulfur dioxide into the stratosphere. The eruptions of Krakatoa (1883) and Mt. Agung (1963) put enough light-scattering particles into the stratosphere to cause readily observable optical effects which persisted for several years. The sulfur released by volcanoes represents a significant fraction of the total emissions to the atmosphere.

Anthropogenic Contaminants

FOSSIL FUEL COMBUSTION FOR POWER PRODUCTION AND SPACE HEATING

Approximately 93% of the current U.S. energy demand is derived from the combustion of fossil fuels, i.e., natural gas, oil, and coal. Of this, about 8% is used as feedstock or raw material for production of products, and 26% is used in transportation. The remaining 59% is burned for space heating (heating of buildings) and to produce electricity. In future years, the consumption of coal will, most likely, increase substantially, primarily for electric power production; oil consumption will increase to a smaller extent, and natural gas consumption will remain the same or possibly decrease slightly.

The fossil fuels contain varying mixtures of hydrocarbons and minerals. Natural gas, as it is commercially distributed, has negligible amounts of sulfur, nitrogen, and noncombustible mineral ash, but may contain noble gases, including radon, and radon daughters. Components of petroleum are classified according to their range of boiling temperatures at atmospheric pressure. The lighter fractions of oil (lower boiling points) will also be low in nitrogen, sulfur, and ash. Coal and the heavier fractions of the oil will have mineral contents depending on their source and the amount of pretreatment, if any, they receive. Coal typically contains about 10% mineral ash.

It is possible to approach complete combustion of fossil fuels in stationary furnaces. However, this becomes increasingly difficult with the increasing viscosity of fuel oil and the increasing content of volatiles (decreasing fixed carbon content) in coal. The difficulty lies in achieving a sufficiently prolonged contact between the oil droplet or coal surface and the oxygen in the flame. If combustion of a particle is not completed within the high temperature zone of the flame, there will be unburned and/or incomplete products of combustion in the effluent.

In an efficient flame, most of the carbon in the fuel will be oxidized to carbon dioxide (CO_2), the hydrogen to water vapor (H_2O), and the sulfur to sulfur dioxide (SO_2). The nitrogen will be released as nitrogen (N_2), nitric oxide (NO), and nitrogen dioxide (NO_2). The amounts of NO and NO_2 are quite variable, and their summed concentration is usually expressed as NO_x. The mineral content of the fuel is converted to an ash composed of mixed oxides. The more volatile materials, such as oxides of lead, cadmium, and mercury, and the radium and polycyclic aromatic hydrocarbons will be released as vapors, but will condense as, or on, fine particulates in the stack. Most of the mass of the ash will fall to the bottom of the furnace (bottom ash), while the fine particles will rise up the stack with the combustion gases (fly ash). Coal-fired utility furnaces generally use various air-cleaning devices to collect most of the fly ash. Table 3–5 presents typical emissions from electric generating plants using the different types of fossil fuels.

There is generally much more NO_x in the furnace effluent than could be accounted for by the nitrogen content of the fuel. The excess is formed in the flame by fixation of the N_2 in the air supply, a process whose rate increases rapidly with rising flame temperature. The reaction between atmospheric N_2 and O_2 to form NO is reversible; if the products of combustion were allowed to cool gradually, the NO would decompose. However, since most large furnaces are designed to permit a rapid extraction of heat for useful work, the gases are rapidly quenched, and the NO does not decompose. Beginning in the furnace and stack, and continuing in the atmosphere, it undergoes conversion, instead, to NO_2, a brownish-colored and much more toxic gas.

The fate and effects of the effluents from fossil fuel combustion depend on how they are released. The large utility furnaces may discharge the effluents through high stacks, diminishing their maximum ground-level

Table 3.5   Typical Emissions from Electric-Power Generating Plants Using Different Types of Fossil Fuels

| Contaminant | Emissions (10^3 metric tons/year for a 1000 MWe plant) | | |
	Coal-fired[a]	Oil-fired[b]	Gas-fired[c]
Carbon monoxide	0.52	0.008	—
Nitrogen oxides	21	21.7	12
Sulfur oxides	139	52.6	0.012
Aldehydes	0.005	0.117	0.031
Hydrocarbons	0.21	0.67	—
Particulates	4.5	0.73	0.46

a. Assumed sulfur content of 3.5% (15% of which remains in the ash), ash content of 9%; plant-pollution-control equipment able to remove 97.5% of the fly ash; no other pollution controls.
b. Assumed 1.6% sulfur content and 0.05% ash; no pollution-control equipment.
c. No pollution-control equipment.

Compiled from: Terrill, J. G. Environmental Aspects of Nuclear and Conventional Power Plants. In: *Selected Materials on Environmental Effects of Producing Electric Power*. Washington, D.C.: U.S. Government Printing Office, 1969.

concentrations, but increasing their residence times in the atmosphere and increasing their dispersion. Space-heating furnaces usually discharge their effluents at or near roof level. The maximum ground-level concentrations can, therefore, be higher, and thus any effects they produce are more likely to be localized.

A potential water contaminant produced by electric power generating plants is heat. Spent steam which leaves the turbine passes through a condenser, where it is cooled and condensed; the water is then returned to the boiler. The simplest cooling method is to allow cold water from a river or stream to pass through pipes in the condenser. This allow heat exchange between the steam and this cooling water. In a once-through system using river water (the cooling water passes through the condenser only once before returning to the waterway), the discharge water can be much warmer than the intake, affecting the ecology of the river. In some cases, there is inadequate river flow, and structures known as cooling towers are used to discharge the heat to the atmosphere via evaporative cooling. The mists generated by such towers may produce fogs, affecting visibility and traffic safety. These fog droplets will also contain traces of the toxic chemicals which are used to prevent algae growth within the tower.

The potential for thermal pollution is greater in nuclear-power electric-generating plants, since they release about 15% more heat per unit of electricity generated than do fossil-fuel facilities.

TRANSPORTATION

In a highly mobile society, such as ours, a great deal of time and energy is consumed in transporting both goods and people. We purchase products which originate in distant states and countries. The manufacturers, in turn, obtain their raw materials from both local and distant sources. We frequently commute to work at locations far from our residences, and many of us travel great distances on business or vacation trips. As a result, a major portion of our total energy consumption can be attributed to that required for movement.

In consuming energy, there is always waste. For transportation, or moving sources, these are heat, fuel spillage, exhaust products, and various by-products.

Fuel combustion which occurs in motor vehicles is generally incomplete, and the exhaust wastes include CO_2, H_2O, SO_2, NO, CO, and a host of organic compounds. They also include the oxidation products of fuel additives, such as tetraethyl lead. Some larger ocean-going ships have nuclear propulsion systems, and electrified land transport often uses energy derived, in part, from nuclear power. Waste products of nuclear-power production include various fission products. This topic is discussed more fully in a later section of this chapter.

The use of transportation systems creates additional environmental contamination. Vehicular motion resuspends settled particles. The vehicles rust and wear, spreading iron oxide, rubber particles from tires, and asbestos fibers from clutch and brake linings, along their paths. They leak or vaporize engine oil and other working fluids, elevate noise levels, and contribute to environmental litter either through the actions of their occupants or by the abandonment of the vehicles or parts thereof.

Finally, transportation vehicles alter the environment in very significant, but less obvious, ways. Land transport vehicles need roads, parking areas, and service stations. Major fractions of our urban land surfaces are covered with concrete and asphalt to accomodate the needs of wheeled transport. The mobility which autos provide has led to the rapid growth of the suburbs which, by and large, are less energy-efficient than cities, and has thereby increased the demand and costs of energy for heat and power usage.

AUTOMOBILE EXHAUST

The greatest single source of air contamination in the United States is automobile exhaust. Prior to 1968, there was no intentional control of tail pipe emissions, except in California. With the constantly increasing numbers of cars in use, the levels of carbon monoxide (CO), hydrocarbons (HC), nitrogen oxides (NO_x), and lead (Pb) in urban air were high enough to cause concern among health authorities. This led to the passage of the Federal Clean Air Act of 1970.

The Act mandated a series of performance goals for automobile engine emissions, with the objective of reducing emissions of CO, HC, and NO_x to the point that they would no longer cause the air quality standards established by the Act to be exceeded. The major auto manufacturers claimed that they could only meet the CO and hydrocarbon emissions limits by the use of catalytic converters in the exhaust system, and their position was accepted by the U.S. Environmental Protection Agency (EPA). Since the catalytic converters are poisoned by lead, which is used as an antiknock additive, they could only be used with unleaded fuel. Thus, lead has been eliminated from a large fraction of the fuel supply.

Catalytic converters increase the extent of oxidation of the hydrocarbons and CO in the fuel; however, they also increase the oxidation state of the sulfur in fuel. The concentration of sulfur in gasoline is relatively low, about 0.03% by weight. Without converters, essentially all of the sulfur is emitted from the tailpipe as SO_2. The total amount of SO_2 so released was not generally considered to be a significant problem, since considerably larger amounts are released by fossil fuel combustion in stationary (nonmoving) sources. In passing through the catalytic converter, some of the SO_2 in the engine exhaust is oxidized to SO_3, which is rapidly hydrolyzed by water vapor in the exhaust to form a sulfuric acid mist (H_2SO_4). Thus, the control device is a source of H_2SO_4, and may be the major source for this highly irritating chemical for passengers in catalyst-equipped cars and pedestrians near heavily travelled roadways.

In modifying the engine designs to reduce emissions, sacrifices had to be made in performance and fuel economy. This is shown in Figures 3–1 and 3–2. Without emission controls, engines generally operated with air/fuel ratios between 12:1 and 15:1, where power is maximal. With increasing ratios, the CO and HC concentrations drop rapidly, but power and fuel economy suffer. Also, emissions of NO_x increase.

In practice, most of the reduction in emissions between 1970 and 1974 was achieved at the expense of fuel economy, as illustrated in Figure 3–2. Energy consumption has, of course, become a factor of increasing concern since 1973, when energy costs rose so sharply, and is exerting a counter pressure affecting the timing and achievement of the original emissions standards.

The preceding discussion applies to the conventional Otto cycle spark-

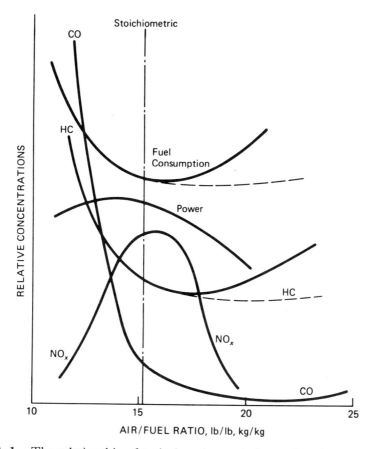

Fig. 3–1   The relationship of typical engine emission and performance to air: fuel ratio. Solid line: conventional engine; dashed line: lean-burn engine.

From: Olson, D. R.: The Control of Motor Vehicle Emissions. Pp. 595–654, in: *Air Pollution*, 3rd Ed., vol. 4, A. C. Stern (ed.), New York: Academic Press, 1977.

ignition internal-combustion engine used in most current automobiles. It has maximum efficiency, and minimal emissions, only when running at constant power. Thus, emissions increase greatly during other operating modes, such as idling, acceleration, and deceleration. While other combustion engine designs could be used, and may be superior with respect to emissions (1), and while electric engines would have essentially no engine exhaust emissions, most of them cannot compete effectively in terms of cost, range, or performance with the conventional Otto cycle engines at the present time. The only engines currently gaining an appreciable share

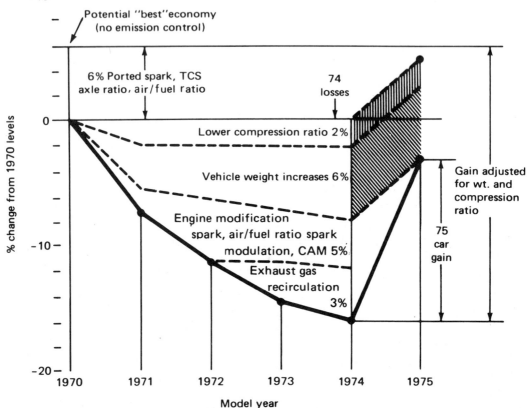

Fig. 3–2 Fuel economy trends [changes (gain/losses) in General Motors city-suburban schedule economy]. Fuel economy has deteriorated as a result of weight increases and emission controls. Use of the catalytic converter for exhaust-emission control reversed that trend for 1975. Analysis of fuel-economy changes: (a) 1975 gain results from engine optimization; (b) 1975 adjusted to 1970 weight and compression ratio approaches 1970 best economy.

From: Olson, D. R. The Control of Motor Vehicle Emissions. pp. 595–654 in *Air Pollution*, 3rd Ed., vol. 4, A. C. Stern (ed.), New York: Academic Press, 1977. Adapted from: Starkman, E. S. Emission Control and Fuel Economy. *Env. Sci. Tech.*9:820–24, 1975.

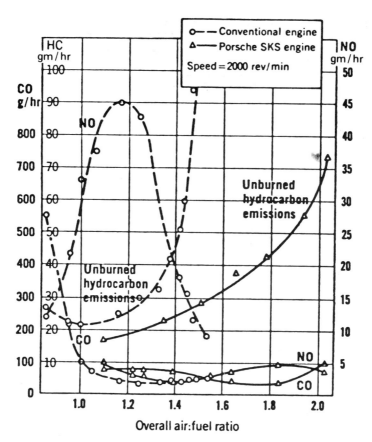

Fig. 3–3 Comparison of combustion products for the Porsche auxiliary chamber stratified charge engine and conventional engines.

From: Olson, D. R. The Control of Motor Vehicle Emissions, pp. 595–654 in *Air Pollution*, 3rd Ed., vol. 4, A. C. Stern (ed.). New York: Academic Press, 1977. Adapted from: Garrett, T. K. Porsche Stratified Charge Engine. *Env. Sci. Tech.*9:826–30, 1975.

of the market are the diesel cycle and the stratified charge version of the Otto cycle, and both are similar in many respects to the conventional Otto cycle type; their emissions are generally lower, but have similar characteristics. Figure 3–3 and Table 3–6 compare emissions from conventional, diesel, and stratified charge engines.

AIRCRAFT EXHAUST

The black plume exhausts of the early commercial jetliners have been essentially eliminated in the newer engines as well as the retrofitted older

Table 3.6   Emission Factors for Transportation Sources*

Contaminant	Source		
	Automobiles	Diesel engines	Jet aircraft
	(mean emission rate in pounds per 1000 gallons of fuel consumed)		(emission rate in pounds per flight at altitude of 35,000 ft)
Carbon monoxide	2300	60	20.6
Nitrogen oxides	113	222	23
Sulfur oxides (as SO_2)	9	40	–
Aldehydes (as HCHO)	4	10	4
Hydrocarbons	200	136	19
Organic acids (acetic)	4	31	–
Particulates	12	110	34

*Emission factor is a parameter commonly used to indicate how much of a contaminant is released for a given amount of fuel consumed.

Compiled from: Office of Air Programs, EPA. *Compilation of Air Pollutant Emission Factors.* Publication No. AP–42, U.S. EPA, Research Triangle Park, N.D., 1972.

engines. However, engine emissions during idling can be a serious problem at airports, as seen in Figure 3–4. Aircraft exhaust products are presented in Table 3–6.

All combustion engines, including aircraft jets, generate nitric oxide. The release of NO by jets in the stratosphere has raised the possibility of significant destruction of ozone by interaction with the NO. The fear of a possible reduction in stratospheric ozone, with a projected subsequent increase in human skin cancer, was a significant factor in the decision of the U.S. Congress not to subsidize the production of a U.S. supersonic transport. The role of NO in O_3 destruction is discussed further in Chapter 7.

WATER CONTAMINATION BY BOATS AND SHIPS

A major problem in aquatic environments is oil contamination. There has been considerable attention paid to oceanic oil spills resulting from the breakup of deep-draft oil tankers. However, the oil contamination of the oceans by more mundane sources such as bilge flushing and leakage may have greater overall effects. Within coastal and inland waterways, leakage and inadvertent spills create most of the problems related to oil. Reported oil spills in U.S. waters alone in 1975 amounted to about 24 billion gallons from 10,500 incidents (2).

Another major problem in inland waters is the sanitary wastes which are

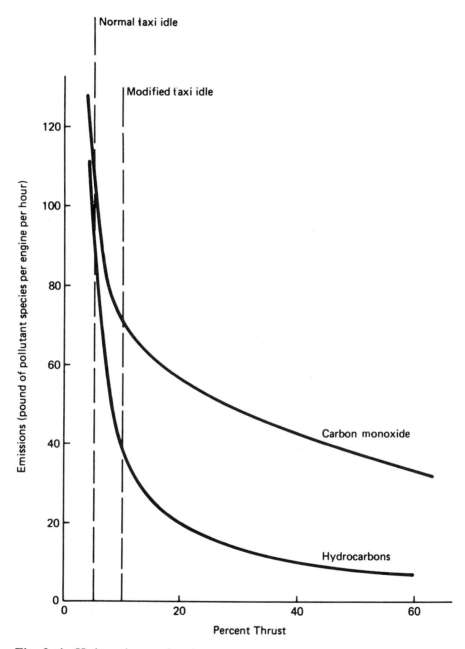

Fig. 3–4 Hydrocarbon and carbon monoxide emissions from a typical aircraft turbine engine.

From: Olson, D. R. The Control of Motor Vehicle Emissions, pp. 595–654 in *Air Pollution*, 3rd Ed., vol. 4, A. C. Stern (ed.), New York: Academic Press, 1977.

discharged directly into the water by commercial and pleasure boats. This problem should diminish through the enforcement of holding tank regulations by the Coast Guard.

INDUSTRIAL OPERATIONS

Industrial operations and processes are extremely varied and complex. They involve a diversity of materials in a variety of forms under a broad range of conditions, making it virtually impossible to provide more than a superficial summary of contaminant releases. Tables 3−7 and 3−8 show the diverse types of air and water contaminants which may result from various industrial operations. The contaminants may evolve from the raw materials or processing agents, or can occur during manufacture, shipping, or storage of raw or processed products.

Detailed information is available on contaminant releases for most manufacturing industries. Specific industries vary greatly in the types and amounts of their effluents. The approach to contamination from industrial operations which has been adopted for this chapter is to illustrate the extent and diversity of the sources through a brief discussion of environmental releases of chemicals from three specific industries.

PETROLEUM REFINING

The petrochemical industry encompasses processes which range from the manufacture of simple compounds to the production of very complex plastics, all of which are potential sources of contaminant release. Our discussion will be limited to the refining of petroleum.

Crude oil, extracted from subsurface wells, is the raw material for a large fraction of our total energy consumption. Most of it is refined into (a) products burned as gasoline or diesel fuel in internal combustion engines, (b) light fuel oils used in domestic and industrial furnaces, or (c) residual fuel oils used in firing large boilers in power plants, industry, and large housing complexes. Approximately 8% is not burned and is converted into petrochemical products, such as plastics, pesticides, and solvents.

Much of the world's crude oil is refined in the United States. In 1974, there were 247 petroleum refineries in the United States, with a total capacity of 14.2 million barrels of crude per day. They represented 27% of the world's total refining capacity.

Crude oil is a mixture of hydrocarbons of varying chain lengths and varying contents of sulfur, oxygen, nitrogen, and trace elements. In its concept and early practice, refining crude oil into commercial products was relatively simple. When oil is heated to a temperature sufficient to

Table 3.7 Major Air Contaminants from Some Industrial Sources

Chemical industry	Synthetic rubber	Construction industry
Ammonia plant	Alkanes	Asphalt roofing
Carbon monoxide	Alkenes	Oil mists
Ammonia	Ethanenitrile	Benzo[a]pyrene
Chlorine plant	Carbonyls	Asbestos
Chlorine gas		Carbon monoxide
Mercury	Metallurgical industry	Brick
Hydrofluoric acid	Aluminum ore reduction	Fluorides
Hydrogen fluoride	Hydrogen fluoride	Sulfur dioxide
Silicon tetrafluoride	Particulate fluoride	Calcium carbide
Sulfur dioxide	Carbon, alumina	Carbon monoxide
Nitric acid manufacture	Copper smelters	Acetylene
Nitric oxide	Carbon monoxide	Sulfur oxides
Nitrogen dioxide	Sulfur oxides	Cement
Paint, varnish	Nitrogen oxides	Various kinds of dust
Aldehydes	Cadmium	Chromium
Ketones	Iron-steel	Ceramic and clay processes
Phenols	Carbon monoxide	Fluorides
Terpenes	Sulfur oxides	Silicates
Glycerines	Iron oxides	Ammonia
Petroleum refinery	Fluorides	Frit (glazing enamel)
Hydrogen sulfide	Nickel carbonate	Fluorides
Selenium	Silicates	Silica
Fluorides	Graphite	Boron
Hydrocarbons	Lead and zinc smelters	Glass
Phosphoric acid	Sulfur dioxides	Chlorine
Silicon tetrafluoride	Fluorides	Fluorides
Hydrogen fluoride	Cadmium	Sulfur oxides
Phthalic anhydride	Magnesium smelters	Nitrogen oxides
Hexane	Fluorides chlorine	Carbon monoxide
Maleic anhydrides	Barium oxide	
Printing ink	Secondary metals industry	
Acrolein	Nitrogen oxides	
Fatty acids	Metal oxides	Food
Phenols	Hydrochloric acid	Coffee roasting
Terpenes	Brass and bronze smelters	Smoke
Sulfuric acid	Zinc oxide	Odors
Sulfur dioxide	Lead oxides	Fish-meal processing
Sulfur oxides	Secondary aluminum smelters	Hydrogen sulfide
Nitrogen oxides	Fluorides, Chlorides, Ozone,	Trimethylamine
	Numerous metals	

Compiled from: Office of Air Programs, EPA, *Compilation of Air Pollutant Emission Factors.* Publication No. AP-42, U.S. EPA, Research Triangle Park, N.C., 1972.

Table 3.8 Major Water Contaminants from Some Industrial Sources

Industry	Origin of contaminants	Components and characteristics of contaminants
Food		
Canning	Fruit and vegetable preparation	Colloidal, dissolved organic matter, suspended solids
Dairy	Whole milk dilutions, buttermilk	Dissolved organic matter (protein, fat, lactose)
Brewing, distilling	Grain, distillation	Dissolved organics, nitrogen fermented starches
Meat, poultry	Slaughtering, rendering of bones and fats, plucking	Dissolved organics, blood, proteins, fats, feathers
Sugar beet	Handling juices, condensates	Dissolved sugar and protein
Yeast	Yeast filtration	Solid organics
Pickles	Lime water, seeds, syrup	Suspended solids, dissolved organics, variable pH
Coffee	Pulping and fermenting beans	Suspended solids
Fish	Pressed fish, wash water	Organic solids, odor
Rice	Soaking, cooking, washing	Suspended and dissolved carbohydrates
Soft drinks	Cleaning, spillage, bottle washing	Suspended solids, low pH
Pharmaceutical		
Antibiotics	Mycelium, filtrate, washing	Suspended and dissolved organics
Clothing		
Textiles	Desizing of fabric	Suspended solids, dyes, alkaline
Leather	Cleaning, soaking, bating	Solids, sulfite, chromium, lime, sodium chloride
Laundry	Washing fabrics	Turbid, alkaline, organic solids
Chemical		
Acids	Wash waters, spillage	Low pH
Detergents	Purifying surfactants	Surfactants
Starch	Evaporation, washing, etc.	Starch
Explosives	Purifying and washing TNT, cartridges	TNT, organic acids, alcohol, acid, oil soaps
Insecticides	Washing, purification	Organics, benzene, acid highly toxic
Phosphate	Washing, condenser wastes	Suspended solids, phosphorus, silica, fluoride, clays, oils, low pH
Formaldehyde	Residues from synthetic resin production and dyeing synthetic fibers	Formaldehyde

Table 3.8 cont.

Industry	Origin of contaminants	Components and characteristics of contaminants
Materials		
Pulp and paper	Refining, washing, screening of pulp	High solids, extremes of pH
Photographic products	Spent developer and fixer	Organic and inorganic reducing agents, alkaline
Steel	Coking, washing blast furnace flue gases	Acid, cyanogen, phenol, coke, oil
Metal plating	Cleaning and plating	Metals, acid
Iron foundry	Various discharges	Sand, clay, coal
Oil	Drilling, refining	Sodium chloride, sulfur, phenol, oil
Rubber	Washing, extracting impurities	Suspended solids, chloride, odor, variable pH
Glass	Polishing, cleaning	Suspended solids

From: Higgins, I. J. and Burns, R. G. *The Chemistry and Microbiology of Pollution*. New York: Academic Press, 1975.

vaporize most of the hydrocarbons, and the vapors are directed into a temperature-controlled column, the vapors will condense at different points along the column according to their individual boiling points. Such a "straight-run" distillation yields products defined by boiling-temperature ranges and, in decreasing order of volatility, are given names such as wet gas, gasoline, kerosine, fuel oil, middle distillate (including diesel fuel), lube distillate, and heavy bottom. A schematic design of such a simple refinery is shown in Figure 3–5. Most of the sulfur and trace elements remain in the residual oil at the bottom of the column.

The "straight-run" distillation products may not match the consumer demand in terms of quantity or quality, and refineries have accordingly become considerably more complex. When a straight-run product is available in excess of its demand, it is reformed into a needed product, and when there is insufficient product, it is synthesized. The industry has developed a variety of techniques for splitting, rearranging, and combining petroleum molecules to form products meeting customer and/or governmental specifications and a product mix giving maximum economic returns. A simplified schematic diagram of a typical modern refinery is shown in Figure 3–6.

The schematic indicates that crude oil is the only input, and that the commercial products listed on the right side of the diagram are the only

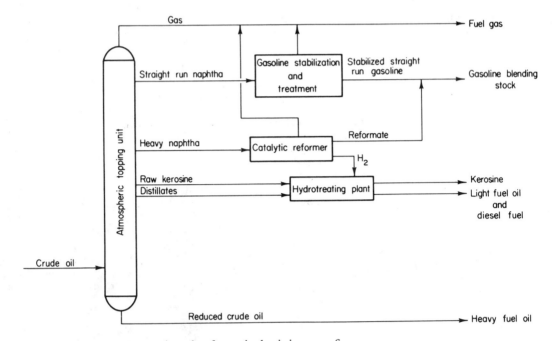

Fig. 3–5   Processing plan for typical minimum refinery.
From: Elkin, H. F. Petroleum Refining, pp. 813–43 in *Air Pollution*, 3rd Ed., vol. 4, A. C. Stern (ed.). New York: Academic Press, 1977.

output. In practice, however, there are additional inputs and many waste products are vented into the atmosphere and discharged in waste water.

A refinery requires large volumes of steam for heating and distillations, and process water for cooling condensers and process extractions. The various treatment steps require the use of a variety of caustics, acids, and special solvents. All of these can, in part, end up in the waste water, along with process spills and leakages of the various hydrocarbon products and intermediates. The process units of refineries are essentially all outdoors, so that storm-water drainage from the refinery becomes contaminated with materials which have been leaked or spilled into exterior surfaces. Storm and waste waters are, however, generally treated to meet applicable effluent standards for BOD, COD, pH, acidity, alkalinity, suspended solids, oil, phenol, and sulfide.

Spills and leaks of liquids also contribute to air contamination if they are volatilized. In addition, there may be leakages of gases and vapors directly into the atmosphere from process equipment. Total hydrocar-

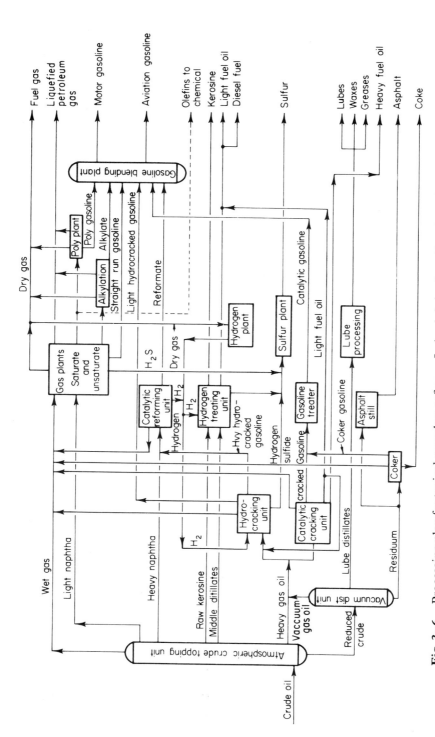

Fig. 3–6 Processing plan for typical complete refinery. Lube: lubricating oil; poly: polymerization.

From: Elkin, H. F. Petroleum Refining, pp. 813–43 in *Air Pollution*, 3rd Ed., vol. 4, A. C. Stern (ed.). New York: Academic Press, 1977.

Table 3.9   Potential Sources of Specific Contaminants from Oil Refineries

Emission	Potential sources
Oxides of sulfur	Boilers, process heaters, catalytic cracking unit regenerators, treating units, H_2S flares, decoking operations
Hydrocarbons	Loading facilities, turnarounds, sampling, storage tanks, waste-water separators, blowdown systems, catalyst regenerators, pumps, valves, blind changing, cooling towers, vacuum jets, barometric condensers, air blowing, high-pressure equipment handling volatile hydrocarbons, process heaters, boilers, compressor engines
Oxides of nitrogen	Process heaters, boilers, compressor engines, catalyst regenerators, flares
Particulate matter	Catalyst regenerators, boilers, process heaters, decoking operations, incinerators
Aldehydes	Catalyst regenerators
Ammonia	Catalyst regenerators
Odors	Treating units (air-blowing, steam-blowing), drains, tank vents, barometric condenser sumps, waste-water separators
Carbon monoxide	Catalyst regeneration, decoking, compressor engines, incinerators

From: Elkin, H. F. Petroleum Refining, pp. 813–43 in *Air Pollution*, vol. 4, A. C. Stern (ed.). New York: Academic Press, 1977.

bon losses may range from about 0.1 to 0.6% of crude throughout, depending on the complexity of the refining and the efforts expended in controlling losses. Some of the crude input ends up within the refinery as fuel to fire boilers and process heaters.

Tables 3–9, 3–10, and 3–11 show the kinds and rates of contaminant emissions into the atmosphere from various sources within refineries.

PRIMARY ALUMINUM PRODUCTION

Various metallurgical processes, especially primary smelting, which is the removal of metal from its ore, have long been associated with the release of chemical contaminants. As shown in Table 3–7, the main contaminants are those associated with the specific ore and smelting process.

Aluminum is the most abundant metallic element in the earth's crust. Its combination of mechanical strength, resistance to corrosion, high electrical and thermal conductivities, and relatively low density make it extremely useful for structural elements, utensils, and electric power lines. Despite its abundance, aluminum has come into widespread use only within the last century. It is not found naturally as the pure metal, and is produced by the electrolytic reduction of alumina (Al_2O_3) in a bath of molten cryolite (Na_3AlF_6).

Table 3.10 Emission Factors Developed from Los Angeles, California Survey

Combustion sources	Units of emission factor Pounds per	Emission factors					
		Hydro-carbons	Aldehydes as HCHO	Carbon monoxide	NOx as NO2	Ammonia as NH3	Particulate matter
Boilers and process heaters	Bbl of fuel oil burned	0.14	0.025	Neg.	2.9	Neg.	0.8
Boilers and process heaters	1000 ft³ of gas burned	0.026	0.0031	Neg.	0.23	Neg.	0.02
Compressor internal combustion engine	1000 ft³ of gas burned	1.2	0.11	Neg.	0.86	0.2	—
Fluid-bed catalytic cracking unit[a]	1000 bbl of fresh feed	220.	19.	13,700	63.	54.	b
Moving-bed catalytic cracking unit[a]	1000 bbl of fresh feed	87.	12.	3,800	5.0	5.0	c

a. Before CO waste heat boiler.
b. With electrical precipitators: 0.0009 % of catalyst circulated; with electrical precipitators: 0.005 % of catalyst circulated.
c. With centrifugal separators: 0.002 % of catalyst circulated.

Compiled from: Elkin, H. F. Petroleum Refining, pp. 813–43 in *Air Pollution*, vol. 4, A. C. Stern (ed.). New York: Academic Press, 1977.

Table 3.11 Hydrocarbon Loss Factors Developed From Los Angeles, California Survey

Evaporation sources	Units of emission factor Pounds of hydrocarbon per	Emission factor
Pipeline valves	Valve per day	0.15
Vessel relief valves	Valve per day	2.4
Pumps seals	Seal per day	4.2
Compressor seals	Seal per day	8.5
Cooling towers	Million gal of cooling water circulated	6.0
Blowdown systems	1000 bbl of refining capacity	5 – 300[a]
Vacuum jets	1000 bbl of vacuum distillation capacity	0 – 130[a]
Process drains	1000 bbl of waste water processed	8 – 210[a]
Other sources	1000 bbl of refinery capacity	10.0

a. Range of values.

Compiled from: Elkin, H. F. Petroleum Refining, pp. 813 – 43 in *Air Pollution*, vol. 4, A. C. Stern (ed.). New York: Academic Press, 1977.

Alumina is produced from bauxite, a naturally occurring ore composed primarily of hydrated oxides of aluminum, with some silica. The bauxite is refined with alkalis to remove the silica. In the presence of the strong base, the alumina is converted to sodium aluminate which forms crystals upon cooling. The crystals are separated by filtration, and washed to yield high-purity alumina.

Alumina is shipped to the primary reduction plant, where it is electrically reduced to aluminum and oxygen. This reduction process is carried out in shallow rectangular cells (pots) made of carbon-lined steel with consumable carbon blocks which are suspended above and extend down into the pot (Figure 3 – 7). The pots and carbon blocks are connected electrically to serve as cathodes and anodes, respectively, for the electrolytical process. Cryolite serves as both an electrolyte and a solvent for alumina. Alumina is added to and dissolves in the molten cryolite. The cells are heated and operated between 950° and 1,000° C, with heat generated by the electrical resistance between the electrodes. During the reduction process, aluminum ions migrate to the cathode, where they are reduced to metallic aluminum. Because of its heavier weight, the aluminum remains as a molten metal layer underneath the cryolite. Oxygen ions migrate to and react with carbon in the anode to form carbon dioxide and carbon monoxide which continually evolve from the cell. Periodically, the molten aluminum is siphoned or "tapped" from beneath the cryolite bath and moved to holding furnaces in the casting area. The product aluminum is held in the molten state until it is cast into billets to await further processing.

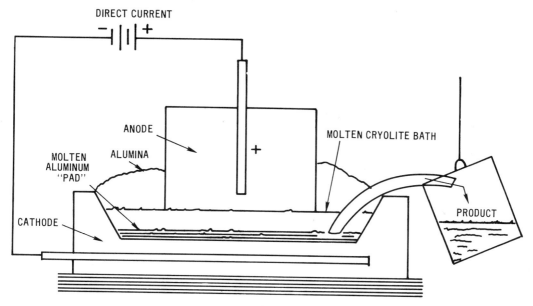

Fig. 3–7   Aluminum reduction cell.

From: Emission Standards and Engineering Div., EPA. Background Information for Standards of Performance: Primary Aluminum Industry, vol. 1: Proposed Standards. EPA-450/2-74–020a, U.S. EPA, Research Triangle Park, N.C., 1974.

Several types of contaminants are emitted during the production of primary aluminum. The major emission source is the reduction cell. Another source is the anode baking facility.

The major contaminant in aluminum smelting is fluoride (F^-), derived from the cryolite. According to the U.S. EPA (3), an uncontrolled primary aluminum plant can emit 40 to 60 pounds of fluoride per ton of aluminum produced. Plant capacities in the United States range from 100 to 750 tons of aluminum per day. In most primary aluminum plants, those emissions that escape the exhaust hoods (the primary control system) exit directly through the roof of the building into the atmosphere (Figure 3-8). Such "uncontrolled" secondary emissions can be several times as large as those which pass through the primary control system.

An uncontrolled plant can also be a significant source of particulates, emitting roughly 112 pounds of particulates per ton of aluminum produced, or over 33 tons each day. Particulates may originate in two ways: simple entrainment in the vent system during periodic additions of alumina and cryolite, and condensation of material vaporized from the molten

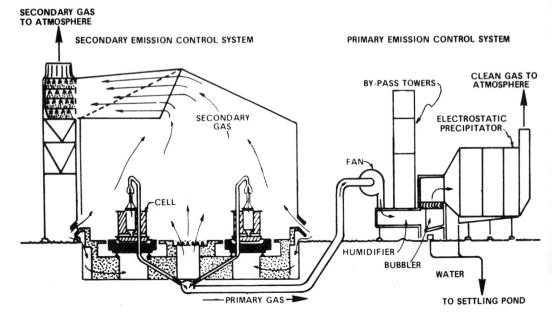

Fig. 3–8   Flow diagram for primary- and secondary-emission control systems in an aluminum plant.

From: Emission Standards and Engineering Div., EPA. *Background Information for Standards of Performance: Primary Aluminum Industry*, vol. 1: *Proposed Standards*. EPA-450/2-74–020a, U.S. EPA, Research Triangle Park, N.C., 1974.

bath and carbon anodes. It is estimated that 25% of the weight of particulates can be complex fluoride compounds such as cryolite (Na_3AlF_6), aluminum fluoride (AlF_3), calcium fluoride (CaF_2), and chiolite ($Na_5Al_3F_4$).

The high temperature of the cell causes emissions of organics (tar-fog) from the anodes at the cells. This fume forms the bluish haze characteristic of aluminum plants.

Concentrations of up to 1% by volume of carbon monoxide (CO) are generated at the reduction cell. Some conversion to carbon dioxide (CO_2) occurs when the hot CO gases contact air. Sulfur dioxide is also emitted as a product of sulfur contamination in the organics from which the anodes are formed; anode bake plants emit from 5 to 80 ppm$_v$(3). For anodes with a sulfur content of 2.5–5.0%, SO_2 emissions of 7–14 tons per day would be generated by a 600 tons/day aluminum plant. Results of a limited number of samples indicate NO_x emissions from primary aluminum plants are very low, only about 5 ppm$_v$.

PULP AND PAPER MANUFACTURE

The pulp and paper industry is one of the largest and fastest growing in the United States economy.

The manufacture of paper can be divided into two phases: (a) pulping the wood and (b) making the final paper product. The raw materials generally used in the pulping phase are wood, cotton or linen rags, straw, hemp, esparto, flax, jute, and waste paper. These materials are reduced to fibers which are subsequently refined, sometimes bleached, and dried.

At the paper mill, which is often integrated in the same plant with the pulping process, the pulps are combined and loaded with fillers and finishes and transformed into sheets. The fillers commonly used are clay, talc, and gypsum. The four main types of pulp used are: (a) groundwood, (b) soda, (c) kraft (sulfate), and (d) sulfite. Fiber industries, therefore, produce two main types of wastes, namely pulp-mill wastes and paper-mill wastes.

The major portion of the contamination from papermaking originates in the pulping processes. Raw materials are reduced to a fibrous pulp by either mechanical or chemical means. Mechanically prepared (groundwood) pulp is made by grinding the wood on large emery or sandstone wheels, and then carrying it by water through screens. This type of pulp is usually highly colored, low-grade, contains relatively short fibers, and is mainly used to manufacture nondurable paper products, such as newspaper.

Chemically prepared pulps are produced by reduction to chips, screening to remove dust, and digestion with chemicals. The various processes differ from one another only in the chemicals used to digest chips. Since each type of pulping produces somewhat different wastes, each should be considered separately. The discussion to follow will be limited to sulfate (kraft) pulping.

Coniferous woods are used in the preparation of kraft pulps. A simplified flow diagram for the kraft pulp process is shown in Figure 3–9. In any given pulp mill, a variety of combinations of the various unit processes might be used depending on the product to be produced, the raw material, yield expected, air quality considerations, and the preferences of the manufacturer.

The kraft process involves the cooking of wood chips in either a batch or continuous digester, under pressure, in the presence of a cooking liquor containing sodium hydroxide and sodium sulfide. The hydroxide is the reagent that dissolves the lignin in the wood chips. During the cooking reaction, the hydroxide is consumed and the sodium sulfide serves to buffer and sustain the cooking reaction. At the same time, small amounts of

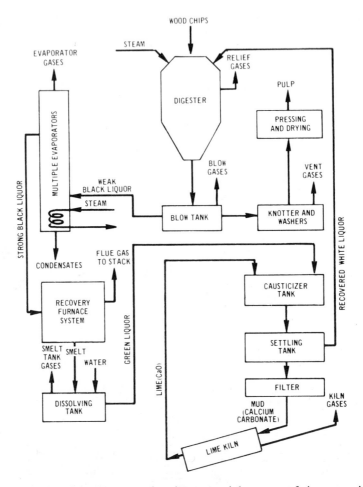

Fig. 3–9   Kraft pulping process, showing potential sources of air contaminants.

From: *Environmental Protection Agency. Atmospheric Emissions for the Pulp and Paper Manufacturing Industry.* EPA-450/1-73-002, U.S. EPA, Research Triangle Park, N.C., 1974.

sulfide react with the lignin giving rise to the odors characteristic of kraft mills (due to hydrogen sulfide and mercaptans).

Upon the completion of the cooking reaction, the residual pressure within the digester is used to discharge the pulp into a blow tank. Gases and flash steam released in the tank are vented through a condenser, where heat is recovered and the condensible vapors removed. The noncondensible gases, which are a source of malodors, are either confined and treated, or released into the atmosphere. At the same time, the pulp in the blow tank is being diluted and pumped to washers where the spent chemicals and the organics from the wood are separated from the fibers.

The spent chemicals and the organics, called black liquor, are then concentrated in multiple-effect evaporators and/or direct-contact evaporators for subsequent burning. The evaporators concentrate the liquor to a solids content of 60–70%, which is a requirement for combustion in the recovery furnace. During evaporation of the black liquor in the multiple-effect evaporators, volatile odorous gases are released. These gases escape when entrained gases and vapors are drawn off by the vacuum system. In order to eliminate the venting of these gases into the atmosphere, they can be confined and destroyed.

The black liquor is generally concentrated further in a direct-contact evaporator using hot flue gases from the recovery furnace. These hot gases, containing carbon dioxide, react with sulfur compounds in the black liquor, leading to the release of gases such as hydrogen sulfide. Prior oxidation of the black liquor will reduce the sulfide content of the liquor and, hence, the amount of hydrogen sulfide released. Some mills, which have gone on stream since about 1969, utilize a type of recovery furnace which eliminates direct contact between the flue gases and the black liquor, and are not sources of malodorous emissions.

The concentrated black liquor is sprayed into a recovery furnace, where its organic content supports combustion. The inorganic compounds, consisting of the cooking chemicals, fall to the bottom of the furnace where

Table 3.12 Comparison of Emissions from an Existing Kraft Mill with No Control Systems and a New Mill Utilizing Best Technology Currently Available

Unit	Total reduced sulfur[a] (TRS)		Particulate matter[b]	
	Existing	New	Existing	New
Recovery furnace	19.5	<0.5	5	<1
Batch digesters	3.7	0	0	0
Multiple-effect evaporators	0.8	0	0	0
Lime kiln	0.8	Trace	10	0.3
Smelt tank	0.05	Trace	1	<0.2
Pulp washers	0.25	0	0	0
Lime slakers	0	0	?	<0.3
	25.1	<0.5	16	<1.8

a. Units are in pounds H_2S per ton of air-dried pulp.
b. Units are in pounds per ton of air-dried pulp.

From: Hendrickson, E. R. The Forest Products Industry, pp. 686–703, in *Air Pollution*, 3rd Ed., vol. 4, A. C. Stern (ed.). New York: Academic Press, 1977.

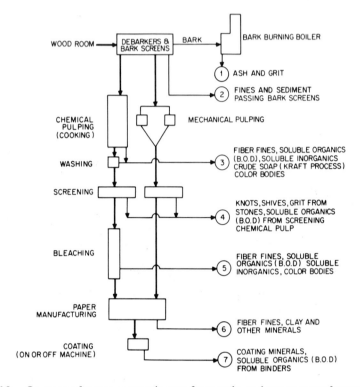

Fig. 3–10   Sources of water contaminants from pulp and paper manufacture.

From: Billings, R. M. and DeHaas, G. G. Pollution Control in the Pulp and Paper Industry, pp. 18.1 – .28 in *Industrial Pollution Control Handbook,* H. F. Lund (ed.). New York: McGraw Hill, 1971.

chemical reactions occur in a reducing atmosphere. The chemicals are withdrawn from the furnace as a molten smelt, containing mostly sodium sulfide and sodium carbonate, which is dissolved in water in a smelt-dissolving tank to form a solution called "green liquor." The green liquor is then pumped from the smelt-dissolving tank, treated with slaked lime (calcium hydroxide) in the causticizer, and then clarified. The resulting liquor, referred to as white liquor, is the cooking liquor used in the digesters. Most kraft mills recover the sludge resulting from causticizing and burn it to lime in a kiln.

Pulp-mill wastes are produced at various stages of the pulping process. These wastes contain pulping liquor, fine pulp, bleaching chemicals, mercaptans, sodium sulfides, carbonates and hydroxides, sizing, casein, clay, ink, dyes, waxes, grease, oils, and fiber.

Particular emissions from the kraft process occur primarily from the recovery furnace, the lime kiln, and the smelt-dissolving tank. They are caused mainly by the carry-over of solids plus the sublimation and condensation of inorganic chemicals. The sublimation and condensation normally produces a plume. Little information is available on the actual range of particle sizes from these sources, especially in the recovery furnace, since agglomeration tends to occur readily. In addition, particulate emssions occur from combination and power boilers. Particulate emissions from an uncontrolled source are summarized in Table 3-12. The newer plants, with better emission controls, discharge much less contaminants into the atmosphere; this is also shown in Table 3-12.

Aqueous paper-mill wastes originate in water which passes through the screen wires, showers, and felts of the paper machines, beaters, regulating and mixing tanks, and screens. The paper machine wastes (white waters) contain fine fibers, sizing, dye, and other loading material.

Figure 3-9 shows the sources of air contaminants in the pulping process, and Figure 3-10 presents a flow sheet of pulp and paper manufacture in terms of water contamination.

CONTAMINANTS RESULTING FROM AGRICULTURAL OPERATIONS

Modern agriculture is similar to mass-production industry in many respects. It relies heavily on highly specialized automated equipment and the economics of high-volume production. In the interests of efficiency, it utilizes man-made fertilizers and pesticides, and concentrates waste products to the point where they create disposal problems. The problems of chemical contamination of the environment by agricultural operations may appear to be remote to urban populations, but the key role of agriculture in the national economy makes it important to understand its total impact on the environment.

AGRICULTURAL USE OF FIRE

Controlling burning of stubble on harvested fields is a cost-effective means of preparing the soil for the next crop, while simultaneously enriching the soil in nutrients. It also produces large amounts of smoke which may be objectionable to downwind populations. In many areas, the controlled burning of agricultural fields is allowed only after a permit has been issued, and only under specified meteorological conditions.

SOIL TILLAGE

Mechanical soil tillage, especially when the soil is dry, can result in substantial quantities of dust immediately, or can make it possible for subsequent wind storms to pick up dust from the tilled surfaces. Marginal dry-land agriculture has resulted in major dust storms and extensive soil erosion.

FERTILIZERS AND SOIL CONDITIONERS

One classic definition of contamination is a resource out of place. Too much inorganic plant nutrient in a waterway is a prime example, since it can cause an undesirable proliferation of rooted aquatic plants and algae, and lead to oxygen depletion and anaerobic decay. The drainage of fertilizers and other soil conditioners from agricultural fields to streams can cause such effects in downstream waters. The main inorganic nutrients used as fertilizers are nitrogen as nitrate, and phosphorous as phosphate.

The application of fertilizers, manures, and lime as powders or sprays can also create local air contamination problems, which may extend beyond the confines of the farm.

PESTICIDES

Pesticides are used because their toxic properties reduce or eliminate species which consume, destroy, or compete with the efficient production of the product of interest. Some important classes of pesticides were listed in Table 2–5.

The ideal pesticide is specific and biodegradable. A pesticide is specific when it destroys the pest, i.e., the insects, fungi, weeds, rats, etc., without affecting nontarget organisms. A pesticide is biodegradable when it is destroyed or degraded by microorganisms within a reasonably short time after it has fulfilled its intended function. The time frame during which the pesticide should persist without biodegradation depends upon the pest. In some cases such as rat control, instantaneous killing is adequate, while for insect control, the pesticide may not be effective unless it persists for the full life cycle of the insect.

Persistent pesticides such as DDT and other organochlorines have practical advantages which led to their widespread use. They have an extremely low order of acute toxicity, and they are relatively inexpensive. Part of their economic advantage rests in their persistence, since they do not require as frequent reapplication as the more degradable pesticides.

Unfortunately, the ability to develop pesticides of sufficient specificity is limited. Frequently, the pesticides used are toxic to species other than the target organisms. The unintended toxicity may be immediate and apparent, such as a fish kill in a farm pond. On the other hand, it may take a long time to develop, and even longer to recognize, such as the effects of the persistent organochlorine pesticides on the reproductive capacities of wild birds in areas removed from the direct applications of the pesticides. Another problem is the potential for human health effects from the ingestion of pesticide residues on or in foods.

In summary, agricultural operations with pesticides are sources of environmental effects off the farm in several ways:

a. The reduction in populations of desirable mobile nontarget species of insects, birds, and mammals because of exposure on the farm.
b. The effects on aquatic organisms resulting from the transport in surface storm drainage.
c. The effects on biota in near and remote downwind regions because of the drift of aerosol sprays or from surface deposits which are volatilized by solar heating and are transported long distances as vapors or condensed fine particles.
d. The effects of residues remaining in or on agricultural produce.

WASTE PRODUCTS

In some agricultural operations, such as fruit growing, only marketable products are harvested and there is relatively little waste to dispose of. In other operations, such as the production of rooted crops, i.e., potatoes, beets, carrots, etc., all of the foliage is waste. In animal husbandry, especially on feedlots, the animals generate large amounts of fecal wastes during their growth. In terms of BOD, cattle produce 16.4 times as much waste per capita as people (4). In the slaughterhouse, large amounts of very rich organic wastes are generated. Such wastes, if not properly processed, can cause significant problems in terms of excessive organic nutrient levels in receiving waters and odor problems in downwind areas.

DOMESTIC ACTIVITIES

The wastes generated in our homes also contribute to the chemical contamination of the environment. Each person disposes of about 500 l. per day of waste water, and approximately 1.6 kg per day of domestic refuse. About half of the refuse is paper, and about 10% metal. Table 3–13 presents the compositions of typical domestic refuse. Table 3–14 shows the average composition of domestic sewage. If buried, the solid wastes can contaminate ground water with dissolved chemicals and oxygen demand. If burned in municipal incinerators of current design, some of the metals, volatile organics, and products of incomplete combustion will be released into the ambient air. It is extremely difficult to maintain efficient combustion with a material as variable in heating value and moisture content as domestic refuse.

Table 3.13 Typical Composition of Municipal Refuse

Material	Percentage
Paper	47
Glass, ceramic, stones	12
Metal	10
Food waste	10
Dirt	6
Grass, leaves	5
Textiles	3
Wood	3
Plastics	3
Rubber, leather	1

Many households minimize the amounts of garbage disposed of as solid wastes by using a garbage grinder on the drain of the kitchen sink. This merely transfers the oxygen demand burden to the sewage treatment facility and/or the receiving waters. The treatment facilities must also cope with the sanitary wastes and laundry wastes from the individual residences, along with the storm waters which pick up some of their waste burdens in passing over the surfaces of residental properties.

MUNICIPAL SERVICES

Municipal governments are generally responsible for disposing of solid and liquid wastes generated by their residents, and, therefore, reduce the volume and strength of the wastes they collect. For example, an efficient sanitary landfill disposes of the solid wastes without significant reentry to the general environment. In sanitary landfilling, the wastes are spread in thin layers, compacted, and covered with soil each day. In some cases,

Table 3.14 Average Composition of Domestic Sewage

State of Solids	Solids, mg/l.			BOD	COD
	Mineral	Organic	Total		
Suspended	65	170	235	110	108
Settleable	40	100	140	50	42
Nonsettleable	25	70	95	60	66
Dissolved	210	210	420	30	42
Total	275	380	655	140	150

From: Fair, G. M., Geyer, J. C., and Okun, D. A. *Water and Wastewater Engineering*, vol. 2. New York: Wiley, 1968.

where wetlands are used as fill sites, there is a negative impact, in the sense that the wetlands environment is destroyed in the landfill process.

Many municipalities use large central incinerators to reduce the mass and volume of solid wastes by about 90%. The residual ash is a relatively clean fill material, and can be used as a cover for the refuse in the landfill. Unfortunately, as discussed, the incinerators put partially burned hydrocarbons, ash, and trace metals into the atmosphere. Community opposition to new incinerators has frequently been sufficiently intense to prevent their construction. There is currently great interest in building and operating incinerators as energy recovery systems, utilizing the heating value of the refuse, but the recent experience with pilot installations has not been encouraging. It has not been possible to operate them as efficiently or economically as had been hoped.

Municipal sewage treatment facilites, are, at best, only partially successful in removing wastes and contaminants from the incoming waters. Some operate only primary treatment facilities, consisting of gravity settlement chambers. These remove the more massive suspended solids, but very little of the finer solids and hardly any of the dissolved solids. Most of the rest of the facilities include secondary treatment involving some kind of biological oxidation. Such plants may remove from 50 to 95% of the oxygen demand from the wastes, but usually have lower efficiencies for the removal of nitrates, phosphates, and trace metals.

The large portion of the oxygen demand removed from the incoming sewage by the treatment plant does not disappear. Some is oxidized to harmless end products in the process, but the rest is simply transferred into sewage sludge, which must itself be disposed of. Sludge disposal is currently a major problem for many large cities.

New York City and other costal cities have, for many years, dumped digested sewage sludge into designated dumping grounds within coastal waters. In New York's case, this is a site about 12 miles out to sea. As a result of alleged hazard to marine life and the public health, the city is under court orders to terminate such dumping by 1981. However, no realistic alternatives to ocean dumping are in sight. There is no adequate technology for land disposal which does not involve significant problems in odor generation and/or ground-water contamination. The sludge could be incinerated, but not without the release of significant amounts of contaminants into the air. Ocean dumping may continue to be the least objectionable alternative for the disposal of the sewage sludge.

Another municipal activity which generates chemical contamination in some regions is the road and street salting done in the winter to improve traffic safety. The salt corrodes metals, damages plants, shrubs, and trees, and increases the salinity of the receiving waters. However, the environmental contamination from street salting appears generally acceptable in comparison to the alternative, i.e., impassable and unsafe roadways.

RADIOLOGICAL CONTAMINATION RESULTING FROM NUCLEAR FISSION REACTIONS

This section discusses the types and sources of anthropogenic radioactive contaminants in the environment derived from military and civilian applications of nuclear fission. The sources for these materials are quite different from those which emit the chemical contaminants described in previous parts of this chapter. In general, the mass concentrations of radionuclides released are almost too low to result in any chemical toxicity following their uptake; the focus of concern is, therefore, on their radiobiological effects. Radiological contamination may be quite pervasive, since radionuclides follow the same environmental and metabolic pathways as do the nonradioactive nuclides of the same element.

While there is no question that the total amount of radioactive material in the world has been increased by man, the radiation dose to the general population from natural sources is much greater than that from anthropogenic environmental releases. For example, in 1970, natural sources accounted for 97% of the per capita annual dose in the United States due to environmental radioactivity (i.e., not including medical diagnostic or therapeutic procedures) (5). However, any exposure above background level is considered deleterious to some degree, since radioactivity can alter genetic materials.

Contaminant radionuclides are generated as fission products and by neutron activation. In the latter process, bombardment by neutrons induces radioactivity in elements in air, water, soil, or in structural materials. Table 3–15 lists some of the important radionuclides produced by nuclear reactions.

To date, the main anthropogenic source of environmental radioactivity has been nuclear-weapons testing. Use of nuclear energy for electric-power generation has added a relatively small increment, and the industrial and medical uses of radionuclides have added still smaller amounts.

NUCLEAR-WEAPONS TESTING

The basic fuel for nuclear weapons is either uranium-235 (^{235}U), which is naturally occurring, or plutonium-239 (^{239}Pu), which is produced from ^{238}U by neutron capture. Contamination due to testing involves the release of: (a) unfissioned ^{235}U or ^{239}Pu, (b) hundreds of different fission-product

Table 3.15 Important Radionuclides Produced by Nuclear Reactions

Nuclide	Half-Life
Fission products	
^{85}Kr	10.4 years
^{89}Sr	50 days
^{90}Sr	28 years
^{95}Zr	65 days
^{131}I	8.05 days
^{133}I	21 hours
^{135}I	6.7 hours
^{137}Cs	30 years
^{140}Ba	12.8 days
^{144}Ce	285 days
Activation products	
^{3}H*	12 years
^{14}C	5730 years
^{54}Mn	314 days
^{55}Fe	2.7 years
^{59}Fe	45.6 days
^{60}Co	5.3 years
^{65}Zr	245 days
^{238}Pu	86.4 years
^{239}Pu	24,000 years

*Hydrogen-3 (tritium) is also produced in fission.

From: Eisenbud, M. The Primary Air Pollutants-Radioactive. Their Occurrence, Sources and Effects, pp. 197–231 in *Air Pollution*, 3rd Ed., vol. 1, A. C. Stern (ed.). New York: Academic Press, 1976.

nuclides having half-lives ranging from fractions of a second to years, the most hazardous of which are iodine-131 (^{131}I), cesium-137 (^{137}Cs), strontium-89 (^{89}Sr), and strontium-90 (^{90}Sr), and (c) products formed by neutron activation of air, water, soil, and the metal casings which surround the weapons.

The most extensive series of atmospheric tests occurred in the 1950's and early 1960's and resulted in the release of about 20 megacuries (MCi) of ^{90}Sr, 30 MCi of ^{137}Cs, 5 MCi of ^{14}C, and 3,000 MCi of ^{3}H into the environment (6). Since the ban on atmospheric testing enacted in 1962, U.S. and U.S.S.R. explosions have been conducted largely underground, with only an occasional release of small amounts of radionuclides on a local basis. Some atmospheric testing has been done by France, China, and India; the contribution since 1972 due to these is about 5% of the amount produced in the earlier test series (6).

A number of proposals have been advanced for the use of peaceful nuclear explosions for various purposes, such as releasing natural gas from underground sources, building harbors, and excavating for canals. If detonated in the atmosphere, these explosions would also release both fission and neutron activation products; the amounts of each would depend upon the particular type of explosive device used and the composition of the soil and rock surrounding the site.

NUCLEAR-POWER REACTORS

In power reactors, controlled fission is used to generate electricity. The first nuclear central power station began operation in 1957 in Pennsylvania. As of January 1, 1976, in the United States, there were 58 operating nuclear-power reactors with a combined total generating capacity of about 40 million kilowatts, and accounting for about 9% of the total U.S. electrical production. In addition, 87 additional plants were under construction, and 93 others were planned. Although continued growth of the nuclear power industry is likely, its exact extent is uncertain at this time.

Nuclear Fuel Cycle   All power reactors currently in use in the U.S. use ^{235}U for fuel. Contaminants may be released into the environment at a number of steps in the fuel cycle. Figure 3–11 presents a flow diagram of the uranium fuel cycle showing typical contaminants associated with each step.

Mining   Uranium is mined both underground and in open pits. In order to protect workers in underground areas, the radioactive radon gas, which diffuses into the mine air, and its daughter products are vented into the atmosphere outside the mine. While locally high radioactivity levels may occasionally result, the releases generally do not contribute significantly to community exposures. Radon levels in discharged air may range from 0.5 to 20 $\mu Ci/min/1000$ ft^3 of air (7). Analogus to the case of ragweed discussed earlier in this chapter, anthropogenic activity results in an increase in what is essentially a natural contaminant. In this case, it is radon.

Milling   Milling involves crushing of the ore and the use of chemical methods to extract, concentrate, and purify the uranium into a semirefined oxide, U_3O_8.

The major contaminant problem in this step is the large amount of residual solids. This waste material, known as tailings, occupies as much volume as the material taken into the mill. These solids contain most of the radium originally present in the ore. By 1970, there were over 80 million metric tons of tailings occupying over 2,100 acres of land in the United States (8). The radiation and radioactive dust from the tailing piles constitute contamination from natural sources which are redistributed and localized in areas which generally are closer and more accessible to human populations than are the ores in the mines. As a result, the radiation levels in cities near piles of tailings may be higher than normal. In addition, in some towns in the western United States, these tailings, which resemble construction sand, have been used for land fill and incorporated into construction materials. Their high radium content results in high radon levels in many buildings.

Milling operations use large amounts of water, which often becomes contaminated with nonradioactive chemicals and with small amounts of uranium, radium, and their decay products. Discharges to streams may result in local contamination problems.

Refining and Conversion   In this step, the uranium concentrate (U_3O_8) is chemically purified. The uranium to be enriched in its ^{235}U content is converted into a volatile hexafluoride, UF_6. Some nonradioactive liquid wastes and acid gases may be released in these chemical conversion processes.

Enrichment   The concentration of ^{235}U in natural uranium is about 0.7%. Most reactors operate more efficiently with higher amounts. Thus, in enrichment, the concentration of ^{235}U in the UF_6 is increased to about 2–4%, by multistage gaseous diffusion or centrifugation. Large amounts of electric power and cooling water are consumed in these processes.

Fuel Fabrication   Enriched UF_6 is converted into a dioxide powder, UO_2, made into pellets and loaded into alloy tubing, which is then formed into individual fuel-rod elements.

Up to this point in the fuel cycle, environmental contamination problems are minimal, and are similar to those of the chemical industry in general. Although dusts and fumes of natural uranium are mildly radioactive, they have been well controlled. The greatest potentials for general environmental contamination are associated with power production and fuel reprocessing.

Power Production   Various types of reactors may be used to convert heat into electrical power. Modern reactors are primarily of two types: (a) pressurized-water reactors (PWR) or (b) boiling-water reactors (BWR). In these, water is used as both the coolant and the moderator of the fission reaction.

In the boiling-water reactor, water is heated as it passes through the reactor core and steam is produced. The steam passes through turbines and is then condensed for return to the reactor.

In the pressurized-water reactor, the water pumped through the core is maintained at a high pressure to prevent it from boiling in the core. Rather, the water, which leaves the core at about 318°C, passes through a steam

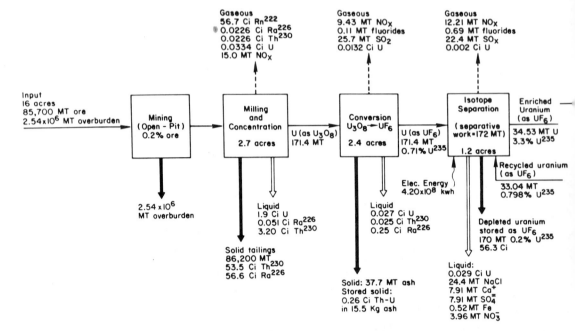

Fig. 3-11 Flowsheet for the fuel cycle of a typical 1000-Mwe uranium-fueled water-reactor power plant (yearly quantities for 100% load factor). These quantities are characteristic of the current generation of water reactors now operating or under construction. When there are significant differences in environmental releases from pressurized-water reactors (PWR) and boiling-water reactors (BWR), the characteristic releases for each of the two types are indicated separately.

From: Pigford, T. H. Environmental Aspects of Nuclear Energy Production, pp. 515–59 in *Annual Review of Nuclear Science*, vol. 24, E. Segré, J. R. Grover, and H. P. Noyes (eds.). Palo Alto, Calif.: Annual Reviews, © 1974. Reproduced with permission of Annual Reviews, Inc.

generator unit, where its heat is transferred to the water-steam loop which passes through the turbines.

The nature and amount of any atmospheric discharges depend upon reactor type, operating history, and condition of the fuel. Reactor products include products of uranium fission and of neutron activation of coolant water in the core. The latter results in short-lived radioactive gases, such as ^{41}Ar, ^{16}N, and ^{19}O. Radioactive noble gases produced from fission, e.g., ^{85}Kr, may diffuse through the fuel-element cladding and be carried by steam to the turbines in a BWR system. Some noble gases may boil off with the steam and escape into the atmosphere via the condensor air ejector and through a stack. Most of this radiation which would escape

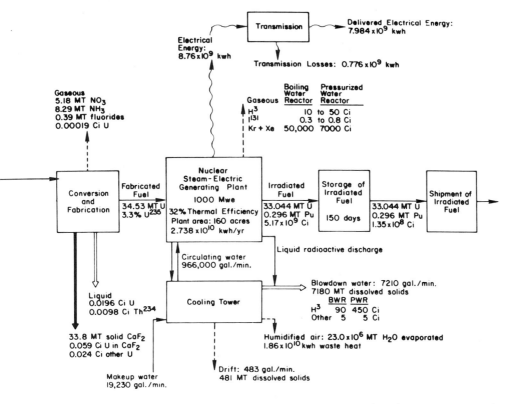

(*continues on next page*)

through the stack is short-lived. Because of this, a 30-minute delay is maintained between the air ejector and the top of the stack to allow for radioactive decay before emission.

Under normal operations, the radionuclide release from power reactors is very low and is not important as a source of public exposure. Radiation doses are less than 1% of the natural background dose and usually much less (5). It is an accidental release of large amounts of radioactive material from these plants that is the focus of public concern.

The only reactor accident which resulted in significant environmental contamination to date occurred in 1957 in England. Various fission products were released. This air-cooled reactor design is not, however, typical of the power reactors in operation today. The track record of existing nuclear power stations has been excellent.

Fuel Reprocessing After some time, the fuel in the reactor must be re-

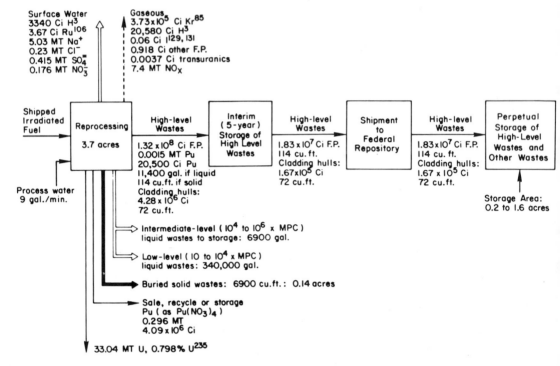

Fig. 3–11   *(continued from previous page).*

processed. This involves recovery of the unused uranium, and also of the [239]Pu which is produced during reactor operation, from the fission wastes. The recovered uranium is recycled for fuel-element fabrication. Reprocessing was generally performed only at government installations and only for military weapons purposes. The only commercial plant to operate in the United States was located in western New York State; it closed in 1972. Table 3–16 shows its main discharges in the year 1969. Future commercial plants can and should be designed with improved emission controls; the main releases, however, will still be [85]Kr and [3]H. Until commercial reprocessing resumes, all power-plant spent fuel is being stored. Fuel reprocessing is the phase of the cycle having the greatest potential for significant radionuclide releases under normal operation.

   Waste Management   Ultimate disposal of radioactive wastes is an unresolved issue. There are low-level (not very radioactive materials) and high-

Table 3.16   Main Discharges of Nuclear Fuels Services
Reprocessing Plant; 1969

Type	Amount (Ci)
Liquid	
^{3}H (tritium)	6000
^{90}Sr	5
Gas	
^{85}Kr	300,000
^{131}I	0.06
Particles	<0.1

From: Harley, J. H. Radiological Contamination of the Environment, pp. 456–80 in *Industrial Pollution*, N. I. Sax (ed.). © 1974 by Litton Educational Publishing, Inc. Reprinted by permission of Van Nostrand Reinhold Co.

level (highly radioactive substances, primarily fission products produced at the power plant and separated during reprocessing) wastes. The former are buried at commercial sites in the United States. The latter are stored in liquid form in underground tanks. Future disposal techniques are a subject of much discussion.

SUMMARY OF PRIMARY ANTHROPOGENIC CONTAMINANTS

In the preceding sections, various anthropogenic activities have been discussed in terms of their contribution to the level of primary contaminants in the environment. In order to put this information into proper perspective, and to clearly show the relative importance of each of these spheres of activity upon specific contaminant burdens, a few additional tables and figures are presented.

Figure 3–12 shows the relative source strengths for each of five significant air contaminants. Table 3–17 presents atmospheric emissions of other selected materials according to the activity which generates them.

It is only within the past few years that nationwide estimates of the amounts of various contaminants discharged into waterways have become available. However, these surveys generally considered only three broad contaminant classes: (a) BOD, (b) total suspended solids (TSS), and (c) inorganic plant nutrients (nitrogen and phosphorus).

Figure 3–13 shows the amounts of BOD and TSS released into U.S. waterways in 1973. Table 3–18 shows the source strengths for nitrogen and phosphorus entering United States waters.

Figure 3–14 shows the production rates for solid and semisolid wastes

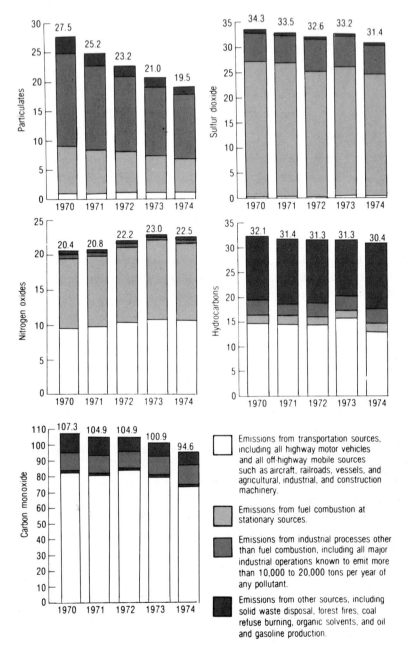

Fig. 3–12 Estimated emissions of major air contaminants, nationwide, 1970–1974 (10⁶ tons/year).

Source: U.S. Environmental Protection Agency.

Table 3.17 Atmospheric Emissions of Certain Materials in the United States (kilotons/year)

Material	Transport	Energy	Mining	Processing	Manufacture	Waste disposal	Use	Totals
Arsenic		0.73		4.6	0.84	0.30	3.01	9.48
Asbestos			5.81		0.49		0.41	6.71
Barium		3.76	0.10	3.70	0.06		0.18	7.80
Benzo(a)pyrene	0.3	0.27		0.11	0.02			3.15
Beryllium		316		21	0.06			337.6
Boron		4.52		2.40		0.54	0.02	6.94
Cadmium				1.68	0.01	0.01	0.01	1.71
Chlorine	21.0	351		47.3	1.82		56.1	477.2
Chromium		1.89		8.29		3.06		13.2
Copper		0.17	0.15	9.68				10.0
Fluorine		20.1		80.9	17.1	1.04	3.30	122
Iron		251		168	0.46	1.82		421
Lead	224	17.4	0.06	10	3.19	5.01		260
Magnesium		50.6	8.48	11.4	38.7			110
Manganese		1.97		14.9	0.48			17.5
Mercury		0.11		0.05		0.18		0.5
Molybdenum		0.66	0.16	0.18		0.01	0.30	1.0
Nickel		3.73	0.14	0.06	0.06	0.04		4.6
Phosphorus		22.5	0.31	13.4	21.7	0.08	0.02	58.0
Selenium		0.72		0.08	0.20			1.0
Silver		0.05		0.11	0.01	0.02	0.02	0.2
Titanium		42.7	0.26	3.29	3.07	0.02		51.1
Vanadium		4.66	0.07	0.17		1.80		4.9
Zinc		4.86	0.05	92.8			2.50	100

From: Barth, D. S., Black, S. C., and Hammerle, J. R. Chemical Agents in Air, pp. 157–66 in *Handbook of Physiology, Section 9, Reactions to Environmental Agents*, D. H. K. Lee, H. L. Falk, and S. D. Murphy (eds.). Bethesda, Md.: American Physiological Society, 1977.

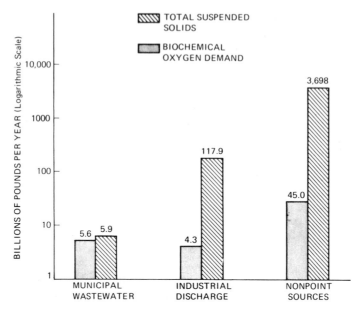

Fig. 3–13 Estimated nationwide water contaminant loadings of suspended solids and oxygen-demanding material. Estimates refer to calendar year 1973. Loadings from municipal and industrial sources tend to produce acute local water-quality problems, more than loadings from nonpoint sources which are more diffuse. Municipal wastewaters include wastes from households; commercial establishments such as garages, laundries and restaurants; certain industrial waste-waters; and the groundwater and runoff from precipitation that enters the municipal sewer system. Nonpoint sources are wastes entering waterways without flowing through sewers, treatment facilities, etc. (e.g., general land runoff, ground-water, and from the atmosphere via precipitation and gravitational settlement).

From: Council on Environmental Quality. *Environmental Quality—1976, Seventh Annual Report.* Washington, D.C.: U.S. Government Printing Office, 1976.

Table 3.18 Sources of Nitrogen and Phosphorus into U.S. Waters

Source	Nitrogen	Phosphorus
	(10³ metric tons/year)	
Natural	470–1900	110–320
Anthropogenic	1810	310–460
Domestic sewage	605	175–200
Urban-land runoff	90	10
Cultivated-land runoff	925	50–175
Runoff from land on which animals are maintained	190	75
Totals	2280–3710	420–780

Compiled from: Ferguson, F. A. A Nonmyopic Approach to the Problem of Excess Algal Growths. *Environ. Sci. Tech.* 2:188–193, 1968.

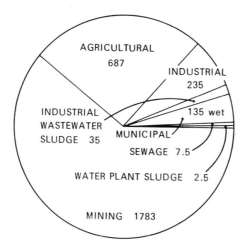

Fig. 3–14 Estimated annual production rates for different types of solid and semisolid wastes, 1970–1974 (dry weight in 10^6 tons/year).
Source: U.S. Environmental Protection Agency.

in the United States; some of these are eventually dumped into the hydrosphere. Table 3–19 presents amounts of vessel-discharged ocean disposal in United States waters in 1975. Table 3–20 shows estimates of petroleum and petroleum products which enter the world's oceans.

SECONDARY CONTAMINANTS

Many of the chemical contaminants in the environment, including some of the most important ones, such as ozone (O_3) in urban air, chloroform

Table 3.19 Vessel Discharged Ocean Disposal in the United States in 1975

Type of Waste	Amount (10^3 tons)
Industrial wastes	3446
Sewage sludge	5040
Construction and demolition	396
Total	8882

Source: U.S. Environmental Protection Agency.

Table 3.20 Petroleum Hydrocarbons* Entering the World's Oceans

Source	Amount (10^6 tons/year)
Natural seepage	0.6
Transportation	2.133
Offshore-oil production	0.08
Coastal refineries	0.2
Coastal municipal wastes	0.3
Coastal, nonrefining industrial wastes	0.3
Atmosphere	0.6
Urban runoff	0.3
River runoff	1.6
Total	6.113

*Includes both crude oil and refined products.

Compiled from: *Petroleum in the Marine Environment*. Washington D.C.: National Academy of Science, 1975.

($CHCl_3$) in drinking water, and aflatoxin in certain foods, are not directly released into the environment by any primary source. Rather, they are formed within the environment by chemical reactions among their precursor compounds, as in O_3 or $CHCl_3$ formation, or by biological processes, as in the formation of aflatoxin by the action of *Aspergillus flavus* mold on stored peanuts and corn.

This section presents some of the important mechanisms for forming secondary environmental contaminants, citing a few of the more notable examples of transformations within the ambient air and surface waters which have affected public health or attracted considerable public attention.

Airborne Reaction Products

Relatively little attention was paid to chemical reactions within the atmosphere until the 1940's, when the characteristic southern California smog, with its visibility reductions and eye-irritation properties, became persistent. It was very different in character from the classic urban air pollution of London or Pittsburgh, which was characterized by its soot and sulfur dioxide content and reducing properties. The Los Angeles smog, by contrast, had oxidizing properties. It was found to contain ozone, which seemed inexplicable initially, since an inventory of known sources did not yield an adequate ozone source. At the time, it seemed unlikely that atmospheric chemistry would account for the ozone. The available information on reaction kinetics suggested that the concentrations of potential reactants in the air were too low. Reactions require molecular collisions, and there are many fewer collisions per unit time in air than, for instance, in water, where the effective density is more than 800 times greater. However, chemical transformations within the atmosphere, while they initially appeared unlikely, do lead to the formation of ozone and a host of other reactive species.

Advances in atmospheric chemistry not only helped to develop an understanding of the formation of the oxidizing atmospheres of southern California, but also were able to help lead to a better appreciation of the transformations undergone by sulfur dioxide within the atmosphere and, therefore, of our understanding of the modes of action and effects produced by the more classic kinds of reducing contaminants. An important ingredient in many of the atmospheric transformations is actinic radiation, i.e., sunlight. Reactions depending on light are known as photochemical reactions.

PHOTOCHEMICAL REACTIONS IN SMOG FORMATION

Solar radiation consists of a broad spectrum of wavelengths and energies. The higher-energy wavelengths dissociate gas molecules in the upper atmosphere. Photodissociation of O_2 produces a mixture of the highly reactive atomic oxygen (O), the stable diatomic molecular oxygen (O_2) and the relatively reactive triatomic ozone (O_3). There is relatively little ozone below the stratospheric ozone layer, except for that formed in urbanized areas. The background tropospheric level of about 0.03 ppm$_v$ is attributable to the small amount of mixing which occurs between the stratosphere and troposphere.

Photochemical reactions are facilitated by the absorption of solar energy, which raises the reactivity of the absorbing molecules to a level enabling them to react with other molecules. Only colored substances can absorb the longer wavelength photons which can penetrate the ozone layer. In plants, the key substance is chlorophyll. In the atmosphere, the key gas is the brownish-colored nitrogen dioxide. Absorption of light quanta having sufficient energy (those with wavelengths up to 0.38 μm) will cause photodissociation:

$$NO_2 + h\nu \rightarrow NO + O \qquad (3.1)$$

where $h\nu$ is an index of the energy of the light, and h = Planck's constant, ν = frequency. Some organic compounds can also undergo photodissociation to form free radicals, e.g.:

$$R - \overset{\displaystyle \overset{O}{\parallel}}{\underset{\displaystyle \underset{H}{\backslash}}{C}} + h\nu \rightarrow \overset{\cdot}{R} + H\overset{\cdot}{C}O \qquad (3.2)$$

The highly reactive atoms and free radicals formed in these dissociations may combine with their original partners but will more likely interact with other molecular species in the atmosphere, utilizing the extra energy received from the solar radiation in their dissociation. Free radicals are so reactive that in aqueous systems their lifetimes are measured in microseconds or nanoseconds. However, in the atmosphere they are present in concentrations of less than 1 part per 100 million, and can have half-lives of minutes to hours. Free radicals can also be formed in nonphotochemical reactions, such as the reactions of ozone with olefins:

$$O_3 + RCH = CHR \rightarrow RCHO + \overset{\cdot}{R}O + H\overset{\cdot}{C}O \qquad (3.3)$$

Solar energy quanta which are not sufficiently energetic to dissociate molecules can excite them, enabling them to react more readily. Oxygen (O_2) absorbs visible light poorly, but there is so much O_2 in the atmosphere that a sufficient number of excited molecules are formed to play a role in reactions with aldehydes and possibly other hydrocarbons.

After the initial steps of photodissociation and excitation, an extremely complex series of secondary and chain reactions can be initiated, depending on the mixture of raw materials available in the atmosphere and on the ambient temperature and humidity. There are so many possible sequences of reactions that the presentation of a particular pathway is highly speculative. Those presented in this discussion are merely some plausible ones.

Some free radicals attach to O_2, forming peroxy radicals ($ROO\cdot$). These can react with NO_2 to form peroxyacetylnitrates (PAN's), which are responsible for much of the eye irritation and plant damage produced by the smog mixtures. The peroxy radicals can also react with other primary and secondary air contaminants to form alcohols, ethers, acids, etc.

Many of the secondary products are relatively unstable and undergo further chemical and photochemical reactions which may generate NO_2, O_3, and a variety of free radicals; these than participate in further chemical and photochemical reactions. The gas-phase photochemistry can continue through many generations of complex chain reactions and has the net effect of converting most of the nitric oxide (NO) in the atmosphere to NO_2, and of increasing the levels of ozone and of light-scattering aerosols of sulfur oxides and highly oxidized hydrocarbons.

The net result of the series of complex reactions is illustrated in Figure 3–15. Levels of the primary contaminants, i.e., the hydrocarbons and NO from auto exhaust, rise through the morning rush hours. However, the buildup of NO is stopped by its conversion to NO_2. Aldehydes are partly a primary contaminant, but their continued rise through the morning, and the late-morning fall in hydrocarbons, indicates that the aldehyde level is augmented by photochemical formation processes. As the NO_2 levels fall, the ozone levels rise (9).

It is interesting that the evening rush hour does not produce the same sequence of events. The better ventilation in the evening limits the accumulation of the primary contaminants. Also, the sunlight is not as strong, nor does it last as long as in the morning. Finally, the residual ozone

Fig. 3–15   Average concentration of contaminants during days of eye irritation in downtown Los Angeles, California. Hydrocarbons, aldehydes, and ozone for 1953–1954. Nitric oxide and nitrogen dioxide for 1958.

From: Haagen-Smit, A. J. and Wayne, L. G. Atmospheric Reactions and Scavenging Processes, pp. 235–88 in *Air Pollution*, 3rd Ed., vol. 1, A. C. Stern (ed.). New York: Academic Press, 1976.

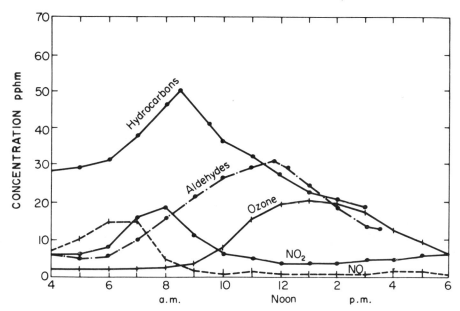

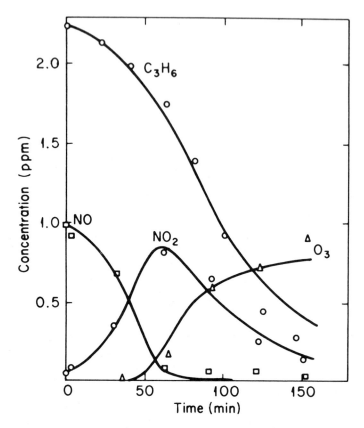

Fig. 3–16 Concentration changes on irradiation of a mixture of NO and NO₂ and C₃H₆; comparison of model predictions with experiment. Solid line: ref. calculated; points: experimental data.

From: Niki, H., Daby, E. E., and Weinstock, B. Mechanisms of Smog Reactions, pp. 16–57 in *Photochemical Smog and Ozone Reactions, Advances in Chemistry Series* #113. Washington, D.C., American Chemical Society, 1972.

reacts with the NO to form more NO₂, accounting for the late-afternoon rise of the latter.

Much of what is known about these complex series of chemical reactions has been learned in smog-chamber tests in which primary contaminants are irradiated by ultraviolet light. The experiment illustrated in Figure 3–16 began with initial concentrations of 2 ppmᵥ of propylene (C₃H₆), 1 ppmᵥ of NO, and 0.05 ppmᵥ of NO₂. The decline in C₃H₆ and NO, and the successive rises of NO₂ and O₃, closely parallel those seen in ambient atmospheres (10).

ATMOSPHERIC TRANSFORMATIONS OF SULFUR DIOXIDE

Sulfur dioxide (SO_2) released to the atmosphere can remain in the air for several days. Some of it, especially when emitted at or near ground level, is taken up by vegetation and materials as SO_2. It can also be taken up as SO_2 by water droplets in the atmosphere. However, most of it will undergo a series of chemical transformations before it reaches the earth's surface.

The first step in the transformation is oxidation to sulfur trioxide (SO_3). This can take place by photochemical oxidation, but the rate is relatively slow and this mechanism does not account for much of the conversion. Most is believed to occur by catalytic oxidation on the surfaces of airborne particles and within droplets. The SO_3 vapor is rapidly hydrolyzed by atmospheric water to form small droplets of sulfuric acid (H_2SO_4). These droplets take on water of hydration, coagulate with other atmospheric particles, and scavenge gas molecules, including more SO_2 and ammonia (NH_3). They thus accumulate trace-metal cations and ammonium ions, and provide a medium for conversion of the absorbed SO_2 to sulfite and sulfate ions.

The end stage of the process is generally an aerosol with a mass median diameter of about 0.5 μm, composed primarily of ammonium sulfate, but also containing traces of other sulfate and nitrate salts. This fine-particle aerosol is very stable in dry conditions and can travel for hundreds and thousands of kilometers before it is removed from the atmosphere, primarily via precipitation.

Chemical Transformations in the Aquatic Environment

The primary contaminants discharged into surface waters can be transformed within the aquatic environment into other species. The possibilities for such transformations are virtually unlimited. The examples to be discussed here illustrate two kinds of transformation. One results from human intervention—this is the unintentional production of chlorinated hydrocarbons during the process of disinfection with chlorine. The other involves the intervention of aquatic bacteria, whereby inorganic mercury compounds are converted into much more toxic organic mercury compounds.

SYNTHESIS OF CHLORINATED HYDROCARBONS IN DRINKING WATER

Surface waters in the United States are all contaminated, to some degree, with organic compounds. Even natural waters in protected watersheds

Table 3.21   Summary of Analytical Results for Volatile Organics

Compound	% of samples giving positive results		Mean concentration (μg/l.)		Maximum concentration (μg/l.)	
	Finished Water	Raw Water	Finished Water	Raw Water	Finished Water	Raw Water
Chloroform	95	27	20	<1	366	94
Bromodichloromethane	78	5	6	<1	51	11
Dibromochloromethane	60	2	1	<1	14	1.4
Bromoform	14	0	<1	<1	7	<1
Carbon Tetrachloride	34*	18	2*	<1	26*	20
Methylene Chloride	8	1	1	<1	7	1
1,2-Dichloroethane	13	14	1	<1	26	15

*11 samples may have been contaminated by exposure to laboratory air containing carbon tetrachloride.

Compiled from: Health Effects Research Laboratory, EPA. *List of Organic Compounds Identified in Drinking Water in the United States.* U.S. EPA, Cincinnati, Ohio, 1976.

will have some contamination from the products and wastes of aquatic life and from the storm-water pickup of land-deposited animal and plant debris and air-contaminant fallout. Many communities produce their drinking water from river water which has been contaminated upstream by agricultural runoff, municipal sewage discharges, industrial wastes and spills. The physical and/or chemical treatments given to these waters to remove their contaminants are never 100% efficient. Thus, when the waters are disinfected, they still contain a variety of organic materials. Some waters are disinfected with ozone, but most are chlorinated. In the process, some of the hydrocarbons react with chlorine to produce chloroform and other trihalomethanes, which are believed to be carcinogenic compounds. As shown in Table 3–21, many drinking-water supplies also have measurable levels of carbon tetrachloride, methylene chloride, and other halogenated hydrocarbons; but some of these do not appear to be formed to any appreciable extent in the treatment process. Of 289 volatile organic compounds identified in drinking water, approximately 38% are halogenated (11).

METHYLATION OF MERCURY BY AQUATIC ORGANISMS

In the natural environment, mercury is found in soil, air, and water. The natural level for mercury in water is 0.06 ppb$_w$ in the northeastern United States.

Mercury can enter the environment, via a host of sources, either directly or indirectly. There are direct mercury releases in the waste effluents from manufacturing processes. The use of fossil fuels, such as coal (containing about 0.5 ppm$_w$ mercury), and the burning of mercury-containing paper constitute indirect means of mercury release. Additionally, accidental misuse must also be considered. The most well-known source of mercury contamination is the chloroalkali industry, which formerly discharged 0.25 – 0.5 lb of mercury per ton of caustic soda produced.

It is important to consider the various forms of mercury present in water. Methyl mercury is the most toxic; other alkyl forms are also toxic but to lesser degrees. Phenyl and inorganic mercury derivatives are also less toxic than the methyl form, but when released into water they can be converted to methyl mercury by a variety of microorganisms.

Various factors influence the rate of bacterial methylation in water. Methylation occurs rapidly in the winter and early spring, but slows down during summer and fall. Waters of low pH contain more methyl mercury than those of high pH. Also, aerobic conditions enhance methylation.

When exposed to similar concentrations of methyl mercury and inorganic mercury, fish are able to absorb methyl mercury from water 100 times as fast as the inorganic mercury and are able to absorb five times as much methyl mercury from food as compared to inorganic mercury (12). Once absorbed, methyl mercury is retained two to five times as long as inorganic mercury. With increased fish size, both the uptake of methyl mercury from the environment and its clearance is decreased. Since methyl mercury is strongly bound to muscle, it accumulates with increased muscle mass and increased duration of exposure. Assuming a steady mercury-containing diet, it takes about a year to establish a balance between mercury intake and mercury excretion.

Very high concentrations of mercury have been found in salmon caught in lakes contaminated by industrial discharges. The FDA limit for mercury in fish is 0.5 ppm$_w$. Among ocean fish, some tuna and most swordfish exceed this limit. However, examination of the tissues of museum specimens captured before the advent of large-scale anthropogenic releases of mercury to the environment show similar levels.

REFERENCES

1. Olson, D. R. The Control of Motor Vehicle Emissions, pp. 595 – 654 in *Air Pollution*, 3rd Ed., vol. 4, A. C. Stern (ed.). New York: Academic Press, 1977.
2. Council on Environmental Quality. *Environmental Quality – 1976; Seventh Annual Report.* Washington, D.C.: U.S. Government Printing Office, 1976.

3. Emission Standards and Engineering Div., EPA. *Background Information for Standards of Performance: Primary Aluminum Industry, vol. 1: Proposed Standards.* EPA-450/2-74-020a, U.S. EPA, Research Triangle Park, N.C., 1974.

4. Wadleigh, C. H. *Wastes in Relation to Agriculture and Forestry.* Misc. Publ. #1065, U.S. Dep't of Agriculture, Washington, D.C., 1968.

5. Eisenbud, M. *Environmental Radioactivity.* New York: Academic Press, 1973.

6. Harley, J. H. Radiological Contamination of the Environment, pp. 456–80 in *Industrial Pollution*, N. I. Sax (ed.). New York: Van Nostrand Reinhold, 1974.

7. Holaday, D. A. In: *Industrial Radioactive Waste Disposal.* Hearing before Joint Committee on Atomic Energy. Washington, D.C.: U.S. Government Printing Office, 1959.

8. Hodges, L. *Environmental Pollution.* New York: Holt, Rinehart and Winston, 1977.

9. Haagen-Smit, A. J. and Wayne, L. G. Atmospheric Reactions and Scavenging Processes, pp. 235–88 in *Air Pollution*, 3rd Ed., vol. 1, A. C. Stern (ed.). New York: Academic Press, 1976.

10. Niki, H., Daby, E. E., and Weinstock, B. Mechanisms of Smog Reactions. *Advan. Chem. Ser. 13*:16–57, 1972.

11. Health Effects Research Laboratory, EPA. *List of Organic Compounds Identified in Drinking Water in the United States.* U.S. EPA, Cincinnati, Ohio, 1976.

12. DeFreitas, A. S. W., Qadri, S. U., and Case, B. E. *Origins and Fate of Mercury Compounds in Fish.* In: Proceedings of the Int. Conf. on Transport of Persistent Chemicals in Aquatic Ecosystems, Ottawa, 1974.

4
Dispersion of Contaminants

INTRODUCTION

The contaminants generated by primary and secondary sources will tend to spread continuously within the environment. In this chapter, we will discuss how the characteristics of the medium affect the dispersion of contaminants within and between the atmosphere, the hydrosphere, the lithosphere, and the biosphere.

CONTAMINANT DISPERSION IN THE ATMOSPHERE

Diffusion in the Atmosphere

The dispersion of contaminants within the atmosphere is generally referred to as diffusion. In still air, the process occurs by molecular diffusion, and can be described by Fick's equation, which was developed in the 1880's for physiological applications. It states that the net rate of flow is directly proportional to the product of the concentration gradient and a constant known as the molecular diffusivity, or coefficient of diffusion. The coefficient decreases with increasing molecular or particle size, since the displacement resulting from a collision of the contaminant molecule or particle with an air molecule decreases as particle size increases.

For practical purposes, however, the dispersion of contaminants within the atmosphere by molecular diffusion is negligible, even for gaseous contaminants of small molecular sizes. This is because the displacements are generally infinitesimal compared to the displacements of the air volumes containing them by the turbulent motions of the air. Thus, atmospheric

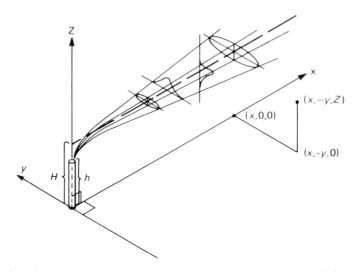

Fig. 4–1   Conventional coordinate system and basic geometry of plume dispersion. h=geometric (actual) stack height, H = effective stack height.

From: Mahoney, J. R. Meteorological Aspects of Air Pollution, pp. 409–55 in *Industrial Pollution*, N. I. Sax (ed.). New York: Van Nostrand Reinhold Co., 1974. © 1974 by Litton Educational Publishing, Inc. Reprinted by permission of Van Nostrand Reinhold Co.

diffusion is synonomous with convective or turbulent diffusion, which may be defined as the random motion and exchange of parcels (small volumes) of air between neighboring parts of the atmosphere. The dispersion of atmospheric contaminants depends primarily on the scale of this turbulence.

Atmospheric Turbulence

Atmospheric turbulence is such a complicated phenomenon that it has defied attempts to describe it in rigorous mathematic terms. In turbulent air, parcels exchange in irregular patterned motions termed eddies. The

Fig. 4–2   (a) Tracing of wind velocity showing fluctuations in the velocity. (b)– (e). The turbulent eddies that cause the fluctuations. The value of u fluctuates about its average, $\bar{u}$, so that $u - \bar{u} = u'$ at any instant in time due to turbulence. The fluctuations are due to the superposition of eddies, each of which has a different frequency and intensity. The contribution due to four eddies is illustrated in (b)– (e). The intensity of each eddy is labeled u (f_n) for turbulent components at frequency $f_n = Tn^{-1}$. For example, u (f_2) = intensity of component at frequency $f_2 = T_2^{-1}$, etc.

From: Williamson, S. J. *Fundamentals of Air Pollution*. Reading, Mass.: Addison-Wesley, © 1973.

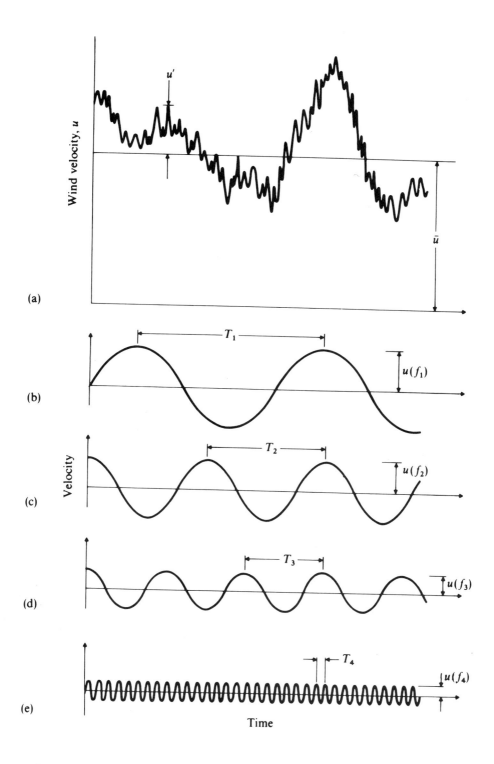

lack of a general theory of turbulence applicable to the prediction of the eddy diffusion of atmospheric contaminants has led to the development of empirical methods. Even the definition of turbulence remains a matter of debate. For our purpose, it is the almost random variation of the velocity of a fluid, in contrast to the constancy in steady-state streamline flow, or the periodicity of wave motion. In a practical sense, fluctuations about the mean or average wind speed are called turbulence. Such variations are associated with complicated eddy patterns of flow in both horizontal and vertical directions.

The characteristics of turbulence can depend upon whether the wind velocity component of interest is directed along the average horizontal wind direction, the transverse direction, or the vertical direction. The conventional coordinate system (Figure 4–1) has the x-axis along the wind direction, the y-axis along the transverse horizontal direction, and the z-axis in the vertical direction. The average wind speed is denoted by $\bar{u}$, and is along the x-axis. The fluctuating components of the wind along the x-, y-, and z-directions are denoted by u', v', and w', respectively.

Actual fluctuations in the wind velocity represent a superposition of a number of different-sized turbulent eddies (Figure 4–2). Some are related to known effects. The intensity of the high-frequency components increases with increasing wind speed and decreases with increasing altitudes, suggesting that they are associated with the friction of the air over the surface terrain. Lower frequency components have been observed to increase with increasing solar flux, suggesting that they are associated with thermal effects. Above about 500 meters, frictional effects from level ground have little effect on the motion of air masses. Over rougher ground, the effects of the surface friction would extend to greater altitudes.

When considering the extent and rate of contaminant dispersion, contaminant sources can be divided into three different categories. These are (a) point sources, such as tall stacks; (b) line sources, such as highways; and (c) area sources, such as whole urban regions. The simplest is an elevated point source, such as a power-plant smokestack. The light-scattering properties of the aerosol in the plume from such a stack, consisting of fly ash and condensed water, enable us to observe plume dispersion with the unaided eye.

Plume Dispersion

The concentration of a contaminant downwind from a stack will fluctuate with time as turbulence distorts the plume. The plume will meander over the ground, as illustrated in Figure 4–3a. At any given instant, the variation of the ambient contaminant concentration along the y-axis may appear as shown in Figure 4–3b. Over a longer period of observation, the plume will wander to either side of the average direction of the wind velocity, with the result that a one-hour average concentration at each point along the y-axis, as illustrated in Figure 4–3c, has a wider spatial extent but a lower maximum concentration. Thus, as a result of atmospheric turbulence, any one location downwind from the source will be subject to an ambient concentration which strongly fluctuates with time, but the average concentration over a long period of time will be considerably less than the peak values measured during short time intervals.

The variation of the plume in the y-direction is, therefore, due both to the spread of the plume and to the shifts in the downwind direction. The plume will also spread vertically along the z-axis. The vertical mixing of air is dependent upon the temperature profile of the atmosphere, i.e., the lapse rate. The immediate ground level concentrations of air contaminants may be reduced by good vertical mixing, since dispersal into higher regions would tend to dilute the contaminant. Poor vertical mixing may allow contaminants released at low altitudes to remain there in relatively concentrated form.

The simplest way to illustrate vertical mixing is in terms of the concept of a parcel of air. The parcel is a small volume of air having a uniform temperature throughout. Furthermore, if some disturbance causes this parcel to rise or fall in the atmosphere, it generally will do so adiabatically, i.e., with no heat exchange between the parcel and its surroundings.

A rising parcel of air expands and, therefore, according to the gas laws, will cool; a descending parcel will contract and become warmer. If the actual lapse rate of the atmosphere is identical to that of the adiabatically rising or falling parcel, the parcel will neither increase nor decrease its rate of ascent or descent. In other words, the parcel will be in equilibrium with the surrounding air. In this case, the atmospheric lapse rate is known as the adiabatic lapse rate. Since vertical motion is neither encouraged nor discouraged, the atmospheric condition is said to be neutral. Mixing only occurs if the parcel continues to be buoyant as it cools. The exact atmospheric lapse rate depends upon the amount of moisture in the atmosphere. The dry adiabatic lapse rate is approximately $10°$ C/km. In moist air, the water vapor has a greater heat capacity than the oxygen, nitrogen, and argon and may also undergo phase changes from vapor to liquid and vice versa, with release or absorption of the latent heat of vaporization. Thus, the actual moist adiabatic lapse rate is always less than the dry rate, rang-

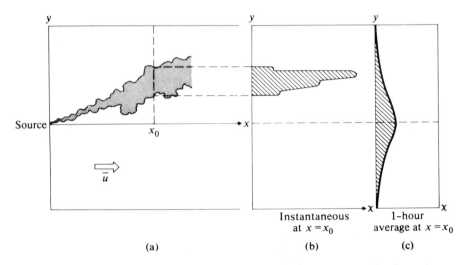

Fig. 4–3   (a) Instantaneous top view of a plume; (b) instantaneous horizontal profile of the plume concentration χ along a transverse direction at some distance
downwind from the source; (c) one-hour average profile for the same downwind
distance.
From: Williamson, S. J. *Fundamentals of Air Pollution.* Reading, Mass.: Addison-Wesley, ©1973.

ing as low as 3.5° C/km depending on the degree of moisture. The average
tropospheric lapse rate is about 6.5° C/km.

Because of the various energy-transport processes occurring in the atmosphere, the actual lapse rate is very often different from the adiabatic
lapse rate.

If the lapse rate is greater than the adiabatic lapse rate, i.e., the ambient
temperature decreases more rapidly with altitude than the adiabatic rate,
the rising air parcel will be warmer, and less dense, than the surrounding
air. It will, therefore, continue to rise, since upward motion is accelerated.
The lapse rate for this situation is called superadiabatic; the atmosphere is
said to be unstable and vertical mixing is facilitated (Figure 4–4).

If the actual lapse rate is less than the adiabatic rate, the parcel of air
will begin to rise until it reaches a point where its temperature equals that
of the surrounding air; upward movement would then cease, opposed by
negative buoyancy. This atmospheric condition leads to a stable condition. The parcel rises (or falls) only a short distance until it reaches a point
of temperature equilibrium (Figure 4–5).

An extreme case of atmospheric stability occurs when the atmospheric
lapse rate is negative, i.e., temperature increases with altitude (Figure

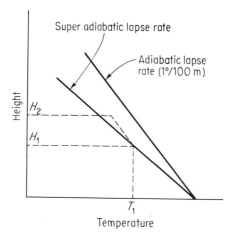

Fig. 4–4   The instability of the superadiabatic atmosphere. A parcel of air raised in height from H₁ to H₂ cools adiabatically, and its rate of rise is accelerated because it becomes warmer and, therefore, less dense than the ambient atmosphere.
From: Eisenbud, M. *Environmental Radioactivity – 2nd Ed.* New York: Academic Press, 1973.

4–6). This condition is known as a temperature inversion. The atmospheric strata within which the inversion occurs is termed an inversion layer. Vertical air movement within inversion layers is essentially nil (low-mixing) and contaminants accumulate within them. These layers are frequently visible because of the light-scattering properties of contaminant aerosols contained within them.

Depending upon how they are formed, inversion layers may have their base on the ground or they may be elevated. Figure 4–7 shows the development of a ground-based inversion, which occurs because the lapse rate near the ground is strongly influenced by radiative heat transfer. This property frequently results in a diurnal variation in lapse rate near ground level.

The effects of lapse rate on the vertical spread of a plume are illustrated in Figure 4–8. In terms of the potential for adverse effects on people, animals, plants, and buildings, we are generally concerned about the ground-level concentrations rather than the concentrations aloft within the plume, and since most effects are concentration dependent, we are most concerned with the maximum ground-level concentration. Thus, for stack discharges, as illustrated in Figure 4–8, the worst condition is fumigation, where the emissions are all retained within the static inversion layer. At the other extreme, the looping that occurs with a very strong lapse can

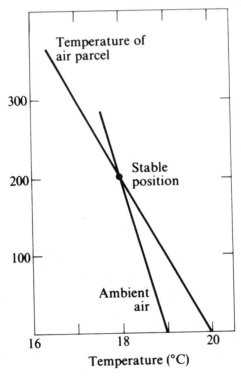

Fig. 4–5   A rising parcel of air will cease its upward motion if at some point its temperature is identical to the ambient temperature and the ambient lapse rate is less than the adiabatic rate.

From: Williamson, S. J. *Fundamentals of Air Pollution*. Reading, Mass.: Addison-Wesley, © 1973.

also bring the plume to the ground close to the stack. However, the time of contact at any one location will be very brief, and the high degree of turbulence will rapidly dilute the contaminant.

Equations have been developed for predicting the downwind concentrations of contaminants discharged from elevated point sources. These are all empirical, since, as discussed earlier, no adequate theory has yet been developed which can cope with the complexity of the atmospheric turbulence.

Even those equations which are based on clearly artificial constraints appear to be formidable to the nonmeteorologist. For example, if we make the simplifying assumption that the average wind speed, $\bar{u}$, does not vary with position, and if contaminants are emitted from an elevated point

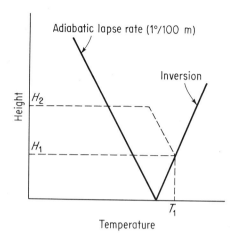

Fig. 4–6 The inherent stability of the inverted temperature gradient. A parcel of air raised in height from H_1 to H_2 cools adiabatically and sinks to its original position because it becomes more dense than the ambient atmosphere.

From: Eisenbud, M. *Environmental Radioactivity – 2nd Ed.* New York: Academic Press, 1973.

Fig. 4–7 Typical diurnal variation of temperatures near the ground. 4 A.M.: Radiation from earth to black sky cools ground lower than air producing a ground-based inversion. 9 A.M.: Ground heated rapidly after sunrise. Slightly subadiabatic. 2 P.M.: Continued heating. Superadiabatic. 4 P.M.: Cooling in the afternoon returns the temperature profile to near adiabatic.

From: Seinfeld, J. H. *Air Pollution: Physical and Chemical Fundamentals.* New York: McGraw Hill, © 1975. Used with permission of McGraw-Hill Book Co.

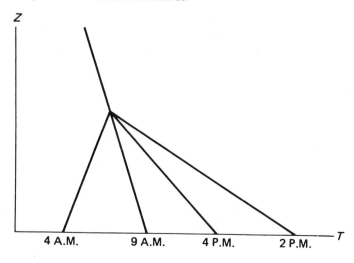

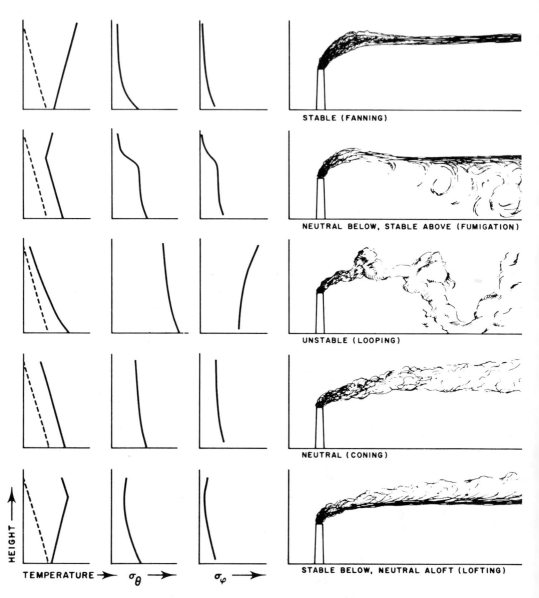

Fig. 4–8 Various types of smoke-plume patterns observed in the atmosphere. The dashed curves in the left-hand column of diagrams show the adiabatic lapse rate, and the solid lines are the observed profiles. The abscissas of the columns for the horizontal and vertical wind-direction standard deviations (σ_θ and σ_ψ) represent a range of about 0° to 25°.

From: Slade, D. H. (ed.). *Meteorology and Atomic Energy*, 1968. U. S. Atomic Energy Commission, TID-24190, Oak Ridge, Tenn., 1968.

source, the downwind ground level $(z = 0)$ concentration can be approximated by:

$$\chi_{(x,\,y,\,0)} = \frac{Q}{\pi\sigma_y\sigma_z\bar{u}}\,e^{-\left(H^2/2\sigma_z^2 + y^2/2\sigma_y^2\right)} \tag{4-1}$$

where

χ = concentration of contaminant in mass per cubic meter at $(x,y,0)$
Q = emission rate from the source in mass per unit time
σ_y = standard deviation of concentration in y-direction
σ_z = standard deviation of concentration in z-direction
H = effective height of discharge above the ground in meters
y = crosswind distance in meters
x = distance downwind in meters

Power-plant stacks emit hot gases with an upward velocity. These emissions are, therefore, known as buoyant plumes, since they tend to rise until their momentum is spent and they reach thermal equilibrium with the surrounding air. This buoyancy produces an effective increment to the actual physical stack height. The effective stack height (H) used in dispersion equations includes both the actual height and the increment due to buoyancy (Figure 4-1).

If one's interest is restricted to the ground-level concentration directly downwind of an elevated source, i.e., where $y = 0$, Eq. $(4-1)$ reduces to:

$$\chi_{(x,\,0,\,0)} = \frac{Q}{\pi\sigma_y\sigma_z\bar{u}}\,e^{-\left(H^2/2\sigma_z^2\right)} \tag{4-2}$$

A feature of Eq. $(4-1)$ and $(4-2)$ is the fact that the maximum concentration depends upon the mass emission rate. For a given release height, H, it is independent of the concentration of the contaminant in the exhaust gas. Thus, emitting a more dilute mixture without decreasing the mass emission rate will not reduce the ground-level concentration. The implicit assumption is that only contaminant—and no air—is emitted from the stack, and that dilution in ambient air after emission is the primary means by which the concentration is reduced. If the ambient concentration downwind from an elevated point source is given by Eq. $(4-1)$ with empirical values for the x-dependence of σ_y and σ_z, then several statements can be made concerning the magnitude and distribution of the ground-level concentration. The location which received the maximum ground-level concentration will be, of course, on the plume line $(y=z=0)$. This location depends upon the empirically determined dependence of the standard deviation σ_z on the value of x. For lack of simpler or more accurate alternatives, a power law dependence is often assumed, as introduced by O. G. Sutton:

$$\sigma_y^2 = x^{(2-n)} C_y^2/2$$

$$\sigma_z^2 = x^{(2-m)} C_z^2/2 \qquad (4\text{-}3)$$

where n and m are appropriate numbers representing atmospheric stability factors; these numbers, together with the constants C_y and C_z, are called "diffusion parameters." Thus, for the case where $n = m$, and $C_y = C_z = C$, so that σ_y/σ_z is independent of x, the point of maximum ground-level concentration is given by

$$x = (H^2/C^2)^{\left(\frac{1}{2-n}\right)} \qquad (4\text{-}4)$$

Under the same conditions, the ambient concentration at this location can be obtained by evaluating Eq. $(4\text{-}1)$, with the result

$$\chi_{max} = \frac{2Q}{e\pi\bar{u}H^2} \qquad (4\text{-}5)$$

Representative values of n and C^2 are shown in Table 4–1.

Under these simplifying assumptions, it can be seen from Eq. $(4-5)$ that the maximum ground-level concentration is directly proportional to the source strength and inversely proportional to the wind speed. It is also inversely proportional to the square of the effective stack height. Thus, tall stacks, i.e., greater than about 100 m, can greatly reduce local ground-level concentrations due to two factors. One is that the maximum ground-level concentration varies as the inverse square of height. The other is that they generally reach above the morning inversion layer and, as shown in Figure 4–8, this results in the condition known as lofting, where the plume theoretically never reaches the ground.

The preceding brief discussion of dispersion from an elevated point source was limited to the simplest, most basic relations and to their application under many simplifying assumptions. They were developed for applications close to the source. The situation becomes much more complex when considering other types of sources. For example, line sources such as roadways are spatially extended, and the wind direction with respect to source geometry influences the downwind dispersion of contaminants. In area sources, contaminants are received not only from sites directly upwind, but also from those to the sides as well.

Table 4.1   Values of n and C^2 for Several Lapse Conditions

Condition	n	C^2			
		at 25 m	at 50 m	at 75 m	at 100 m
Unstable	0.20	0.043	0.030	0.024	0.015
Neutral	0.25	0.014	0.010	0.008	0.005
Moderate inversion	0.33	0.006	0.004	0.003	0.002
Large inversion	0.50	0.004	0.003	0.002	0.001

For more detailed and accurate equations, and for equations which can be used to predict concentrations more than a few kilometers from the source, there are a number of good texts (1, 2, 3) and reference manuals (4, 5, 6) available.

Effects of Surface Features

In addition to the imperfections in all of the available empirical relationships for predicting downwind concentrations of airborne effluents, there are a number of other complicating factors which limit their general application.

One key assumption is that there is a free flow field around the top of the stack. This is frequently not the case. Figure 4–9 shows a common complication. Many power plants and factories are built within river valleys because of their need for large volumes of cooling water and/or waterborne transportation (7). In deep valleys, even a tall stack may not extend above the valley walls. Table 4–2 describes some of the effects of natural surface features upon plume dispersion.

The ability of a stack to disperse contaminants can also be diminished by the presence of a large building in the vicinity. The plume becomes distorted, even if it does not actually contact the building. This effect occurs because the plume is carried in an air stream that accommodates itself to the shape of the building. If the airflow is disturbed locally, that portion of the plume which penetrates the disturbed flow region will also become distorted. Plume distortions near buildings are controlled by the local air motions.

Figure 4–10 shows characteristic flow zones around a sharp-edged cubical building oriented with one wall normal to the wind direction. The background flow, or the flow which would have existed in the absence of the building, is shown at the left, where the streamlines are horizontal. The mean velocity increases upward from zero at the ground, rapidly at first

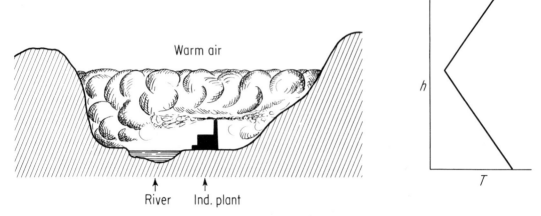

Fig. 4-9 Fumigation of a valley floor caused by an inversion layer that restricts diffusion from a stack.

From: Eisenbud, M. *Environmental Radioactivity - 2nd Ed.* New York: Academic Press, 1973.

Table 4.2 Effects of Terrain Features Upon Plume Dispersion

Topographic feature	Effects
1. Elevated regions	a. Increased wind speed (and increased ventilation) over hill tops. b. Occasional impacting of elevated plumes on ground level.
2. Deep valleys	a. Channeling of wind flow along the valley axis, resulting in higher average concentrations in the valley. b. Development of stable drainage winds during calm, nighttime conditions, resulting in higher concentrations along the valley floor.
3. Undulating regions	a. Increased atmospheric turbulence near the ground level during times of moderate or strong winds. This results in lower pollutant concentrations at locations near sources. b. Accumulation of pollutants in low spots during calm, nighttime conditions (i.e., localized drainage-wind conditions).
4. Regions of tree cover	a. Enhanced turbulence near the ground during moderate or strong winds, resulting in lower concentrations for locations near sources. b. In fully covered regions, blockage of elevated plumes, resulting in lower concentrations at ground level.
5. Bodies of water	a. Increased moisture content in the local atmosphere, favoring fog formation at low-lying spots, and affecting the removal rate of SO_2 and other pollutants from the atmosphere. b. For larger bodies of water, formation of local circulation (lake and sea breezes) which can cause ground-level fumigation on the landward side of sources, during sunny daytime conditions.

From: Mahoney, J. R. Meteorological Aspects of Air Pollution, pp. 409–55 in *Industrial Pollution*, N. I. Sax (ed.). New York: Van Nostrand Reinhold, 1974. © 1974 by Litton Educational Publishing, Inc. Reprinted by permission of Van Nostrand Reinhold Co.

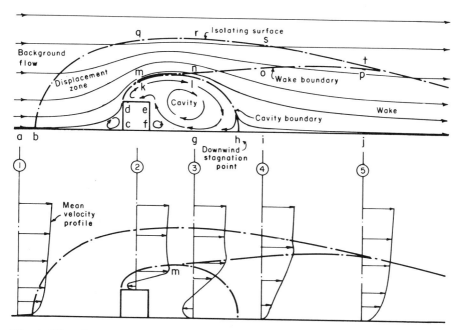

Fig. 4–10   General arrangement of flow zones near a sharp-edged building.

From: Slade, D. H. (ed.). *Meteorology and Atomic Energy*, 1968. U. S. Atomic Energy Commission, TID-24190, Oak Ridge, Tenn., 1968.

and more slowly at high elevations. In the center of the figure, the building creates a disturbance in the flow, whose main characteristic is the highly turbulent wake.

Within the wake, adjacent to the ground and walls and roof of the building, there exists a region called a cavity; in the cavity, the mean flow is in the direction of the background flow in the outer portion and opposite to the background flow near the axis. Changes in building shape and orientation to the wind affect the cavity dimensions and flow. Rounded buildings have smaller displacement zones and wakes.

Contaminant Removal to Surfaces

All of the preceding discussion was based on the behavior of the inert gas within the plume. It does not take into account removal of contaminant by precipitation or by contact with the surface of the ground or vegetation. It also does not take the gravitational sedimentation of particles into account, although the sedimentation rate of particles smaller than 20 μm

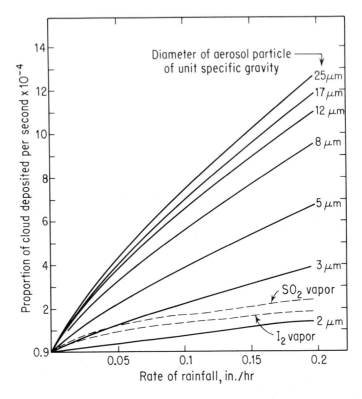

Fig. 4–11 Percentage removal of particles according to particle size and rate of rainfall.

From: Eisenbud, M. *Environmental Radioactivity – 2nd Ed.* New York: Academic Press, 1973.

diameter is so low that there is little significant error introduced when applying the empirical equations to the turbulent diffusion of small particles within the air.

Empirical correction factors for the diffusion formulas have been developed to account for the capture of particles smaller than 20 μm diameter and reactive gases by ground-level surfaces. They employ the concept of deposition velocity (v_g), which is determined experimentally and defined as follows:

$$v_g = \frac{\text{mass deposited per unit area per unit time}}{\text{mass concentration above the surface}}$$

The product of the mass concentration given by the diffusion equation and v_g in m/sec will be the deposition in mass/m²/sec.

Table 4.3   Summary of Effects of Meteorological Conditions Upon
Plume Behavior

Condition	Effect
Wind speed	Determines initial dilution
Wind direction	Determines downwind geometry
Atmospheric stability category	Determines plume spread associated with turbulent motions in the atmosphere
Mixing layer depth (or height of inversion base)	Determines limit to the vertical spread of the plume – important for downwind flow distances greater than 1,000 m
Humidity	High humidity associated with visibility decreases for water-vapor plumes
Surface temperature	Determines possibility of ice formation from water-vapor plumes
Precipitation	Determines possibility of washout of contaminants near the source

The effect of precipitation on the removal of particles and soluble gases can be accounted for using a washout factor rather than a deposition velocity. The washout factor is defined as the fraction deposited per unit time, and some experimental values are illustrated in Figure 4–11.

Table 4–3 provides a summary of the effects of various meteorological factors upon plume dispersion.

Dispersion from Ground-Level Sources

In contrast to elevated sources, most of which can be considered discrete point sources whose impacts on the environment are at distant sites, most ground-level sources have their major effects immediately downwind. Furthermore, since such emissions occur within the air most affected by surface roughness, they rapidly become well mixed into the surface air, a layer of varying height which normally extends from about 100 m to perhaps 2,000 m. As illustrated in Figure 4–10, the effluent from a short stack or roof vent will be influenced by the aerodynamic flow pattern around the building(s) with which it is associated. In estimating the downwind concentration patterns caused by this phenomenon, the most conservative approach is to assume that the effluent is trapped in the wake and brought quickly to the ground without being enveloped and mixed in the cavity directly behind the structure.

If the effluent is trapped in the cavity, it will be mixed rapidly into a volume determined essentially by the cross-sectional area of the structure and the wind speed. An appropriate formula or the downwind concentration is:

$$\chi = \frac{Q}{(\pi\sigma_y\sigma_z + cA)\bar{u}} \, e^{-[y^2/2\sigma_y^2]} \tag{4-6}$$

where A is the cross-sectional area of the building normal to the wind and c is a shape factor, ranging from $1/2$ for a relatively streamlined shape to 2 for a blunt building. This formula gives a maximum concentration directly downwind of the building (σ_y, $\sigma_z = 0$) of $Q/cA\bar{u}$, a simple volume approximation.

In urban areas where there are many buildings with short stacks and roof vents, the situation becomes more complex, but the likelihood of thorough mixing of the various effluents is generally quite high. There is usually relatively little variation of concentration with height for CO and Pb emitted at street level, or for SO_2 emitted from rooftops of private homes and apartment houses.

Long-Range Transport in the Troposphere

While elevated discharges are an effective means of reducing maximum ground-level concentrations, they obviously do nothing to diminish the total amounts of contaminants emitted into the atmosphere. In fact, by minimizing contact between the ground-level surfaces and the contaminants, they actually act to increase the residence times and total atmospheric burdens of the contaminants. The development of a persistent haze layer over the eastern third of the continental United States in recent decades appears to be due in large part to the sulfate particles in the air which are transported in the prevailing winds over many hundreds of kilometers. These in turn were formed in the atmosphere from the airborne oxidation and hydrolysis of SO_2, much of which was discharged at elevated levels by the tall stacks of utility boilers. Even in the vast continental atmosphere, dilution may no longer be the solution to pollution.

The spread of environmental contaminants via the air depends upon their physical and chemical forms. Nonreactive and water-insoluble gases, vapors, and submicrometer particles can remain airborne for weeks within the troposphere and can, therefore, spread throughout the world. Evidence for such transport of chemical contaminants via the atmosphere is the presence of lead, nuclear weapons test debris, and various organic pesticides in samples of polar ice collected at locations which are remote from known sources of such contaminants.

Tropospheric-Stratospheric Interchange

The transport of contaminants from near-surface sources covers time scales measured in hours and days. When contaminants are generated by thermonuclear explosions, they can spread throughout the atmosphere on a global scale, following injection of some of the debris into the stratosphere.

Our knowledge of stratospheric dispersion has been derived from gas and dust samples obtained by aircraft or balloons at an altitude of about 35 km. On two occasions, nuclear weapons were exploded at unusually high altitudes, and unique tracers were intentionally added to facilitate the dispersion studies. ^{102}Rh was injected into the stratosphere by an explosion at an altitude of 43 km above Johnston Island in the Pacific in August, 1958. The cloud from this nuclear explosion was believed to have risen to about 100 km. In July, 1962, an explosion conducted about 400 km above Johnston Island contained a known amount of ^{109}Cd.

Stewart et al. developed a model of stratospheric-tropospheric exchange that is consistent with the observed pattern of nuclear weapons fallout (8). According to this model, air enters the stratosphere in the tropical regions, where it is heated and rises to an altitude of about 30 km, at which level it begins to move toward the poles. As shown in Figure 2–6, the tropopause is lower in the polar regions than at the equator, and tropopause discontinuities in the temperate regions facilitate transfer from the stratosphere to the troposphere. The jet streams, with velocities of 100 to 300 km/hr, occur at these discontinuities. The rate of transfer from the lower stratosphere is most rapid in the winter and early spring.

CONTAMINANT DISPERSION IN THE HYDROSPHERE

The aquatic environment is considerably more varied and complex than the atmosphere with respect to contaminant dispersion. There are large differences in dilution volumes, mixing characteristics, and transport rates between rivers, lakes, estuaries, coastal waters, and oceans, making a generalized approach to the dispersion of contaminants introduced into bodies of water not possible with the current state of knowledge. Also, there is a much more active interchange of contaminants between the hydrosphere and the biosphere than between the atmosphere and biosphere.

While this section will be concerned primarily with contaminant transport within the abiotic hydrosphere and a subsequent section will be con-

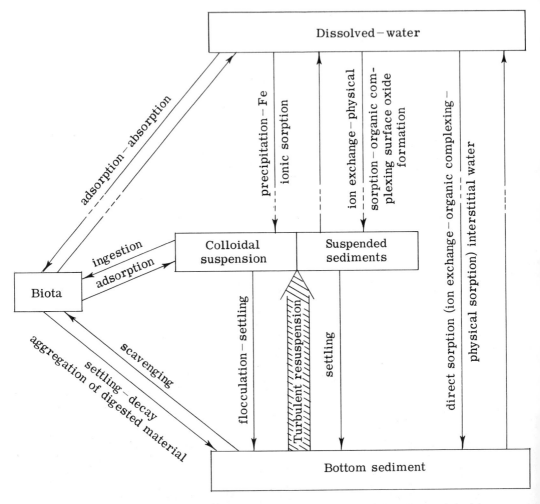

Fig. 4–12 Routes of contaminant transfer between the hydrosphere and the biosphere.

From: Eisenbud, M. *Environmental Radioactivity – 2nd Ed.* New York: Academic Press, 1973.

cerned primarily with transport within the biosphere, it is not really possible to treat either in isolation. Some of the possible routes of contaminant transfer between the hydrosphere and the biosphere are shown in Figure 4–12.

Precipitation and Surface Runoff

The atmosphere is a source of chemical contaminants for the surface of the earth, for example, by gravitational settlement and via precipitation. Precipitation on the land becomes surface runoff, which picks up additional contamination by dissolution and suspension of contaminants deposited on the surface since the last rain. On the other hand, the land surface also acts as a filter and chemical buffer to the surface runoff, reducing part of its contaminant load. During light precipitation, the runoff may become relatively concentrated with dissolved contaminants which do not react with nor are readily sorbed by surface soil; at the same time, it would be relatively well filtered of suspended solids. On the other hand, during heavy precipitation the soluble contaminants would be very much diluted, while the turbulent flow would stir up relatively high concentrations of suspended solids.

Streams, Rivers, and Lakes

Rivers and streams vary so much in bed structure, depth, and flow rate that it is not possible to generalize dispersion rates for the contaminants discharged into them. Further complications are added for "managed" rivers. Most of the larger rivers in the United States have been extensively dammed for purposes of flood control, hydroelectric power generation, and navigation, and large stretches of them are really a series of shallow lakes rather than a free-flowing river. There are also major withdrawals of river water for drinking-water supplies, industrial and utility usage, and irrigation of agricultural lands. Some of these waters, especially the industrial and utility process and cooling waters, are largely returned to the river, albeit with added contaminant loads. On the other hand, the irrigation waters are largely lost to evaporation, ground-water recharge, and possibly to drainage into another watershed. In an extreme case, such as the Colorado River where withdrawals by the United States and Mexico are equal to the total flow, none of the water leaves the mouth of the river.

Lakes have the additional component of greater depth of water. As discussed in Chapter 2, the water can become thermally stratified, with the lower portion (hypolimnion) becoming isolated from oxygen sources. In nutrient-rich lakes, this results in complete oxygen depletion and anaerobic decay within the hypolimnion. The spring and fall turnovers of the waters in the lake then spread the products of this decay throughout the lake.

The natural dispersion of wastewater effluents in lakes is often poor. In the absence of wind and tide there is little turbulent mixing. Lake waters are normally colder and, therefore, heavier than wastewater effluents. Effluents discharged at or near the surface of denser receiving waters are likely to overrun them. Discharged at some depth below the surface, they rise like smoke plumes and, on reaching the surface, fan out radially. Because chemical diffusion is slow, natural dispersion or mixing of the unlike fluids is mainly a function of winds and currents.

Estuaries

The spread of contaminants within estuarine waters depends heavily on the structure of the estuary. In some, the sea water forms a distinct wedge beneath the fresh water. More typically, there is sufficient turbulent mixing to create a more gradual change from fresh to salt water. At the other extreme, there is so much turbulent mixing that both the horizontal and vertical concentration gradients disappear.

Estuary flow is so complicated that there is no generalized dispersion model; the physical characteristics of each estuary must be studied on an individual basis. For example, in a study of diffusion and convection in the Delaware River Basin, Parker et al. (9) constructed a scale model of the Delaware Basin at the United States Army Waterways Experiment Station at Vicksburg, Mississippi. The model was 750 ft long and 130 ft wide, and the expected dilution was studied using dyes. Figure 4–13 illustrates the type of information obtained from a study of the dilution of dye over a period of 58 tidal cycles (about one month). In this particular study, which involved instantaneous injection of a given dose, the concentration at the end of the 58 cycles remained at approximately 1% of the maximum concentration during the initial tidal cycle. The conditions of the experiment were conservative: There were no losses due to sedimentation or biological uptake. Thus, only the diminution owing to mixing was measured.

The Oceans

Dispersion of contaminants in ocean waters depends primarily on the characterisitic physical features discussed in Chapter 2. These are the

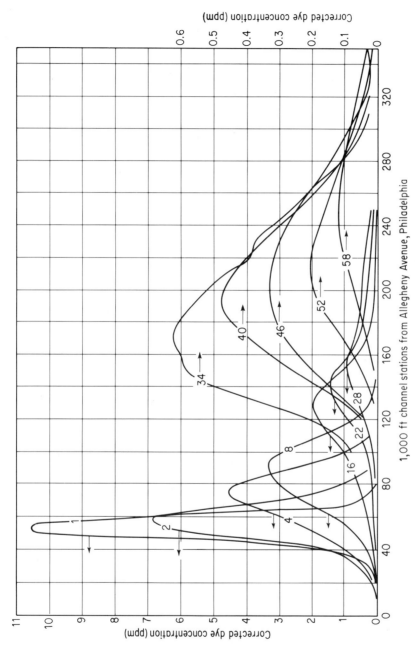

Fig. 4–13 Longitudinal distribution of contaminants after a designated number of tidal cycles in the Delaware River near Philadelphia.

From: Parker, F. L., Schmidt, G. D., Cottrell, W. B. and Mann, L. A. Dispersion of Radiocontaminants in an Estuary. *Health Physics* 6:66, 1961. By permission of the Health Physics Society.

temperature, density, and salinity gradients with depth, and the surface currents.

Vertical mixing is vigorous within the surface zone due to wind action, but is very poor within the lower parts of the intermediate zone. This is illustrated in Figure 4–14, which shows the distribution of radioactive debris at 6, 28, and 48 hours after it was deposited on the surface of the water.

Since material deposited on the surface of the ocean mixes very slowly into the deep water, it is possible to find measurable amounts of residual

Fig. 4–14   The vertical distribution of radioactivity in the ocean at selected times after fallout of debris from a nuclear explosion.

From: Lowman, F. G. Marine Biological Investigations at the Eniwetok Test Site, pp. 105–38 in *Proc. Disposal Radioactive Wastes*, IAEA, Vienna, 1960.

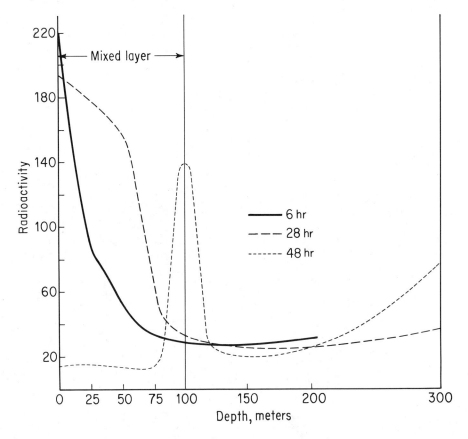

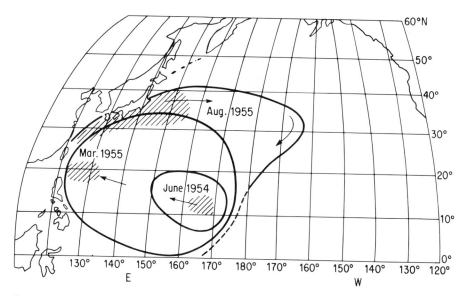

Fig. 4–15   The horizontal dispersion of nuclear weapons debris in the north Pacific Ocean after tests by the United States in the Marshall Islands, 1951. Striped regions indicate areas of maximum contamination.

From: Miyake, Y. and Saruhashi, K. Vertical and Horizontal Mixing Rates of Radioactive Material in the Ocean, pp. 167–73 in *Proc. Disposal of Radioactive Wastes*. Vienna: IAEA, 1960.

material on the surface at a considerable time following deposition. For example, Figure 4–15 shows the first year's path for weapons test debris from a June, 1954 test. Comparison of this figure with Figure 2–11 shows that the debris followed the normal circulation of ocean surface water.

While the rate of movement of the bottom water has not been mapped, Koczy (10) estimated rates of vertical diffusion from measurements of radium and other substances dissolved in ocean waters. Dissolved ^{238}U gives rise by radioactive decay to ^{230}Th, which precipitates rapidly to the bottom sediments. The decay of ^{230}Th results in the formation of ^{226}Ra, which tends to return into solution at the ocean bottom. The vertical gradient of dissolved radium is a measure of the rate of movement of bottom water toward the surface. Koczy's model of vertical diffusion is shown in Figure 4–16. Dissolved substances released from the ocean floor diffuse slowly through a friction layer 20 to 50 m in depth, where diffusion rates are on the order of molecular diffusion. Mixing is most rapid (3–30 cm²/sec) just above the friction layer and decreases rapidly with height above the ocean

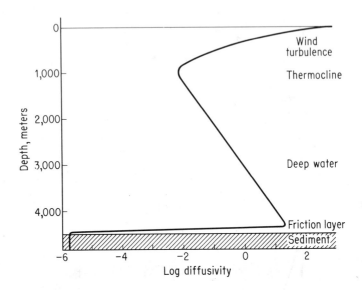

Fig. 4-16   Vertical diffusion in the oceans.

From: Koczy, F. F. The Distribution of Elements in the Sea, pp. 189-97 in *Proc. Disposal of Radioactive Wastes*, Vienna: IAEA, 1960.

floor to a level about 1,000 m below the surface, where a secondary minimum (10^{-2} cm²/sec) is thought to exist. Diffusion rates then increase as one approaches the mixed layer, where diffusion coefficients ranging from 50 to 500 cm²/sec are found.

Koczy's estimate of vertical transport in the Atlantic Ocean is between 0.5 and 2 m/year at depths ranging from 750 to 1,750 m. If these values apply at greater depths, a solution placed at a depth of 3,000 m would not appear in the surface water for more than 1,000 years.

Limited areas of the oceans are characterized by upwellings which carry water from deep regions to the surface via vertical currents having a velocity of a few meters per day. These upwellings could facilitate the vertical transport of contaminants.

CONTAMINANT DISPERSION IN SOIL

Like the hydrosphere, the soil environment is quite complex with regard to contaminant dispersion. The different soil types have individual characteristics which affect the movement of chemicals introduced, intentionally or otherwise, into or onto the soil.

Contaminant dispersion in soil depends upon the specific nature of the chemical, the soil type, and other soil factors such as mositure, pH, and temperature. Except for gases, which readily diffuse through the air channels, most chemical contaminants do not readily move once they enter the soil. The main processes which affect movement of contaminants through soil are adsorption, leaching, and diffusion.

Chemicals may become dispersed in soils via movement with the soil water. The major direction of movement is downward to lower horizons, but it may also be upward, due to evaporation of water, or even laterally, due to surface runoff. Downward movement is, of course, the major direction in soils through which there is rapid percolation of water. As the rate of percolation slows in deeper layers, chemicals may diffuse into the pores between soil particles, and be further dispersed and redistributed.

Many chemicals become adsorbed to soil particles via various processes; these are discussed more fully in Chapter 5. However, since adsorption affects subsequent leaching and diffusion, it may restrict the distribution of a contaminant. Chemicals which are tightly bound to soil particles are poorly leached by percolating water. Many high molecular weight halogenated organic compounds, such as DDT and PCB, show little movement, even after several years of exposure to water percolating through the soil. In this case, their low degree of water solubility is also a factor. As other examples, ammonium and phosphate ions are strongly bound to soil particles, while nitrate is not and readily leaches through the soil. Thus, strongly adsorbed contaminants may be found only in the upper few inches of the soil, while those less strongly bound readily travel through lower horizons. These are, of course, generalizations, since binding and subsequent leaching of any contaminant is highly dependent upon the specific soil type.

Little is known of the rate at which chemical contaminants migrate through different types of soils under different environmental conditions, i.e., degree of drainage, pH, degree of moisture. The migration of radionuclides has been examined to some extent. For example, Spitsyn et al. (11), analyzed the movement of ^{90}Sr in soil having a moderately high ion exchange capacity and which was permeated with ground water. The annual movement downward was found to be less than 5 m. Thus, most ^{90}Sr from fallout would tend to be held tightly in the upper soil layers.

CONTAMINANT DISPERSION IN THE BIOSPHERE: TRANSPORT THROUGH FOOD CHAINS

In the biosphere, essential nutrients and energy are tranferred from organism to organism along pathways called food chains. Very often, single chains are interconnected, resulting in a more complex interlocking pattern termed a food web.

Solar energy is utilized by organisms known as producers, for the assimilation of simple, inorganic chemicals (nutrients) into energy-rich compounds. The primary producers are green plants, whose chlorophyll absorbs solar radiation, and which produce organic compounds from CO_2 and H_2O via photosynthesis. The potential food energy and nutrients within the plants are then transferred to the primary consumers, or herbivores, i.e., organisms that eat plants. By eating the herbivores, carnivorous secondary consumers obtain energy and nutrients indirectly from plants. Food webs can also involve omnivores, organisms intermediate between the primary and secondary consumers, who eat both plants and animals. Carnivorous tertiary consumers feed on secondary consumers, etc., with the number of levels of carnivores varying with the specific biotic community. The organic compounds in waste products and in dead plant and animal bodies provide the substrate for the decomposers, largely bacteria, which break down these compounds, releasing inorganic nutrients. A generalized food chain is shown in Figure 4–17.

Organisms which obtain their nutrients from plants by the same number of steps along a food chain are said to belong to the same trophic level. Green plants are the first level, herbivores the second, and so on. Trophic structure is often graphically represented as an ecological pyramid, with producers comprising the base of the pyramid, and successive trophic levels making up the tiers. Pyramids may be based upon the total number of organisms in each level, their biomass, or their energy content. One such pyramid is shown in Figure 4–18. Man's place in the ecological pyramid is generally at the apex, i.e., at the end of food chains. Thus, man is very dependent upon the natural environment for his nutrient needs.

Transport via food chains is an important route by which chemical contaminants may reach man. In addition, damage to the environment may also occur along the transport chain.

Contamination of food chains may occur in the aquatic or terrestrial environments. There is, however, never a sharp delineation between these two, and transport of chemicals often occurs between them. For example, carnivorous birds may consume contaminated marine organisms, then

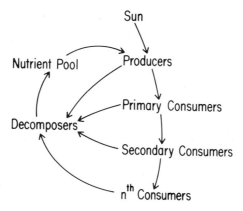

Fig. 4–17   A generalized food chain.

deposit their droppings on land. Man, a terrestrial organism, often consumes marine creatures directly.

In aquatic environments, contaminants may become adsorbed on solids suspended in the water. The solids may eventually settle to the bottom, deposit on the surfaces of aquatic plants, or be directly taken up by filter-feeding marine organisms. Bottom sediments receive the settling remains of dead organisms or deposits of contaminated excreta. Dissolved chemicals may also be directly taken up by plants or animals.

The sediments often provide the organic substrate which supports the

Fig. 4–18   A trophic pyramid, arranged according to biomass, in an aquatic ecosystem in Silver Springs, Florida. The numbers represent gm of dry biomass per m^2. P = Producers; C_1 = Primary Consumers (herbivores); C_2 = Secondary Consumers; C_3 = Tertiary Consumers.

From: Odum, E. P. *Fundamentals of Ecology*. Philadelphia: W. B. Saunders, 1971.

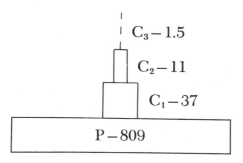

benthic organisms. Contaminants taken up by the latter may then be passed to higher trophic levels, and then back to the sediments, or possibly to man via his consumption of aquatic biota. This chain of contaminant uptake bypasses the primary trophic level, since chemicals may directly enter the food chain at a consumer level.

As in aquatic environments, contaminants generally enter terrestrial food chains through the producers. Incorporation into green plants may occur by root uptake from the soil or by absorption following deposition upon foliar surfaces. Uptake from soil depends upon the amount of time the contaminant remains within the root zone and, of course, whether it is in a form usable by the plant. The structure of a plant is a prime determinant of its ability to act as an efficient trap for contaminant deposition.

In some biota, the concentrations of particular chemicals may be greater than those in the surrounding environment. The process by which this occurs is called biological magnification or bioconcentration. The degree of increase is often indexed by a concentration factor (CF); this is the ratio of the contaminant concentration in the organism to that in its environment. The effects of bioconcentration may lead to reduced populations among members of a food chain. However, when the acute toxicity of a contaminant is low, harmful effects may be delayed until chronic symptoms develop. By this time, damage to the biota may persist long after emissions are curtailed. Bioconcentration poses the greatest threat to those consumers, such as man, in the upper trophic levels.

In some cases, a process of biological discrimination may occur, i.e., an organism actually has a lower concentration of a contaminant than is present in its environment. For example, carrots, potatoes, and alfalfa which were grown in soils having $0.48-8.36$ ppm$_w$ of aldrin residues were found to contain only $0.009-0.32$ ppm$_w$ of the pesticide (12).

Examples of Food-Chain Transport of Contaminants

RADIONUCLIDES

Most of the basic chemical elements required by plants are obtained from the soil in terrestrial environments, or from the water and sediments in aquatic environments. As these nutrients are taken up, so may contaminants. This is especially true of many radionuclides.

The property of radiation itself does not usually affect the uptake of an element, although once absorbed, tissue damage may ensue. A biological system will incorporate material based solely upon its biochemical properties, without discrimination between radioactive and nonradioactive isotopes. The advanced technology available for isotope identification and quantitation has made it possible to do tracer studies with radionuclides, which have provided some of the best information on food-chain transport of contaminants.

Studies in the Pacific Ocean atolls performed following nuclear testing in the 1950's showed there existed a difference between the types of radionuclides which enter aquatic and terrestrial food chains. Marine organisms contained highest amounts of radionuclides which either formed strong complexes with organic matter (e.g., ^{54}Mn, ^{59}Fe, ^{65}Zn, ^{60}Co) or which occurred in particulate or colloidal form (e.g., ^{95}Zr, ^{106}Rh, ^{144}Ce, ^{144}Pr). Soluble fission products, e.g., ^{90}Sr and ^{137}Cs, were found in highest concentrations in land biota (13, 14).

In general, the relative proportion of radionuclides which enter food chains, as well as their concentration factors, are greater in nutrient-poor environments. For example, concentration factors tend to be greater in fresh water than in sea water, since the former usually contains lower concentrations of mineral nutrients than does the latter. Radionuclides often act as sources of enrichment for nutrient-deficient soils. For example, the concentration of ^{137}Cs in deer was greater in the sandy, low coastal plains of the southeastern United States than in adjacent Piedmont regions, where the soil had better drainage and a high clay content, even though rainfall was comparable in both regions (15). Concentration factors in aquatic environments, however, tend to be greater than in terrestrial environments, largely due to the more rapid nutrient cycling which occurs in the former.

Some of the earliest data which demonstrated the bioconcentration of contaminants in both aquatic and terrestrial food chains were obtained from studies at the AEC Hanford reactor on the Columbia River in the state of Washington (16, 17, 18). Trace levels of both induced radionuclides, such as ^{32}P, and fission products, such as ^{90}Sr, ^{137}Cs, and ^{131}I, were released into the Columbia River, holding ponds, and the atmosphere. The river water was found to contain ^{32}P concentrations around 3×10^{-5} mg/gm H_2O. However, yolks from eggs of game birds were found to contain about 6 mg/gm; the average concentration factor was about 200,000. Concentration factors for ^{137}Cs of 250 in the muscle of waterbirds and for ^{90}Sr of 500 in their bones, in relation to concentrations in the water of holding ponds where the birds fed, were also observed. Thyroid glands of rabbits were found to have concentration factors for ^{131}I of about 500 compared to desert vegetation.

Table 4.4 Typical Concentration Factors of Elements for Various Classes of
Aquatic Organisms

Biota	CF (concentration organism/concentration water)					
	Co	Cs	I	K	Sr	Zn
Fresh water plants	6,760	907	69	–	200	3,155
Fresh water molluscs	32,408	–	320	–	–	33,544
Fresh water crustacea	–	–	–	–	–	1,800
Fresh water fish	1,615	3,608	9	4,400	14	1,744
Marine plants	553	51	1,065	13	21	900
Marine molluscs	166	15	5,010	8	1.7	47,000
Marine crustacea	1,700	18	31	12	0.6	5,300
Marine fish	650	48	10	16	0.43	3,400

Compiled from: Eisenbud, M. *Environmental Radioactivity – 2nd Ed.* New York: Academic Press, 1973.

A potential danger to man is posed by the selective concentration of radionuclides from the soil by food-crop plants. In one analysis (19), for
example, relative concentration factor values (ppm$_w$ in dry plant materi-
al/ppm$_w$ in dry soil) of particular elements were examined, and found to be
10 to 1,000 (strongly concentrated) for K, Rb, P, Na, Li; 1–100 (slightly
concentrated) for Mg, Ca, Sr, B, Zn; smaller than 0.01 (strongly excluded)
for Pb, Pu, Zr, Y. Table 4–4 presents average concentration factors of
some elements for the main groups of edible organisms in aquatic environments.

A major route of food-chain contamination by radionuclides is foliar
contamination via fallout. For example, during the atomic bomb testing
periods in the 1950's and 1960's, about 80% of the total ^{90}Sr in British
milk was due to foliar contamination (20). The degree of danger from foliar contamination depends partly upon the half-life of the nuclide. Short-
lived nuclides (^{131}I) could produce immediate contamination of the food
chain following foliar deposition, since the process of uptake from soil
would be slow relative to the half-life.

A classic example of food chain concentration involves the Finnish
Laplanders and Alaskan Eskimos. The matlike vegetation of the tundra
provides a highly efficient collection surface for fallout. Caribou, which
feed upon this vegetation, are a staple in the diet of these people. The Eskimos were found to have 2 to 3 times the level of ^{137}Cs present in the caribou (21, 22).

The potential danger to man of the bioconcentration of any contaminant
is dependent upon whether the particular site of sequestration in the plant

or animal is eaten. To use radionuclides as the example, ^{90}Sr is concentrated in shells and bones, while ^{60}Co and ^{137}Cs concentrate in the edible tissues of organisms. In addition, other factors also affect the amounts eaten by man. Food processing and cooking may result in the reduction of the amounts of many contaminants from levels found in the raw food.

PESTICIDES

Together with radionuclides, pesticides have provided good examples of contaminant transport along food chains, with the best examples of pesticide transport having been provided by DDT. Although pesticide residues may reach man directly via his water supply, the greatest danger to both man and the environment is due to bioconcentration along food chains in aquatic environments.

Because it is fat-soluble, DDT is concentrated in the fatty tissue at each link of the food chain. Residues may thus be passed to large game fish and carnivorous birds. For example, a Long Island marsh had been sprayed with DDT for 20 years to control mosquitoes; plankton were found to contain 0.04 ppm$_w$, minnows 2 ppm$_w$ and carnivorous gulls 75 ppm$_w$ (whole body, wet weight) in their tissues (22). Eastern U.S. oysters have been found to concentrate DDT up to as much as 70,000 times (23).

POLYCHLORINATED BIPHENYLS

PCB's undergo bioconcentration along the aquatic food chain, being concentrated in fatty tissue because of their high lipid solubility. Shrimp exposed to 10 ppb$_w$ for only 48 hours accumulated 1,300 ppb$_w$ (24); large concentrations subsequently occur in fish and carnivorous birds. Peregrine falcons from the coast of California were found to have up to 2,000 ppm$_w$ in their fatty tissue (24).

MERCURY

Mercury compounds can undergo bioconcentration in aquatic food chains by up to as much as 10,000 times the concentration in the sea water. The greatest degree of concentration of mercury occurs for its alkyl compounds, such as methyl mercury. In one stream in central Sweden, water was found to contain 0.13 μg/gm H$_2$O, while a one-kilogram pike had 300 μg/gm body weight (25). Some degree of bioconcentration may occur in terrestrial food chains. For example, some seed-eating birds were found to have concentration factors of mercury in their livers of only 2 to 3; however, the factors for predatory birds were in the 100's or 1,000's (26).

REFERENCES

1. Williamson, S. J. *Fundamentals of Air Pollution*. Reading, Mass.: Addison-Wesley, 1973.
2. Strom, G. H. Transport and Diffusion of Stack Effluents, pp. 401–501 in *Air Pollution—3rd Ed., vol. 1*, A. C. Stern (ed.). New York: Academic Press, 1976.
3. Pasquill, F. *Atmospheric Diffusion*. London: Van Nostrand, 1962.
4. Smith, M. (ed.). *Recommended Guide for the Prediction of the Dispersion of Airborne Effluents*. New York: American Soc. of Mechanical Engineers, 1968.
5. Briggs, G. A. *Plume Rise*, TID-25075. Springfield, Va.: Clearinghouse for Federal Scientific and Technical Information (CFSTI), NBS, 1969.
6. Slade, D. H. (ed.). *Meteorology and Atomic Energy—1968*, TID-24190. Springfield, Va.: CFSTI, NBS, 1968.
7. Eisenbud, M. *Environmental Radioactivity—2nd Ed.* New York: Academic Press, 1973.
8. Stewart, N. G. et al. *World-Wide Deposition of Long-Lived Fission Products from Nuclear Test Explosions*. U.K. Atomic Energy Auth. Res. Grp. Rep. MP/R 2354.
9. Parker, F. L., Schmidt, G. D., Cottrell, W. B., and Mann, L. A. Dispersion of Radiocontaminants in an Estuary. *Health Physics* 6:66, 1961.
10. Koczy, F. F. *The Distribution of Elements in the Sea*. Proc. Disposal of Radioactive Wastes, IAEA, Vienna, 1960.
11. Spitsyn, V. I. Sorption Regularities in Behavior of Fission Product Elements during Filtration of their Solutions through Ground. Proc. Disposal of Radioactive Wastes, IAEA, Vienna, 1959.
12. Edwards, C. *Persistent Pesticides in the Environment*. Cleveland, CRC Press, 1970.
13. Seymour, A. H. The Distribution of Radioisotopes Among Marine Organisms in the Western Pacific. *Publ. della Stazione Zool. Napoli* 3(Suppl.):25–33, 1959.
14. Palumbo, R. F. The Difference in Uptake of Radioisotopes by Marine and Terrestrial Organisms, pp. 1367–72 in *Recent Advances in Botany*. Toronto: University of Toronto, 1961.
15. Odum, E. P. *Fundamentals of Ecology*. Philadelphia: W. B. Saunders, 1971.
16. Foster, R. F. and Rostenbach, R. E. Distribution of Radioisotopes in the Columbia River. *J. Amer. Water Works Assoc.* 46:563–640, 1954.
17. Hanson, W. C. and Kornberg, H. A. Radioactivity in Terrestrial Animals Near an Atomic Energy Site. *Proc. Int. Conf. Peaceful Uses Atomic Energy, Geneva* 13:385–88, 1956.
18. Davis, J. J. and Foster, R. F. Bioaccumulation of Radioisotope through Aquatic Food Chains. *Ecology* 39:530–35, 1958.
19. Menzel, R. G. Soil-Plant Relationships of Radioactive Elements. *Health Phys.* 11:1325, 1965.
20. Russell, R. S. Interception and Retention of Airborne Material on Plants. *Health Phys.* 11:1305–15, 1965.

21. Miettinen, J. K. Enrichment of Radioactivity by Arctic Ecosystems in Finnish Lapland. Radioecol. Proc. Nat. Symp. 2nd, USAEC CONF-670503, 1969.

22. Woodwell, G. M. Toxic Substances and Ecological Cycles. *Sci. Amer. 216*: 24–31, 1967.

23. Pimental, D. Evolutionary and Environmental Impact of Pesticides. *BioScience 21*:109, 1971.

24. Gustafson, C. G. PCB's-Prevalent and Persistent. *Env. Sci. Technol. 4*: 814–19, 1970.

25. Johnels, A. G., Westermark, T., Berg, W., Persson, P. I., and Sjostrand, B. Pike (*Esox lucius L.*) and Some Other Aquatic Organisms in Sweden as Indicators of Mercury Contamination in the Environment. *Oikos 18*:323–33, 1967.

26. Borg, K., Wanntorp, H., Erne, K., and Hanko, E. Alkyl Mercury Poisoning in Terrestrial Swedish Wildlife. *Viltrevy 6*:301–79, 1969.

5

Fate of Contaminants: Transformations and Sinks

INTRODUCTION

Following their release, chemical contaminants may be converted to different forms and/or transferred within and between the various compartments of the environment, i.e., the atmosphere, the hydrosphere, the lithosphere, and the biosphere. Unfortunately, our knowledge of the transformations and ultimate fate of many contaminants is still quite inadequate.

The transformation, degradation, and sequestration of chemicals in the environment may occur via three processes: (a) chemical, e.g., atmospheric oxidation by O_2, and its subclass, photochemical reaction; (b) biological, the degradation due to the metabolic action of microorganisms, primarily bacteria, and which occurs mainly in soil and aquatic sediments; and (c) physical, e.g., solubility and gravitational settlement.

The ultimate disposition site or removal mechanism for a chemical contaminant is known as a sink. Ideally, the concentration of a contaminant in the environment would be determined by the dynamic steady-state existing between all of its sources, intermediate forms, and sinks. However, many chemicals are not normally in a steady-state, largely because their source strengths vary too rapidly due to economic factors, seasonal cycles, and climatic and technological changes.

The times between contaminant releases and their transformations or entrapment are quite variable, and depend upon the physical and chemical characteristics of the materials released, the characteristics of the environmental compartment into which they are released, and the degree to which they cross between compartments. There may be several time constants for retention within the various component regions of each compartment. For example, within the atmosphere, the average residence times may be very different in the stratosphere and troposphere, while in the biosphere, with contaminants interacting among numerous organisms, the complications increase enormously.

FATE OF AIR CONTAMINANTS

Contaminants emitted into the atmosphere are removed by various natural mechanisms, either in the primary form in which they were originally released, or following conversion to other forms. To a large extent, the fate of an airborne chemical depends upon whether it is in the particulate or gas phase.

Particles

Particulates may be emitted directly, or be produced in the atmosphere as a secondary contaminant. Regardless of the initial mode of production, they are ultimately removed primarily via natural processes, often after undergoing alterations in size, number, or chemical composition.

The principal processes for removal of particulates from the atmosphere are: (a) gravitational settlement, (b) impaction on and interception by earth-surface objects, and (c) rainout and washout. Certain of the mechanisms are more effective for particulates of one size regime than for those of another, and for particles in specific regions of the atmosphere.

Gravitational settlement, or sedimentation, increases with the aerodynamic diameter of the particle and is important for those with diameters greater than about 20 μm. Below this size, the particle's settling velocity is insignificant compared with its displacement by vertical motions of the atmosphere.

Impaction and interception onto surface objects (buildings, vegetation, etc.) are often effective removal processes for particles suspended near ground level; these particles may be transported to the surface by atmospheric motions. Impaction occurs when the momentum of a particle moves it across flow streamlines, causing it to collide with a surface. Interception refers to contact with a surface because the flow streamline coincident with the center of the particle is less than one particle radius away from the surface. The efficiency of removal by these processes is dependent upon such factors as the rate at which the particulates are supplied to the surface and the collection efficiency of the surface features. The former is affected by wind and general stability of the atmosphere, the latter by the shape, size, texture, and degree of wetness of the particular surface. Submicrometer particles, which have high diffusivity, may be removed via diffusion to surface objects.

Sedimentation and surface impaction, which are also known as dry deposition, are the principal removal mechanisms in the lower troposphere. At altitudes in the troposphere above 100 m, the primary removal mechanisms are the precipitation scavenging processes of rainout and washout, also known as wet deposition.

Rainout refers to the scavenging and incorporation of contaminants into precipitation elements within clouds, and their subsequent removal via precipitation, while washout refers to the incorporation by raindrops (or snowflakes) below the clouds, i.e., direct removal by collision with falling precipitation. The washout mechanism is effective for particles greater than about 2 μm diameter, while rainout is most effective for those with diameters less than this size.

Physical processes which occur in the atmosphere may affect the particulate removal efficiency. Hydroscopic particles may increase in size due to the accumulation of water from the vapor phase in the atmosphere. This growth may enhance their removal by sedimentation and washout.

As an aerosol ages, the constituent particles may grow by coagulation with smaller and larger particles, forming even larger particles which may have more appreciable settling velocities. The primary mechanism responsible for coagulation is Brownian motion, and results in the formation of droplets and solid aggregate particles up to about 0.7 μm in diameter. Particles with diameters greater than 0.3–0.5 μm have small diffusivities, and act as acceptors of the more rapidly diffusing, smaller particles. Particulate growth may also occur due to electrostatic forces. Very small particles, those below 0.1 μm, act as condensation nuclei, which induce the accumulation of vapors into aerosols; these subsequently coagulate with other particulates and are thereby eventually removed from the atmosphere.

Because of the large variation in the particle size of atmospheric aerosols, and the dependence of removal mechanisms upon size, no precise values for particulate residence times in the troposphere are possible. Furthermore, residence times are highly dependent upon the amount of precipitation in any particular area. Nevertheless, average residence times in the lower troposphere have been estimated to range from about five days to two weeks, while particulates in the upper troposphere may remain as long as 30 days (1, 2, 3). Particle residence times in the lower stratosphere are estimated at from four months to one year, and for those in the upper stratosphere from one to five years (1, 2). The stratospheric residence time depends upon latitude, with the shortest times occurring over the poles, and the longest over the equator. Residence times in the mesosphere are estimated at between five to ten years (1).

Gases and Vapors

The main processes for removal of gases and vapors from the troposphere are: (a) precipitation scavenging (with or without chemical reaction), (b) absorption by or reaction on earth-surface objects, and (c) chemical reactions in the atmosphere and conversion to other gases or particles. In the last case, the level of one contaminant may decline, while that of another increases. Another removal mechanism is escape through the tropopause and destruction in the stratosphere; the stratospheric sink may be significant for gases with long tropospheric residence times. The high energy, short wavelength radiation in the stratosphere degrades many molecules by disruption of their chemical bonds.

The precipitant scavenging of gaseous contaminants is not as well understood as for particulates. Whereas particulate scavenging may be thought of as a collisional phenomenon, gaseous scavenging is a diffusional phenomenon, which is dependent upon the concentration gradients of the particular gas across a liquid-air interface. Very often, chemical reactions occur following incorporation of the gas into a water droplet.

Absorption and/or reaction at the surface of the earth is a sink for some gaseous contaminants. As with surface impaction of particles, the efficiency of surface absorption of gases is dependent upon factors such as meteorologic conditions, which affect the rate at which a gas is applied to a surface, properties of the gas, and the nature of the particular surface. Certain surfaces are excellent sinks for specific gases, irreversibly absorbing them.

Atmospheric chemical reactions are an important removal process for many gaseous contaminants in the troposphere by producing stable gaseous end products or a particulate form easily removed from the atmosphere. Some particulates may accelerate atmospheric reactions by providing surface catalysis sites for both simple and complex reactions, or by sorption of gases from dilute gaseous concentrations, thereby providing discrete sites of higher concentration.

The relative importance of any particular removal mechanism for gases depends upon the specific chemical and environmental conditions. Removal can be described in terms of sink strength, the rate at which the contaminant is removed from the atmosphere via a particular route. There is generally a broad range of reported sink strengths, largely because their estimation is quite difficult. Rates of removal by various routes are very often extrapolated from limited studies and from models where contaminant levels are unrealistically high, or where other factors are not adequate simulations of conditions in the atmosphere. Thus, global sink strength estimates are only rough numbers.

In the following sections, the sinks of the major gaseous air contaminants are discussed.

NITROGEN-CONTAINING GASES AND VAPORS

The main gaseous compounds of nitrogen in the atmosphere are N_2O (nitrous oxide), NO (nitric oxide), NO_2 (nitrogen dioxide), and NH_3 (ammonia). NO_2 is in equilibrium with its dimer, N_2O_4 (nitrogen tetroxide).

NITROUS OXIDE

The most abundant nitrogen compound in the troposphere is N_2O. However, this gas is emitted almost totally by natural sources, is very stable chemically, and, thus, plays an insignificant role in low-level air pollution chemical reactions. Most of the N_2O is removed via its only known atmospheric reaction, which is photolysis occurring in the upper troposphere and stratosphere and resulting in the production of NO and N_2. Some N_2O is also believed to be removed from the lower troposphere by soils, plants, and bodies of water.

NITRIC OXIDE AND NITROGEN DIOXIDE

Nitric oxide is removed largely via oxidation to NO_2. Although this may occur by reaction with O_2, ozone oxidizes NO much more rapidly. For example, at atmospheric concentrations of 1 ppm$_v$ of NO, about 100 hours are necessary for conversion of 50% of the NO to NO_2; with ozone, the half-life of NO is only 1.8 seconds (4, 5). In the upper troposphere and stratosphere, photolysis of NO results in the production of atomic nitrogen (N), which may then form N_2 following reaction with other molecules of NO.

The primary mechanism for removal of NO_2 is dissolution in cloud and rain droplets. Although numerous schemes have been proposed for subsequent reactions of NO_2, the end result is always production of nitrous acid (HNO_2) and nitric acid (HNO_3), which may then be converted to nitrite and nitrate salt aerosols by reaction with ammonia or other atmospheric bases, or become adsorbed onto other particulates. These acid or salt particles are then removed by wet or dry deposition mechanisms. Some nitrogen dioxide may react with O_3, also producing particulate nitrites and nitrates. Both HNO_2 and HNO_3 may be photochemically decomposed, resulting in the reformation of NO and NO_2.

Minor sinks for NO_2 and NO are vegetation and soils, the latter partly by biological means. However, the NO_2 which is absorbed by soils is eventually oxidized to nitrate which, upon oxidative decomposition, produces NO_2 once again.

Nitrogen dioxide is important because of its involvement in atmospheric photochemical reactions. It absorbs solar energy over the entire visible and UV range of the spectrum in the troposphere. At wavelengths less than 0.38 μm, NO_2 is photodissociated into NO and O; the latter reacts with O_2 to produce O_3. The photochemical reactions of NO_2 have been discussed in detail in Chapter 3.

In the stratosphere, NO and NO_2 form nitrogen pentoxide (N_2O_5), which combines with water vapor to produce HNO_3 vapor.

AMMONIA

Although most sources of NH_3 are natural, it plays a major role in the formation of particulates in the atmosphere. In fact, the main sink for NH_3 is aerosol formation. This occurs by reaction with acids or acid-forming oxides in the gaseous phase or after dissolution in rain water; the latter process also enhances the oxidation of dissolved SO_2 and NO_2. The main aerosol species produced are ammonium bisulfate (NH_4HSO_4), ammonium sulfate [$(NH_4)_2SO_4$] and ammonium nitrate (NH_4NO_3).

Because of its high water solubility, NH_3 may be removed from the troposphere by absorption onto wet surfaces such as vegetation, bodies of water, and soils. Some ammonia may also be removed by reaction with hydroxyl radicals to form NO.

Estimates of tropospheric residence times for gaseous nitrogen com-

Table 5.1 Mean Tropospheric Residence Times for Nitrogen Gases

Chemical	Residence time
N_2O	4 years
NO	4–5 days
NO_2	3–5 days
NH_3	7–14 days

Compiled from: Seinfeld, J. H. *Air Pollution: Physical and Chemical Fundamentals.* New York: McGraw Hill, 1975. Hidy, G. M. Removal Processes of Gaseous and Particulate Pollutants, pp. 121–76 in *Chemistry of the Lower Atmosphere*, S. I. Rasool (ed.). New York: Plenum Press, 1973. Robinson, E. R. and Robbins, R. C. Emissions, Concentrations and Fate of Gaseous Atmospheric Pollutants, pp. 1–93 in *Air Pollution Control Pt. II*, W. Strauss (ed.). New York: Wiley, 1972.

Table 5.2   Sink Strengths for Nitrogen Compounds

Sink	Sink strength (10^{10} kgN/year)
N_2O, NO, NO_2	
Rainout/washout	2 – 7.5
Dry deposition	1.9 – 7
Stratosphere	0.03 – 0.2
Gaseous deposition on surfaces	4.5
NH_3	
Rainout/washout	3 – 18.6
Dry deposition	4.9 – 7
Gaseous Deposition	74.9
Oxidation in troposphere	7
Stratosphere	0.04

Compiled from: Rasmussen, K. H., Taheri, M., and Kabel, R. L. Global Emissions and Natural Processes for Removal of Gaseous Pollutants. *Water, Air and Soil Pollution 4*:33– 64, 1975.

pounds are given in Table 5 – 1. Sink strength estimates are presented in Table 5 – 2.

CARBON-CONTAINING GASES

The main carbon-containing gases in the atmosphere are carbon monoxide (CO) and carbon dioxide (CO_2).

CARBON MONOXIDE

Except for carbon dioxide, more carbon monoxide is released into the atmosphere than any other contaminant. The major removal pathways of CO are via the soil and via gas-phase reactions in the troposphere and stratosphere.

Many soils have been found to be able to remove CO from the atmosphere, releasing CO_2. Greatest activity was shown by organically rich tropical soils, and organically poor desert soils had the least. Although the exact mechanism is not clear, CO oxidation may be due to biological activity, since removal was inhibited in soil that had been sterilized (6).

There are two atmospheric sinks for CO. One involves the reaction of CO with hydroxyl ($\cdot OH$) and hydroperoxyl ($\cdot OOH$) radicals in the troposphere, producing CO_2 and H atoms. The other involves reaction with hydroxyl radicals in the stratosphere, following CO migration through the tropopause. The importance of the stratosphere as a sink is dependent upon ambient CO levels, rate of transport into the stratosphere and hydroxyl levels in the stratosphere. Based upon a highly uncertain flux estimate for

Table 5.3   Sink Strengths for CO

Sink	Sink strength (10⁹ kg/year)
Soil	67–1400
Gas phase oxidation in stratosphere	52–71

Compiled from: Rasmussen, K. H., Taheri, M., and Kabel, R. L. Global Emissions and Natural Processes for Removal of Gaseous Pollutants. *Water, Air and Soil Pollution 4:33–64*, 1975.

CO movement into the stratosphere, it has been estimated that 11–15% of the total tropospheric CO is removed via this route each year (7).

The residence time for CO in the troposphere is estimated at 0.09–2.7 years; the uncertainty is fairly high (8–13). Sink strengths for CO are presented in Table 5–3.

CARBON DIOXIDE

About 30–50% of the CO_2 released into the atmosphere, largely due to combustion of fossil fuels, remains. The remaining 50–70% enters sinks in the hydrosphere and biosphere.

Green plants may be a temporary sink for CO_2; although green plants take it up in photosynthesis, the CO_2 ultimately returns to the atmosphere during respiration and upon oxidative decomposition of dead plants.

The largest natural influence upon CO_2 levels in the atmosphere is the exchange with the oceans, which contain about 60 times as much CO_2 as the atmosphere. The average residence time of CO_2 in the atmosphere before it transfers to ocean waters has been estimated to be around 10 years (14), but the actual approach to equilibrium levels between atmosphere and oceans may be much slower, inasmuch as atmospheric CO_2 levels are increasing.

Carbon dioxide is destroyed by photochemical decomposition in the mesosphere.

SULFUR-CONTAINING GASES AND VAPORS

The primary gaseous compounds of sulfur present in the atmosphere are hydrogen sulfide (H_2S) and sulfur dioxide (SO_2). Another sulfur species of importance is sulfur trioxide (SO_3).

HYDROGEN SULFIDE

Hydrogen sulfide is produced primarily from natural sources and is not an important general air contaminant in terms of anthropogenic emissions. Its only significant mode of removal from the atmosphere is oxidation to

SO_2. Although the reaction proceeds fairly slowly in the gas phase, it may be catalyzed by the presence of particulate matter, which provides reaction surfaces.

Most of the oxidation of H_2S occurs by O_3. Reaction with atomic oxygen may be significant in the stratosphere and under conditions of photochemical smog, where significant quantities of atomic oxygen are produced by the photolysis of ozone. The reaction system in this latter case results in the production not only of SO_2, but also SO_3 and H_2SO_4.

The estimates of tropospheric residence time for H_2S range from a few hours to two days (15, 16, 17).

SULFUR DIOXIDE

The sinks for SO_2 are precipitant scavenging, diffusion to and absorption onto earth-surface features, and chemical conversion into other sulfur species.

Washout and rainout are the major mechanisms for removal of atmospheric SO_2, largely because SO_2 is very soluble in water. Scavenged SO_2 undergoes a series of chemical reactions, ultimately forming sulfuric acid (H_2SO_4) and sulfate salts.

Under conditions of high relative humidity and in the presence of catalytic surfaces on particulates, SO_2 may undergo oxidation. Salts of certain metals, e.g., iron and manganese, serve as the catalysts, either by acting as condensation nuclei or by undergoing hydration; both actions result in the production of liquid droplets. Both SO_2 and O_2 readily dissolve in these droplets, and oxidation proceeds quite rapidly in the liquid phase, producing sulfite (SO_3^{-2}) and eventually H_2SO_4 droplets. The H_2SO_4 may react with other atmospheric constituents to produce ammonium, sodium, and calcium salts, and numerous other sulfate species.

The presence of NH_3 in water droplets also enhances the oxidation of SO_2 in solution. As a droplet becomes highly acidic, the rate of SO_2 oxidation decreases, due to the decreased solubility of SO_2 in acid solutions. Ammonia dissolved in the droplet increases SO_2 solubility, and if sufficient levels of NH_3 are present, oxidation is not impeded by accumulation of H_2SO_4, since conversion to $(NH_4)_2SO_4$ occurs. Calcite dust and other airborne alkalis besides ammonia are also involved in the rapid oxidation of SO_2 to sulfate.

The rate of catalytic oxidation of SO_2 decreases if the water concentration in the atmosphere falls below the level necessary to maintain catalyst droplets. The critical point seems to be 70% relative humidity, for above this level the rate of catalytic oxidation increases dramatically (18).

Table 5.4 Sink Strengths for Sulfur Compounds

Sink	Sink strength (10^9 kgS/year)
Gaseous absorption, SO_2	
Vegetation	15–75
Ocean	25–100
Washout/rainout (land)	70–86
Dry deposition (land)	10–20
Wet and dry deposition, SO_4^{-2}	71–100
(oceans)	

Compiled from: Rasmussen, K. H., Taheri, M., and Kabel, R. L. Global Emissions and Natural Processes for Removal of Gaseous Pollutants. *Water, Air and Soil Pollution* 4:33–64, 1975.

In clean, dry air, SO_2 is only very slowly oxidized via homogeneous reactions (gas phase) to SO_3 vapor. The most important transformation of SO_2 under low-humidity conditions is photochemical oxidation. Photochemical reactions have been discussed in Chapter 3, and only a general description will be presented here.

In the initial photochemical event, the absorption of light results in activation of a molecule of SO_2 which then proceeds to react with O_2 or O_3 at a faster rate than would unactivated SO_2 molecules. The resultant SO_3 reacts almost immediately with water vapor to produce H_2SO_4.

The rate of photochemical oxidation of SO_2 is fairly slow; estimates range from 0.006%/min to a theoretical maximum of 0.03%/min (19, 20). However, the rate may be greatly accelerated by the copresence in the air of certain reactive intermediate species in the photooxidation of hydrocarbons under the influence of NO_x, e.g., NO and olefins.

In the stratosphere, SO_2 reacts with atomic oxygen, resulting in the formation of SO_3 vapor, which reacts with water vapor to form H_2SO_4 vapor. The latter combines with more water to produce H_2SO_4 droplets, which are removed by precipitation.

Aside from atmospheric sink reactions, SO_2 may also be removed by other mechanisms. Plants may absorb SO_2. In fact, this is actually a method to obtain nutrient sulfur in some areas where the soil has a sulfur deficiency (21). Soils are also able to absorb significant amounts of SO_2; the absorbed SO_2 may then be oxidized to sulfate. Other sinks are absorption of SO_2 by the hydrosphere, and the uptake by carbonate stones, e.g., buildings, in moist atmospheres.

The estimated tropospheric residence time for SO_2 ranges from 20 minutes to 7 days (17, 22). Sink-strength estimates for sulfur compounds are presented in Table 5–4.

HYDROCARBON GASES AND VAPORS

A broad range of gaseous hydrocarbons are emitted into the atmosphere. Certain of these may be taken up via soil microbial action. This has been shown to be a sink for ethylene and acetylene (23), and, to some extent, for polycyclic aromatic hydrocarbons (24).

Most organic molecules are oxidized only very slowly in clean atmospheres. The mechanism involves reaction chains which are initiated by removal of a hydrogen atom from the molecule, producing a radical which combines with an oxygen molecule to form a peroxide. The latter, being reactive, may go on to remove a hydrogen from another organic molecule.

In the presence of light sensitizers or initiators, hydrocarbons may be degraded by photochemical reactions. These initiators provide the energy necessary for radical production from the organic molecule. Photochemical reactions are the primary degradation processes for reactive hydrocarbons, e.g., olefins and polycyclic aromatic hydrocarbons. These molecules rapidly undergo transformations under the influence of sunlight and in the presence of atomic oxygen, O_3, SO_2, and NO_2. The result is conversion to other organic molecules. The exact reaction rates for various hydrocarbons are quite different, and there are few data on the rates for many species or for the products formed.

Up to about 10% by weight of the reactive hydrocarbons emitted into the atmosphere eventually is converted to or adsorbed onto particulates which are then removed by wet or dry deposition processes (16). The remainder of the hydrocarbons eventually undergo chemical degradation, becoming oxidized to other hydrocarbons, and if oxidation is complete, to CO_2 and H_2O. For substituted hydrocarbons, the noncarbon constituents are found in various salts.

Methane (CH_4) is the predominant atmospheric hydrocarbon; however, the paraffin group, of which CH_4 is a member, is not significantly involved in photochemical reaction processes. The main sink for CH_4 is oxidation to form CO and CO_2, through the action of hydroxyl radicals in the troposphere.

Except for CH_4, the lack of precise estimates of emissions and ambient levels of hydrocarbons prevents an accurate estimation of atmospheric residence times. The residence time of CH_4 in the troposphere is estimated at 1.5–2 years; that for higher-molecular-weight hydrocarbons is on the order of days to months (15, 25).

In the stratosphere, many hydrocarbons react with atomic oxygen, and may eventually be oxidized to CO_2 and H_2O.

OZONE

Ozone is a constituent of photochemical smog atmospheres. A major O_3 sink is the surface of the earth, primarily soils and vegetation; absorption by oceans may also occur to some extent.

Ozone is a powerful oxidizing agent. It is involved — especially in contaminated atmospheres — in many reactions, such as photooxidation of hydrocarbons in atmospheres containing NO_2.

FATE OF WATER CONTAMINANTS

Chemical transformations and physical or biological removal of contaminants from the atmosphere may be thought of as natural purification processes which produce nonreactive (both physiologically and chemically) products and/or otherwise result in removal from the atmosphere of the chemical agent. Analogous self-purification processes occur in aquatic environments. However, sinks for aquatic contaminants and chemical reactions of contaminants in natural aquatic environments have not been as extensively studied.

The self-purification processes by which natural waters tend to rid themselves of contaminants involve physical, chemical, and biological mechanisms which occur simultaneously and interact with each other. At one time, these processes were sufficient to prevent permanent effects on many bodies of water, but now this is not always the case.

Self-purification processes in aquatic environments are quite complex, especially due to the interaction of the biota. Each specific waterway, be it estuary, stream, river, lake, or ocean, has its own capacity for self-purification and recovery due to its unique combination of physical, geochemical, and biological characteristics. Nevertheless, the main processes in all environments are basically the same. This section will deal with these general processes.

Physical Self-Purification

The physical mechanisms involved in self-purification of water are the processes of dilution, mixing, sorption (adsorption and absorption) and sedimentation. Dilution of chemicals occurs due to diffusion, advection (movement of one parcel of water with respect to another), and turbulence (eddy diffusion). Mixing is primarily due to turbulence, which results from temperature gradients in large bodies of water, and from bed friction in rivers and streams. In coastal areas and estuaries, the action of tides also helps in the mixing of contaminants.

Sorption of chemicals may alter their behavior and often results in a purification of the water. Particles are quite ubiquitous in natural waters, occurring in a wide assortment of sizes, shapes, chemical compositions, and concentrations. In terms of sorption and binding of chemicals, the colloidal particles are the most important, with clay being the most significant class of common mineral which occurs as colloidal matter in natural waters. Because of their strong tendency to sorb chemicals from the water, clays may effectively immobilize dissolved chemicals, purifying the water. In addition, some microbiological degradation processes for organic wastes may also occur at the clay-particle surfaces, giving clay a role in biological purification.

A wide variety of dissolved chemical effluents may be sorbed and bound by suspended particles. This includes various organic chemicals, such as herbicides and nonvolatile organics, as well as dissolved inorganic materials, such as trace metals.

Particulates, either natural or those discharged into water, along with any sorbed chemicals, may eventually be removed from suspension by sedimentation. This process depends upon the particle size, shape, and density, and the mixing and flow characteristics of the waterway. In turbulent streams, suspended particles may be carried for long distances before they settle.

The sedimentation of the small colloidal particles may be enhanced by aggregation, which involves the processes of coagulation and flocculation. Coagulation is the aggregation of colloidal particles of the same material. Flocculation is dependent upon the presence of bridging compounds which form chemically bonded links between the colloidal particles and enmesh the particles in large floc networks.

Movement of suspended particles into different waters may affect their sedimentation. The size of colloidal particles may change due to coagulation when they travel from fresh water into salt water; the salt ions act to reduce the normal electrostatic repulsion between the particles. This process may be accompanied by the uptake and/or loss of some of the sorbed contaminant ions.

Particulates which deposit on the sediments may remain there, undergoing no removal processes. However, very often accumulation and precipitation processes are reversible. Bottom sediments undergo a continuous process of leaching and ion exchange with the overlying water.

For example, trace metal ions sorbed on particles deposited in the sediments may be released by ion-exchange mechanisms. Sediments which move from aerobic to anaerobic environments may undergo changes in sorbed chemical species.

Nevertheless, bottom sediments are often the ultimate sinks for many contaminants. The deposition may occur in the waterway into which the contaminant discharge occurred, or in downstream areas. In this regard, oceans are an enormous sink, often representing the ultimate "downstream" locale for flowing waterways. Rivers deliver billions of tons of dissolved and suspended matter each year to the oceans and seas; much of this is subsequently deposited in the sediments, often in coastal regions.

Chemical Self-Purification

Specifics of the chemical reactions of many contaminants in aquatic environments are unknown; the nature of chemicals is often influenced by the biota. In addition, chemical reactions are also affected by water temperatures and pH. It is beyond the scope of this book to discuss the fate of specific chemicals in water. However, certain types of chemical reactions are often involved in the self-purification of inorganic contaminants, and these will be presented. Purification of many organic contaminants involves biological processes, and these are discussed in a subsequent section.

Acid-base reactions are important in the assimilation of acid and alkaline wastes discharged into water. Acids may react with numerous minerals, such as those in crystalline rocks, clays and other silicates, limestone and other carbonate rocks. The capacity of a waterway to neutralize acid is primarily a function of the bicarbonate, carbonate, and hydroxide ions. Alkaline contaminants are purified mainly via reactions with silica, bicarbonate, and free carbonic acid in the water, resulting in the production of silicate and carbonate salts. The rates of acid-base reactions differ in different aquatic systems, and are important determinants of the area of the waterway over which unassimilated acids and alkaline wastes traverse.

Nonbiologically mediated oxidation-reduction reactions are also of great significance in the purification of aquatic environments; many hydrocarbons and heavy metals undergo these types of reactions.

Metal ions may be removed from solution by the formation of insoluble complexes known as coordination compounds. Some naturally occurring organic complexing agents, e.g., humic and fulvic acid, are found in certain aquatic environments, and these agents strongly bind heavy metals.

Synthetic chelating agents (e.g., NTA, EDTA) are used in detergents and industrial processes, and are often discharged into surface waters where they may form complexes with metals. On the other hand, metal ions may also be solubilized by formation of complexes.

Some dissolved chemicals may be precipitated under the influence of other agents in the water (coprecipitation), while some may leave by volatization. Some chemicals may be hydrolyzed in the aquatic environment. For example, hydrolysis of esters results in products which may then be degraded by biological processes. Some organic chemicals may undergo light-induced transformation reactions in water.

In aquatic chemistry, interactions occurring solely in solution are of less significance than those which occur at solid-water and gas-water interfaces. In this regard, colloidal solids are of great significance in chemical interactions in the aquatic environment.

Although some chemicals may undergo a limited number of reactions which may serve to remove them from the water, others, e.g., nitrate (NO_3^{-2}), undergo few reactions without the action of microorganisms.

Biological Self-Purification

Biological purification processes generally involve the assimilation of putrescible organic wastes entering waters. The process involves the transformation of the chemicals into innocuous products due to action of the aquatic biota, primarily microorganisms; it usually occurs at the expense of dissolved oxygen and of some species which normally are present in the biota of the uncontaminated waterway.

Because of the multitude of microbial types which exist in aquatic environments, the overall process is quite complex. However, most aquatic environments show, to some extent, common biological reactions which may vary due to differences in flow and mixing characteristics and according to the concentration and composition of the specific contaminant. The commonality occurs because all living systems respond quite similarly to insult.

As an example of biological self-purification, consider the case of a discharge of organic matter into a stream. The discharge initially results in an increase in the BOD; the levels of DO in the stream below the point of discharge, therefore, decrease. Depending upon the initial waste load, a certain time and distance of flow downstream are necessary before stream reaeration processes return the DO to near air-saturation levels. By considering factors of BOD loading, reaeration processes and rate of flow, an oxygen sag curve may be developed (Figure 5 – 1).

Suspended organic solids eventually settle out to the bottom. These may be subject to microbial decomposition, although the processes differ from those occurring in flowing water. Decomposition in bottom sediments varies, with the depth of deposit, from aerobic to anaerobic.

Dissolved solids are also removed or diluted as the stream flows. As the various nutrient cycles occur, organic compounds are broken down and levels of inorganic nutrients (ammonia, nitrate, phosphate) are increased. This results in changes in biota, since species differ in their ability to tolerate increased nutrients (some are favored and others are disadvantaged). Bacteria and sewage fungi predominate in the zone where oxygen becomes a severely limiting factor. Further downstream, these decline and algae may bloom, benefiting from the increase in mineral nutrients. Eventually, these also decline. Low oxygen levels cause reduction in the number of species of larger invertebrates; highly specialized sewage organisms, e.g., tubificid worms, midge larvae, and other undesirable forms which can tolerate these conditions increase in number due to the large supply of food and little interspecies competition. Increasing dilution occurs with flow downstream and, as the river oxygen and minerals return to normal, the number of species able to survive increases and the strong numerical predominance of particular species tends to disappear. A gradual return to the community found upstream of the discharge occurs. The water has purified itself and has recovered from the discharge. The entire scheme of biological purification is shown in Figure 5 – 1.

Natural microbial purification is not a fast process, and heavily contaminated streams may have to travel fairly long distances before any significant degree of purification occurs. Exact rates are quite difficult to estimate because of the compositional and mixing complexities of most waterways and the interaction of biological, chemical, and physical processes.

There are generally significant differences in biological reactions between flowing waterways and enclosed bodies of water. Flow and mixing in the former systems make them less susceptible to permanent damage by a given concentration of organic wastes. For example, an increase in the decomposition of sewage acts to reduce DO in the bottom waters of a lake. In flowing waters, the constant mixing acts to hinder establishment of such distinct surface-to-bottom gradients in the entire waterway. Under severe conditions, however, streams can be completely changed, returning

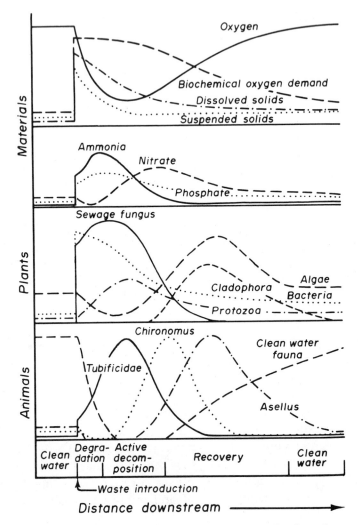

Fig. 5–1 Natural purification in a river following introduction of sewage. Note the changes in water quality and species of plants and animals with distance from the source of the sewage.

From: Warren, C. E. *Biology and Water Pollution Control.* Philadelphia: W. B. Saunders, 1971. After Hynes, H. B. N. *The Biology of Polluted Waters.* Liverpool: Univ. of Liverpool Press, 1960.

to normal only after long distances downstream from the point of contaminant discharge.

Ground water also has some capacity for self-purification, although the mechanisms differ from that in surface waters. Because of certain limiting factors, such as darkness, the variety of microbes in ground waters is restricted. However, the reduction in biological self-purification is offset by the process of physical purification via filtration through the soil and rocks. Soils may remove particulates by physical means and chemicals by various means, as discussed in a subsequent section. But there is a limited ability to remove many chemical compounds that commonly occur in industrial waste waters.

Aside from sewage, other organic contaminants may be subject to microbial attack. For example, the degradation of many hydrocarbons, such as those in fuel oils, proceeds primarily via microbial utilization of the constituent alkanes.

Residence Times in the Hydrosphere

Residence times for chemical contaminants in the hydrosphere are extremely difficult to determine because of the various types of aquatic environments and the complex physical, chemical, and biological factors which determine the fate of the contaminant and which differ to some degree in these different environments.

Since the oceans do often represent the ultimate sink for contaminants contained in flowing waterways, various estimates of the residence time of chemicals in the marine environment have been made (26, 27).

Unless they are removed into the atmosphere, the ultimate fate of chemical contaminants entering the oceans is usually removal to the sea floor. Thus, the residence time in the marine environment is generally considered to be the time the contaminant remains in the ocean water between its introduction and its incorporation into the sediments.

Residence time depends upon the degree of reactivity of the specific contaminant, a factor which also determines the region in the marine environment in which the chemical reaches its sink. As a generalization, chemically reactive elements, e.g., iron and aluminum, and those involved in biological cycles, e.g., phosphorus and nitrogen, have shorter residence times then do those contaminants which tend to be less reactive or inert in solution, e.g., alkali metal ions (26, 28). The first two groups tend to become associated with particulate matter and/or biota prevalent in the coastal ocean regions, and are largely removed there by deposition processes. The latter group tends to reach the open ocean waters where they accumulate until they eventually undergo downward transport in the waters. Specific oceanic residence times, for example, vary from a century for iron to hundreds of millions of years for sodium (28). Organic contaminants which do not undergo biological degradation may have residence half-times in the ocean of thousands of years.

FATE OF CONTAMINANTS IN SOIL

Soil may be a sink for numerous chemicals. The fate and persistence of chemicals in soil is a complex function of physical, chemical, and biological factors, the main ones being the specific chemical and its formulation, soil type and specific physicochemical properties, type of cover vegetation, degree of soil cultivation, and nature of the microbial population.

The major routes of removal of contaminants from soil are: (a) degradation via microbial action, which is discussed in the next section, (b) chemical degradation, e.g., hydrolysis, (c) evaporation and volatization from the surface, and (d) uptake by vegetation. Most contaminants are not effectively removed by leaching.

Uptake by plants may only be a temporary sink, since the contaminant may return to the soil in plant litter. Temporary removal may also be afforded by microbial assimilation without degradation, which may immobilize the contaminant until the death of the microorganism. Chemicals may also be immobilized in soil via various adsorption processes.

Colloidal soil particles may adsorb various chemicals. Neutral organic molecules, for example, may be adsorbed by physical mechanisms, such as via van der Waals forces. Polar or polarizable organic ions, especially cations, and inorganic metallic ions, such as Pb^{+2}, Hg^{+2}, $^{90}Sr^{+2}$, may be adsorbed via ion exchange. Other processes implicated in adsorption of contaminants, particularly pesticides, are hydrogen bonding and coordination, the latter being formation of a complex between the chemical and an exchangeable ion on the soil particle. Some metals, such as Cu and Zn, may become tightly bound to soil organic matter by chelation. Other soluble inorganic chemicals may precipitate out of solution in soil water as oxides or hydroxides.

Thus, a wide variety of immobilization processes may involve all types of chemicals. Immobilization affects the subsequent fate of the contaminant in the soil in any number of ways. It may hinder volatization and leaching, act to catalyze chemical degradation, prevent biological degradation by sequestering the contaminant from the microbes, enhance biological decay by concentrating the chemical near microbes, etc.

Specific components of certain soils may affect the fate of contaminants. For example, part of the humic fraction in many agricultural soils is composed of lipids; contaminant-lipid interactions in these soils often serve to immobilize the fat-soluble organochlorine pesticides.

There are few reliable data on the residence times of chemicals in soils, since these depend upon so many physicochemical and biological factors. Type of soil is, of course, a major factor. For example, many pesticides tend to persist longer in soils with higher organic matter than in sandy soils, since the former tend to bind the residues much tighter.

FATE OF CONTAMINANTS IN THE BIOSPHERE

The discussion in this chapter has concentrated on the physicochemical processes by which contaminants are removed from air, water, and soil environments. In this section, we shall discuss microbially-mediated transformations and their relation to the biodegradability of chemicals.

Biogeochemical Cycles

The biological cycles for carbon, nitrogen, and sulfur are discussed in the following sections. Bear in mind that they are not independent of each other, but, as with most environmental processes, they interact.

CARBON CYCLE

There are two main carbon reservoirs: the atmosphere and the hydrosphere. In both, carbon naturally exists in the form of CO_2; interchange occurs via diffusion, evaporation, and precipitation.

The atmospheric or hydrospheric CO_2 is fixed into biomass via photosynthesis, primarily by green plants on land and by phytoplankton in the sea. The carbon then moves through various trophic levels and is eventually released back to the reservoir as CO_2, via the respiratory activity of plants and animals, in the processing of waste products and remains of dead biota by decomposers, and via the process of combustion.

Some carbon is trapped in deposits of the remains of plants and animals and becomes peat, coal, and oil, some in the shells of aquatic organisms,

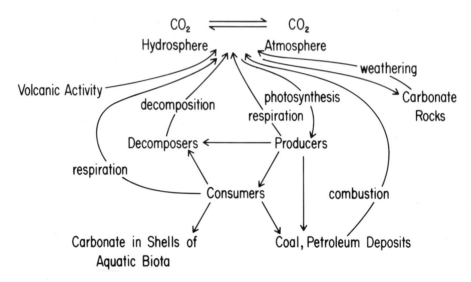

Fig. 5–2   The biological carbon cycle.

and some in formation of carbonate rocks in the oceans. The weathering of these rocks and the combustion of oil and coal release this formerly trapped carbon into the cycle once again. The biological carbon cycle is shown in Figure 5 – 2.

NITROGEN CYCLE

Although nitrogen is the dominant gas in the atmosphere, very few living organisms are able to use gaseous nitrogen. Rather, nitrogen must first be fixed into an inorganic compound for utilization by the producers in biological processes. The primary compound for such utilization is nitrate, and the primary mechanism for fixation is biological. Various groups of free-living aerobic and anaerobic microorganisms, primarily bacteria, and also certain symbiotic microorganisms, e.g., the root-nodule bacteria of legumes, are able to fix atmospheric nitrogen. Fixation occurs primarily in terrestrial ecosystems, although certain organisms in aquatic ecosystems, e.g., blue-green algae, are also capable of nitrogen fixation. Physicochemical processes in the atmosphere may also result in fixation of nitrogen; these include electrification (lightning) and cosmic radiation.

The nitrogen contained within the bodies of plants and animals exists as ammonium ions or in amino-compounds, such as proteins and nucleic

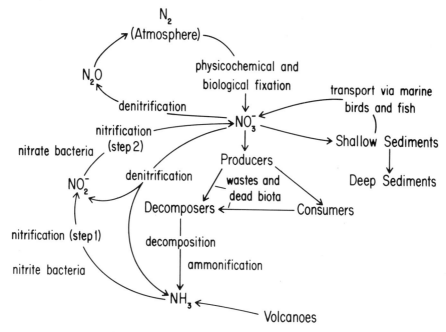

Fig. 5-3   The biological nitrogen cycle.

acids. Mineralization of this nitrogen occurs via the processes of ammonification and nitrification.

In the process of ammonification, numerous aerobic and anaerobic microorganisms metabolize bound nitrogen, releasing NH_3 or ammonium
(NH_4^+) compounds. The subsequent conversion of the NH_3 and NH_4^+
salts back to nitrate (NO_3^-) is called nitrification. This occurs in two
stages, and involves two classes of aerobic bacteria. The first step is oxidation of ammonia to nitrite (NO_2^-) by nitrite bacteria; the nitrite is then
converted to nitrate (NO_3^-) by nitrate bacteria.

Under total or partial anaerobic conditions in soils, various microorganisms convert nitrate into nitrite, various nitrogen oxides, NH_3, and gaseous N_2 in processes collectively termed denitrification. Some of these
products remain in the soil and others are released into the atmosphere.
Under certain soil conditions, such as high acidity, chemical denitrification of nitrate may also occur, resulting in production of NO_2, N_2O and,
eventually, HNO_3.

Some nitrogen is trapped in the deep ocean sediments and in sedimentary rocks. The biological nitrogen cycle is shown in Figure 5 – 3.

SULFUR CYCLE

Elemental sulfur is not utilized by producer organisms. Rather, the principle form of sulfur used in biological systems is inorganic sulfate (SO_4^{-2}), although some organisms may obtain their sulfur from certain organic compounds. Microbial decomposition results in mineralization of sulfur bound in biomass.

Under anaerobic conditions, especially in swamps, marshes, and soils, various bacteria reduce sulfate, producing sulfides, such as H_2S, or elemental sulfur. In aerobic environments, microorganisms oxidize H_2S to elemental sulfur, H_2S to SO_4^{-2}, or elemental sulfur to SO_4^{-2}.

Sulfur may be removed from active cycling by precipitation in neutral or alkaline water under anaerobic conditions, by formation of ferrous and ferric sulfide. The sulfur may be trapped to the limits of the amount of available iron present. The biological sulfur cycle is shown in Figure 5–4.

Fig. 5–4   The biological sulfur cycle.

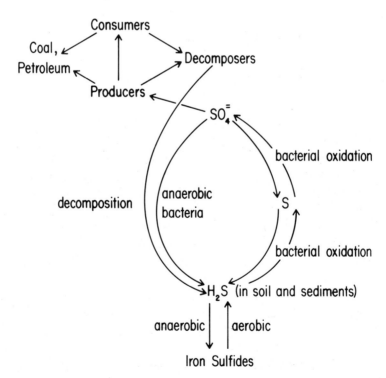

Biodegradability of Chemical Contaminants

It should be clear from the previous discussion that the ultimate minerali-zation and degradation of a chemical in any biogeochemical cycle is de-pendent upon microorganisms, primarily bacteria. Microbial metabolism is, therefore, responsible for the mobility, or lack of it, of chemicals in the biosphere. These microbes purify streams into which organic contami-nants are dumped, degrade pesticides in soil, decompose sewage sludge, and even have a role in the degradation of oil spilled into waterways. Con-taminants which may be degraded by microbes are termed biodegradable. Although these chemicals are primarily organic in nature, they yield inor-ganic end products. It should, however, be mentioned that microbial con-version may result in products which are more hazardous than the original material.

On the other hand, certain chemicals released into the environment are resistant to microbial decomposition or else are degraded at such slow rates that they build up in the environment. These types of contaminants are termed nonbiodegradable, refractory, or persistent, and include many chemicals. Notable examples are organochlorine insecticides, e.g., DDT, and numerous synthetic polymers, such as PCB, PVC, teflon, polyeth-elenes, and mylars. The residence times in soil for selected biodegradable and persistent pesticides are presented in Table 5–5.

Various reasons for resistance of chemicals to microbial degradation have been proposed. They include the inability to be attacked by a microbe having an enzyme capable of producing degradation, or the complete ab-sence of an essential microbial enzyme.

Table 5.5   Residence Times of Selected Pesticides in Soil

Designation	Chemical group	Residence half-life (years)
Degradable	Organophosphate insecticides	0.02–0.2
	Carbamate insecticides	0.02
Moderately degradable	Urea herbicides	0.3–0.8
	2,4-D; 2,4,5-T herbicides	0.1–0.4
Persistent	Organochlorine insecticides	2–5
Permanent	Lead, arsenic, copper pesticides	10–30

Compiled from: Salvato, J. A., Jr. *Environmental Engineering and Sanitation.* New York: Wiley-Intersci-ence, 1972.

CONTAMINANT RESIDUES

Following their release, contaminants may be dispersed throughout the environment until they reach a sink. The dispersal may result in global distribution, as well as high levels on a local or regional scale.

The actual concentration of contaminant residues in the air, water, soil, and biota varies greatly in different parts of the world due to differences in dispersion patterns and source strengths and characteristics. Based upon the discussions in this and the previous chapters, it should be clear that the distribution and, therefore, resultant residue levels are a complex function of physical, chemical, and biological processes acting on the contaminant and within the components of the environment. Thus, it is difficult to generalize as to global residue levels. However, in order to appreciate actual levels of contaminants in the environment, a number of tables and figures are presented.

Residues in the Atmosphere

Contaminants may exist as particulates or gases in the ambient air.

PARTICLES

Trace metals, with the exception of mercury, are associated almost entirely with particulate matter in the atmosphere. Table 5–6 presents concentrations of selected metals in urban air. Often, the particle size differs for the various metals. For example, the mass median size of the lead contain-

Table 5.6   Selected Trace Metals Found in Air

Metal	Approximate concentration range, U.S. cities ($\mu g/m^3$)	
Arsenic	0–0.75	(mean: 0.02)
Beryllium	0–0.008	(mean: <0.0005)
Cadmium	0–0.3	(mean: 0.002)
Chromium	0–0.35	(mean: 0.015)
Lead	0.1–5	(higher in traffic)
Manganese	0.01–10.0	(mean: 0.10)
Nickel	0.01–0.20	(mean: 0.032)
Vanadium	0–1.4	

Compiled from: *Environmental Engineers Handbook*, vol. 2. Air Pollution. B. G. Lipták (ed.). Radnor, Pa.: Chilton, 1974.

Table 5.7 Selected Pesticide Levels in Air

Pesticide	Range of maximum concentration (ng/m³)	Number of cities with positive samples*
DDT	2.7–1560	9
BHC	4.4–9.9	4
Lindane	0.1–2.6	3
Heptachlor	2.3–19.2	2
Methyl parathion	5.4–129	3

*Air samples were measured in nine cities across the United States.

Compiled from: Stanley, C. W., Barney, J. E. II, Helton, M. R., and Yobs, A. R. Measurement of Atmospheric Levels of Pesticides. *Env. Sci. Technol.* 5:430–35, 1971.

Fig. 5–5 Trends in levels of total suspended particulates for five cities.
Source: U.S. Environmental Protection Agency.

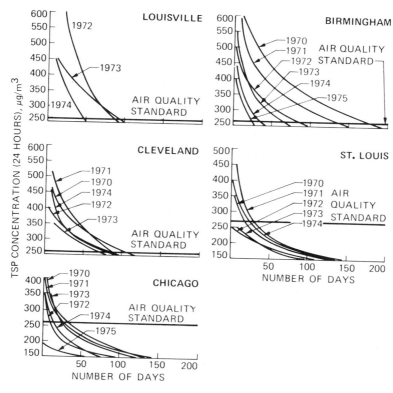

ing aerosol is generally in the 0.2–0.4 μm diameter range, while cadmium is associated with particles larger than 2 μm. These differences generally reflect various sources for the contaminants.

Considerable interest has been expressed in the levels of pesticides in ambient air. In most cases, these are measured as particulate matter. Table 5–7 presents some maximum pesticide concentrations obtained in a survey of nine cities in the United States.

A summary of daily total suspended particulate (TSP) observations made in 1970–1975 for five cities in the United States is presented in Figure 5–5. Total suspended particulate measurements include any solid or liquid particle dispersed in the air, and do not distinguish among the vari-

Fig. 5–6   Distributions of 24-hour SO₂ levels for four cities (1974 data).
Source: U.S. Environmental Protection Agency.

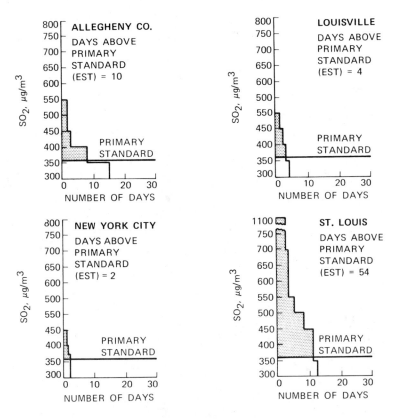

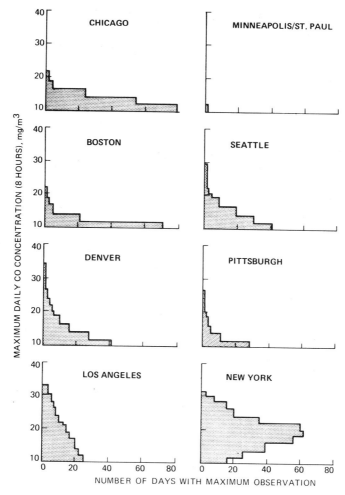

Fig. 5-7 Distribution of CO levels for eight cities.
Source: U.S. Environmental Protection Agency.

Table 5.8 Levels of NO_2 in Selected Urban Areas

Urban region	Annual average concentration, $\mu g/m^3$
Los Angeles	166
Denver	98
New York	121
Philadelphia	94
Chicago	133
Baltimore	120
Canton, Ohio	113
Atlanta	90
Springfield, Mass.	111
Youngstown, Ohio	96
Steubenville, Ohio	98
Boston	128

Source: U.S. Environmental Protection Agency.

ous particle sizes or between natural and anthropogenic aerosols. A discussion of the air-quality standard shown in the figure is presented in Chapter 8.

GASES

There are many data on the concentrations of various gaseous contaminants in ambient air at various locations.

The distribution of daily levels of SO_2 in four urban areas of the United States in 1974 is shown in Figure 5–6; there are relatively few data on SO_2 levels in rural regions.

Figure 5–7 shows the analysis of 1974–1975 data for levels of CO in eight regions. The "street-canyon" effect caused by buildings and heavy continuous traffic congestion results in relatively high levels at the New York City sampling sites.

Levels of NO_2 in selected United States urban areas are shown in Table 5–8. Figure 5–8 presents levels of O_3 in eight cities. Note the differences in O_3 levels between those cities in southern California and the other regions measured. Although maximum levels may be similar in all cases, these occur much more frequently in southern California.

Residues in the Hydrosphere

Trace amounts of many chemicals may be found in surface waters, including potable water. What is actually observed is very often limited by the

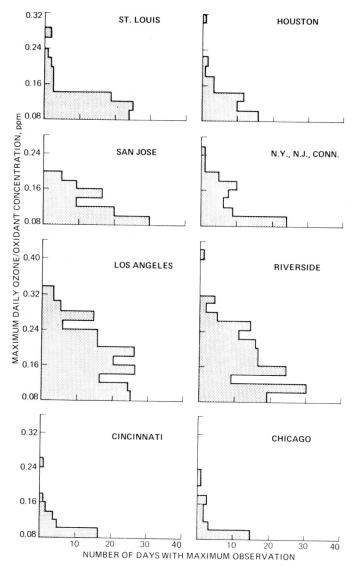

Fig. 5–8 Distribution of O$_3$ levels for eight cities.
Source: U.S. Environmental Protection Agency.

Table 5.9 Trace Metals in Surface Waters in the United States

Metal	Mean detectable* concentration ($\mu g/l$)	Percentage of samples containing metal
Aluminum	74	31
Arsenic	34–308	6
Barium	43	~100
Boron	101	98
Cadmium	10	3
Chromium (VI)	≤25	5–56
Cobalt	1–48	3
Copper	15	74
Iron	19–173	59–99
Lead	23	20
Manganese	20–600	51
Molybdenum	68	33
Nickel	19	16
Silver	2.6	7
Strontium	<200	99
Vanadium	105	3.4
Zinc	64	76

*These values depend on the sensitivity of the method of detection. The smallest detectable level varied from 0.1 $\mu g/l$. for beryllium to 100 $\mu g/l$. for lead. A range of values indicates the data were presented separately for various river basins.

Compiled from: Kopp, J. F. and Kroner, R. C. *Trace Metals in Waters of the United States*. U.S. Dept. of the Interior, Federal Water Pollution Control Administration. Division of Pollution Surveillance, Cincinnati, Ohio, 1962–67.

sensitivity of the detection methods. The increasing application of more specific and sensitive techniques for water-quality analysis has resulted in the identification and quantification of an increasing range of chemical contaminants in water.

A number of surveys of raw and finished water supplies in the United States have been performed. Table 5–9 presents some of the results of a five-year (1962–1967) study aimed at measuring the concentration of trace metals in surface waters of the United States. A total of 1,500 samples were examined.

Contamination of surface waters is frequently reflected in contamination of municipal water supplies. Table 5–10 presents the results of a study of drinking water supplies conducted in 1969. A total of 2,595 samples were examined.

One problem in any survey is determining the amount of a chemical in the water which is due directly to human activity. Since "natural" waters contain many of the chemicals of interest, it is very often exceedingly diffi-

Table 5.10   Selected Chemicals in U.S. Drinking Water Supplies

Chemical	Detectable concentration (mg/l.)*	
	Mean	Range
Chloride (Cl⁻)	27.6	<1.0 – 1,950
Cyanide (CN⁻)	0.00009	<0.1 – 0.008
Fluoride (F⁻)	0.32	<0.2 – 4.40
Nitrate (NO₃⁻)	6.3	<0.1 – 127
Sulfate (SO₄⁻²)	46	<1.0 – 770
Radium (²²⁶Ra)	2.2 pCi/l.	0 – 135.9 pCi/l.
Strontium (⁹⁰Sr)	<1.0 pCi/l.	0 – 2.0 pCi/l.

*Occurrence in 2,595 samples taken in 1969.

Compiled from: *J. Amer. Water Works Assn. 63*:728, 1971.

cult to assign source strengths to anthropogenic emissions. It is, however, of interest to compare Tables 5 – 9 and 5 – 10 to the "background" concentrations of some of these same materials presented in Tables 2 – 9a and 2 – 9b.

Recently, considerable concern has been expressed about organic carcinogens present in many drinking-water supplies in the United States. Table 5 – 11 presents a list of some recognized and suspected carcinogens found in such supplies in the United States. Most of these are also found in drinking water in other areas of the world, although the measured levels of the various chemicals may be expected to differ.

Table 5.11   Some Recognized and Suspected Carcinogens in U.S. Drinking Water Supplies

Chemical	Concentration (μg/l.)
Aldrin	5.4
Benzene	50.0
Benzo(a)pyrene	0.0002 – 0.002
Bis(2-chloroethyl)ether	0.42
Carbon tetrachloride	2.0 – 3.0
Chlordane	0.1
Chloroform	0.1 – 3.11
Dieldrin	8.0
Endrin	0.004
Vinyl chloride	10.0

Compiled from: Kraybill, H. F. Global Distribution of Carcinogenic Pollutants in Water. *Ann. N.Y. Acad. Sci.* 298:80 – 89, 1977; and *U.S. Environmental Protection Agency Report to Congress on Preliminary Assessment of Suspected Carcinogens in Drinking Water.* Washington, D.C., 1975.

Table 5.12   Some Organic Residues in Marine Animals

Species	Source	Chemical (mg/kg fresh tissue)		
		Dieldrin	DDT	PCB
Shellfish				
mussel	Mediterranean	0.001–0.030	0.01–0.65	0.08–0.18
	Baltic	–	0.005–0.07	0.01–0.33
Fish (mg/kg muscle)				
herring	North Sea	<0.001–0.034	0.035–0.17	<0.001–0.48
sardine	Mediterranean	<0.001–0.14	0.001–0.63	0.03–6.9
cod	North Sea	<0.001–0.023	<0.003–0.052	<0.001–0.099
cod	Newfoundland	–	0.011	0.038
(liver)		–	2.7	22.
haddock	Newfoundland	–	0.002	0.030
shark (liver)	North Atlantic	–	4.8	5.8
Mammals (mg/kg blubber)				
grey seal	Scotland	0.46–1.7	8.5–36.3	12–88
common seal	Netherlands	<0.02–0.09	9.5	385–2530
porpoise	Scotland	3.1–4.5	16.8–37.4	31–68
sea lion	California	–	41–2678	–
various cetaceae	Atlantic Ocean	<0.1–3.0	1.1–268	0.7–114
Birds				
white tailed eagle (muscle)	Baltic Sea	–	290–400	150–240
cormorant (fat tissue)	Arctic Ocean	–	6.5–15	14–47
various (liver)	North Atlantic	–	–	0.02–311

Note: Some areas are considered particularly heavily contaminated with DDT and/or PCB. These are the Baltic, W. Mediterranean, southern area of the North Sea, west coast of the United Kingdom, and coast of California.

Compiled from: Whittle, K. J., Hardy, R., Holden, A. V., Johnston, R., and Pentreath, R. J. Occurrence and Fate of Organic and Inorganic Contaminants in Marine Animals. *Ann. N.Y. Acad. Sci. 298*:47–79, 1977.

Residues in the Biosphere

Contaminants present in the abiotic components of the environment may eventually occur as residues in the biosphere. In addition, because of bio-concentration processes, the levels of some chemicals in biota may be orders of magnitude greater than the levels in the abiotic environment in which they live.

The global distribution of many contaminants is made quite evident by analysis of marine biota in remote areas. Since many marine creatures are of commercial value as foods, they become a potential direct source of exposure to man.

Table 5–12 provides some examples of the common organochlorine

Table 5.13 Estimates of Petroleum-Derived Hydrocarbons in Marine Animals

Organism	Type of area collected	Hydrocarbon type	Hydrocarbon content[*] (μg/gm wet wt)
Molluscs			
mussel	single oil spill	No. 2 fuel oil	218
clam	single oil spill	No. 2 fuel oil	26
scallop	single oil spill	No. 2 fuel oil	7
oyster	chronic coastal contamination	polycyclic aromatics	1
Fish			
minnow	single oil spill	No. 2 fuel oil	75
smelt	chronic harbor contamination	benzo(a)pyrene	0–5 (dry wt)
flying fish	oceanic	C_{14}–C_{20}	0.3
pipefish	oceanic	C_{14}–C_{20}	8.8
Bird			
herring gull	single oil spill	No. 2 fuel oil	535

[*]Corrected for natural hydrocarbon levels in biota, i.e., represents estimate of levels due solely to exposure in environment.

Compiled from: National Academy of Science, *Petroleum in the Marine Environment*, Washington, D.C., 1975.

Table 5.14 Concentrations of Some Metals in Marine Animals of Commercial Value

Species	Metal (μg/gm wet wt)				
	Arsenic	Cadmium	Chromium	Lead	Mercury
Shellfish					
Clams, all species	3.13 ± 3.85[*]	0.22 ± 0.17	0.33 ± 0.12	1.60 ± 2.28	0.08 ± 0.05
Crab, blue	12.16 ± 11.79	3.62 ± 6.31	0.14 ± 0.06	0.54 ± 0.18	0.16 ± 0.15
Oysters	9.12 ± 15.59	1.88 ± 1.29	0.12 ± 0.04	24.92 ± 38.97	0.10 ± 0.06
Shrimp, brown	8.71 ± 7.63	0.09 ± 0.06	0.07 ± 0.05	0.46 ± 0.06	0.70 ± 2.02
Fish					
Cod	2.14 ± 1.26	0.04 ± 0.02	–	0.28 ± 0.07	–
Flounder	2.45 ± 1.92	0.03 ± 0.01	0.09 ± 0.08	0.36 ± 0.17	0.22 ± 0.20
Haddock	2.56 ± 2.77	0.05 ± 0.03	0.11 ± 0.12	0.33 ± 0.17	0.07 ± 0.06
Tuna	–	–	–	–	0.51 ± 0.30

[*]Mean $\pm$ standard deviation.

Compiled from: Whittle, K. J., Hardy, R., Holden, A. V., Johnston, R., and Pentreath, R. J. Occurrence and Fate of Organic and Inorganic Contaminants in Marine Animals. *Ann. N.Y. Acad. Sci.* 298:47–49, 1977.

Table 5.15   Average Body Burden in Man of Selected Chemical Elements*

Chemical	Body burden (mg/70 kg)	Remarks
Antimony	7	nonessential
Arsenic	18	nonessential
Cadmium	30	nonessential
Chromium	2	Chromium (III) is an essential nutrient.
Cobalt	1	essential nutrient
Lead	150	nonessential
Mercury	13	nonessential
Uranium	0.02	nonessential

*These are average values (for the U.S.); some specific tissues may show variations in burden for certain metals.

compounds found in various marine animals in different parts of the world. Levels of PCB in other types of food have already been presented (Table 2–4). Estimates of petroleum hydrocarbons in marine organisms due to anthropogenic sources, including oil spills, are provided in Table 5–13.

Many metals are essential to life, but excessive exposure may produce adverse effects; other metals are not known to be needed, and any increase in their level is undesirable. Table 5–14 shows residue values for some potentially toxic metals in marine species used as food sources for man. The residue levels for some metals in humans are shown in Table 5–15. Pesticide residues in man were presented in Table 2–6; numerous other exogenous organic compounds have also been identified in human tissue (29).

Residues in Remote Locations

The long-range transport of chemical contaminants often results in their presence in regions far removed from any known anthropogenic source. This is especially true for those refractory chemicals which do not degrade and may, therefore, travel long distances before reaching some ultimate sink.

For example, DDT has been found in Antarctic snow, and one estimate is that about 2.4×10^6 kg have collected in the snow over a 22-year period (30). These residues were probably transported from Europe and North America via northeasterly trade winds and were removed by wet deposition (31).

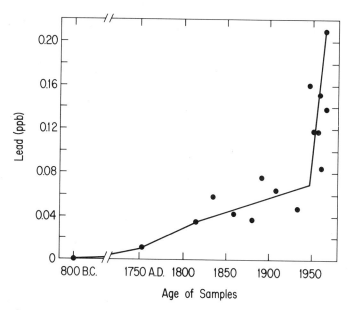

Fig. 5–9   Lead concentration in dated levels of a Greenland glacier.
From: Murozumi, M., Chow, T. J. and Patterson, C. Chemical Concentrations of Pollutant Lead Aerosols, Terrestrial Dusts, and Sea Salts in Greenland and Antarctic Snow Strata. *Geochim. Cosmochim. Acta* 33:1247–94, 1969. Reprinted by permission of Pergamon Press.

Another example is lead, which has been analyzed in chronologic layers of snow strata in Greenland (32). Annual ice layers from the interior of northern Greenland show a rise in lead concentration (Figure 5–9). The layer corresponding to 1750 represents the lead concentration at the beginning of the Industrial Revolution. The concentration at that date was already 25 times greater than the natural level. In the early 1970's, the lead concentration in Greenland snow was about 400 times the natural level.

Additional evidence reflects increasing lead concentrations in other regions. For example, the preindustrial lead content for marine water is estimated at 0.02–0.04 μg/kg (33). However, ocean surface waters in many areas away from anthropogenic sources show lead concentrations as high as 0.20–0.35 μg/kg, and only deep waters (below 1,000 m) remain relatively uncontaminated.

REFERENCES

1. *Man's Impact on the Global Environment. Report of the Study of Critical Environment Problems.* Cambridge, Mass: Mass. Inst. of Technology Press, 1970.

2. Newell, R. E. The Global Circulation of Atmospheric Pollutants *Sci. Amer.* *224*:32–42, 1971.

3. Moore, H. E., Poet, S. E., and Martell, E. A. [222]Rn, ^{210}Pb, [210]Bi and [210]P Profiles and Aerosol Residence Times Versus Altitude. *J. Geophys. Res.* 78:7065–75, 1973.

4. Johnston, H. S. and Yost, D. M. The Kinetics of the Rapid Gas Reaction Between Ozone and Nitrogen Dioxide. J. Chem. Phys. 17:386–92, 1949.

5. Johnston, H. S. and Crosby, H. J. Rapid Gas-Phase Reaction Between Nitric Oxide and Ozone. *J. Chem. Phys. 19*:799, 1951.

6. Inman, R. E., Ingersoll, R. B., and Levy E. A. Soil: A Natural Sink for Carbon Monoxide. *Science 172*:1229–31, 1971.

7. Pressman, J. and Warneck, P. Stratosphere as a Chemical Sink for Carbon Monoxide. *J. Atmos. Sci. 27*:155–63, 1970.

8. Weinstock, B. Carbon Monoxide: Residence Times in the Atmosphere *Science 166*:224–25, 1969.

9. Weinstock, B. and Niki, H. Carbon Monoxide Balance in Nature *Science* 176: 290–92, 1972.

10. Robinson, E. and Robbins, R. C. *Sources, Abundance and Fate of Gaseous Atmospheric Pollutants*, Final Report of Project PR-6755. Menlo Park, Calif.: Stanford Research Institute, 1968.

11. Maugh, T. H., III. Carbon Monoxide: Natural Sources Dwarf Man's Output. *Science 177*:338–39, 1972.

12. Weinstock, B. and Niki, H. Carbon Monoxide Balance in Nature *Science 176*: 290–92, 1972.

13. McConnell, J. C., McElroy, M. B., and Wofsy, S. C. Natural Sources of Atmospheric CO. *Nature 233*:187–88, 1971.

14. Haagen-Smit, A. J. and Wayne, L. G. Atmospheric Reactions and Scavenging Processes, pp. 235–88 in *Air Pollution vol. 1*, A. C. Stern (ed.). New York: Academic Press, 1976.

15. Seinfeld, J. H. *Air Pollution: Physical and Chemical Fundamentals*. New York: McGraw-Hill, 1975.

16. Hidy, G. M. Removal Processes of Gaseous and Particulate Pollutants, pp. 121–76 in *Chemistry of the Lower Atmosphere*, S. I. Rasool (ed.) . New York: Plenum Press, 1973.

17. Robinson, E. R. and Robbins, R. C. Emissions, Concentrations and Fate of Gaseous Atmospheric Pollutants, pp. 1–93 in *Air Pollution Control Pt. II*, W. Strauss (ed.). New York: Wiley, 1972.

18. Cheng, R. T., Corn, M., and Frohliger, J. O. Contribution to the Reaction Kinetics of Water Soluble Aerosols and SO_2 in Air at ppm Concentrations. *Atmos. Environ. 5*:987–1008, 1971.

19. Gerhard, E. R. and Johnstone, H. F. Photochemical Oxidation of Sulfur Dioxide in Air. *Ind. Eng. Chem. 47*:972–6, 1955.

20. Sidebottom, H. W., Badcock, C. C., Jackson, G. E., Calvert, J. G., Reinhardt, G. W., and Damon, E. K. Photooxidation of Sulfur Dioxide. *Environ. Sci. Technol. 6*:72–9, 1970.

21. Fried, M. The Absorption of Sulfur Dioxide by Plants as Shown by the Use of Radioactive Sulfur. *Soil Sci. Soc. Am. Proc. 13*:135–38, 1948.

22. Nordo, F. J. *Proc. of 3rd Int. Clean Air Congress*. Düsseldorf: VDL-Verlag, Gmblt, B-105.

23. Smith, K. A., Bremner, J. M., and Tabatabai, M. A. Sorption of Gaseous Atmospheric Pollutants by Soils. *Soil Sci. 116*:313 – 19.

24. Suess, M. J. The Environmental Load and Cycle of Polycyclic Aromatic Hydrocarbons. *Sci. Total Env. 6*:239 – 50, 1976.

25. Levy, H., II. Tropospheric Budgets for Methane, Carbon Monoxide, and Related Species. *J. Geophys. Res. 78*:5325 – 32, 1973.

26. Goldberg, E. D., Broecker, W. S., Gross, M. G., and Turekian, K. K. Marine Chemistry, pp. 137 – 46 in *Radioactivity in the Marine Environment*, Washington, D.C.: National Academy of Sciences, 1971.

27. Bewers, J. M. and Yeates, P. A. Oceanic Residence Times of Trace Metals. *Nature 268*:595 – 98, 1977

28. *Principles for Evaluating Chemicals in the Environment*. Washington, D.C.: National Academy of Sciences, 1975.

29. Laseter, J. L. and Dowty, B. J. Association of Biorefractories in Drinking Water and Body Burden in People. *Ann. N.Y. Acad. Sci. 298*:547 – 56, 1977.

30. Peterle, T. J. DDT in Antarctic Snow. *Nature 224*:620, 1969.

31. Risebrough, R. W., Huggett, R. J., Griffin, J. S., and Goldberg, E. D. Pesticides: Transatlantic Movements in the Northeast Trades. *Science 159*: 1233 – 36, 1968.

32. Murozumi, M. T., Chow, J., and Patterson, C. C. Chemical Concentrations of Pollutant Lead Aerosols, Terrestrial Dusts and Sea Salts in Greenland and Antarctic Snow Strata. *Geochim. Cosmochim. Acta. 33*:1247 – 94, 1969.

33. Chow, T. J. Isotope Analysis of Seawater by Mass Spectrometry. *J. Water Poll. Control Fed. 40*:399 – 411, 1968.

6

Effects of Contaminants
on Human Health

INTRODUCTION

The first indications that chemicals in the environment could affect human health were associations between certain diseases and specific occupations. It is, however, now clear that many chemicals in the ambient environment may influence the health of the general population as well.

This chapter presents a broad overview of the health effects of chemical contaminants in the environment. The emphasis will be on general aspects of environmental toxicology. For specific consideration of the effects due to individual contaminants, the reader is referred to the toxicology books listed in the Appendix.

Some of the chemicals cited in this chapter are of primary concern in the occupational environment. However, the effluents from industrial operations can also pose similar hazards to the general community. There have been numerous incidences of disease in neighborhoods downwind or downstream from certain industries.

Human exposures to chemicals are very often mediated by complex environmental pathways, and uptake may occur via more than one route. People may be exposed to a chemical via the air they breathe, the water and food they ingest, or by skin contact with air, water, or solid surfaces. As an example, Figure 6–1 shows a diagrammatic representation of the environmental pathways for lead, leading to contact with man and, when exposures are great enough, ultimate effects upon human health

Depending upon its specific nature, a chemical contaminant may exert its toxic action at various sites in the body. However, the first contact is at a portal of entry: the respiratory tract, gastrointestinal tract, or skin. At the

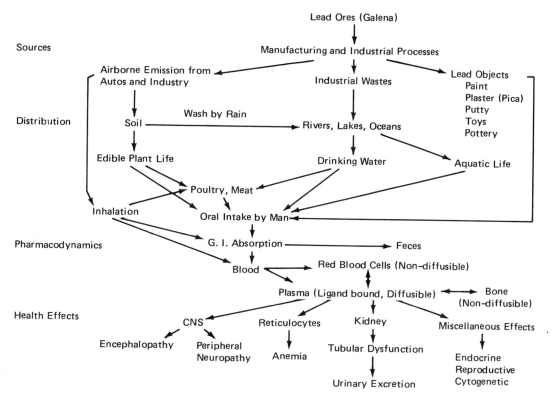

Fig. 6–1   Pathways of lead in the environment and its effects in man.

From: Goyer, R. A. and Chisolm, J. J. Lead, in *Metallic Contaminants and Human Health*, D. H. K. Lee (ed.). New York: Academic Press, 1972.

portal, a chemical may have a topical effect. However, for actions at sites other than the portal, the agent must be absorbed through one or more body membranes and enter the general circulation, from which it may become available to affect internal tissues (including the blood itself). The ultimate distribution of any chemical contaminant in the body is, therefore, highly dependent upon its ability to traverse biological membranes. There are two main types of processes by which this occurs: passive transport and active transport.

Passive transport is absorption according to purely physical processes, such as osmosis; the cell has no active role in transfer across the membrane. Since biological membranes contain lipids, they are highly permea-

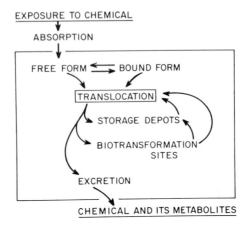

Fig. 6-2   Schematic representation of the pathways through which a chemical agent may pass during its presence in the body.

From: Loomis, T. A. *Essentials of Toxicology*, 2nd Ed. Philadelphia: Lea and Febiger, 1974.

ble to lipid-soluble, nonpolar, or nonionized agents, and less so to lipid-insoluble, polar, or ionized materials. Many chemicals may exist in both lipid-soluble and -insoluble forms; the former is the prime determinant of the passive permeability properties for the specic agent.

Active transport involves specialized mechanisms, and the cell actively participates in transfer across the membrane. These mechanisms include carrier systems within the membrane and active processes of cellular ingestion, i.e., phagocytosis and pinocytosis. Phagocytosis is the ingestion of solid particles, while pinocytosis refers to the ingestion of fluid containing no visible solid material. Lipid insoluble materials are often taken up by active-transport processes. Although some of these mechanisms are highly specific, if the chemical structure of a contaminant is similar to that of an endogenous substrate, the former may be transported as well.

In addition to its lipid-solubility characteristics, the distribution of a chemical contaminant is also dependent upon its affinity for specific tissues or tissue components. Internal distribution may vary with time after exposure. For example, immediately following absorption into the blood, inorganic lead is found to localize in the liver, the kidney, and in red blood cells. Two hours later, about 50% is in the liver. A month later, approximately 90% of the remaining lead is localized in bone (1, 2).

Once in the general circulation, a contaminant may be translocated

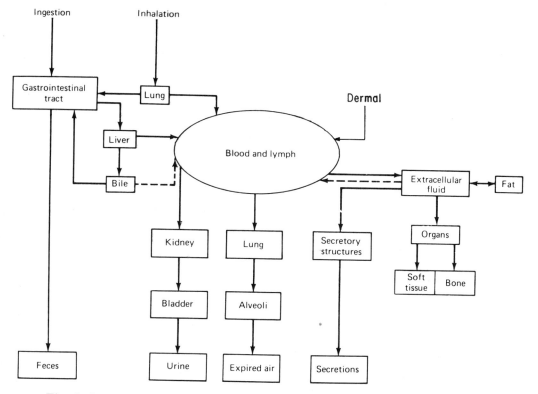

Fig. 6-3   Routes of absorption, distribution, and excretion of toxicants in the body.

From: Klaassen, C. D. Absorption, Distribution and Excretion of Toxicants. pp. 26-44 in *Toxicology: The Basic Science of Poisons*, L. J. Casarett and J. Doull (eds.). New York: Macmillan, 1975.

throughout the body. In this process it may: (a) become bound to macromolecules, (b) undergo metabolic transformation (biotransformation) in certain organs, (c) be deposited for storage in depots which may or may not be the sites of its toxic action, or (d) be excreted. Toxic effects may occur at any of several sites.

The biological action of a contaminant may be terminated by storage, metabolic transformation, or excretion, the latter being the most permanent form of removal.

Figure 6-2 presents a generalized diagram of the possible fate of chemicals in the body. Figure 6-3 shows the specific routes of absorption, distribution, and excretion of toxicants.

ENTRY OF ENVIRONMENTAL CHEMICALS
INTO THE BODY

The portal of entry presents the first possible site of action of, and first line of defense against, environmental chemicals. Some environmental agents have multiple portals of entry, with the specific physicochemical properties in the exposure environment determining the significant entry path. The specific portal may be an important factor in determining the hazard from exposure. For example, more inorganic lead is absorbed into the general circulation following inhalation than would be following ingestion (3).

Respiratory Tract

The respiratory tract is the prime portal of entry for airborne chemicals. It may be separated into two functional zones: the conducting airways and respiratory airways (Figure 6-4). The former extends from the upper respiratory tract (nasopharynx, pharynx, and larynx) through the terminal bronchioles, and serves as a conduit system for the transport of air into and out of the lungs. The latter zone, often called the alveolar region, consists of respiratory bronchioles, alveolar ducts, alveolar sacs, and alveoli, and is involved in the mutual gas exchange between air and the pulmonary circulation. The very thin cellular layer between blood and air in the alveoli reduces the effectiveness of the lungs as a barrier to the systemic uptake of many inhaled chemicals.

The fate of a specific contaminant within the tract is dependent upon the form in which it exists, i.e., gaseous or particulate.

PARTICLES

Deposition is a main factor in determining the fate of inhaled particles which are not subsequently exhaled. Current knowledge of particle deposition patterns in the respiratory tract is by no means complete. A variety of physical mechanisms act to remove suspended particles in various regions of the tract. The main processes are inertial impaction, sedimentation, Brownian diffusion, interception, and electrostatic attraction. The effectiveness of these various mechanisms depends upon airway geometry, breathing rate and pattern, and certain characteristics of the particle itself, such as size, shape, density, hygroscopicity, and charge. Particles with aerodynamic diameters greater than $10-20$ μm are trapped in the nasal passages if breathed through the nose, or retained in the mouth,

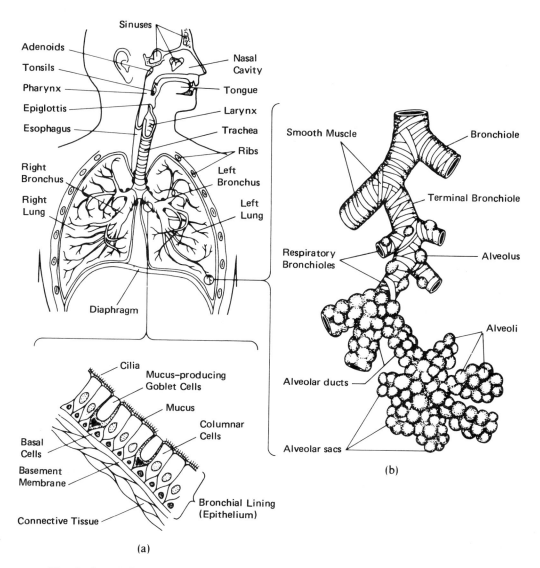

Fig. 6-4 (a) Conductive airways of the respiratory system, with details of the ciliated epithelial lining. (b) Detailed drawing of the components of the alveolar region.

From: Williamson, S. J. *Fundamentals of Air Pollution*. Reading, Mass.: Addison-Wesley, © 1973.

pharynx, and larynx if breathed through the mouth. Those from 2 to 10 μm are largely trapped in the nasal passages and/or bronchial tree, while those less than 2 μm penetrate extensively to the bronchioles and alveoli. Deposition of particles in the bronchial tree is an important protective mechanism for those environmental agents which produce health effects only when they penetrate to and deposit in the alveoli.

The conductive airways are lined with a ciliated epithelium, overlaid by a thin fluid layer composed largely of water and glycoproteins, and formed by secretions of specialized glands and individual cells (goblet cells) (Figure 6–4). This "mucous blanket" is propelled by the cilia towards the pharynx, where it may be swallowed or expectorated. A similar blanket in the nasal passages and sinuses also moves toward the pharynx. The rate of mucous transport varies in different regions of the tracheobronchial tree, and the composition and/or consistency of the fluid lining also differs in different regions.

Insoluble particles which deposit on the nasal passages and tracheo-bronchial tree may be cleared from the respiratory tract via this mucociliary system, with residence times ranging from minutes up to about 24 hours. Frequently, the rapidity and degree of clearance plays a major role in determining the risk from particulate exposure. Extended contact time between deposited toxic particulates and the epithelium may lead to local damage or an increased degree of absorption and resultant systemic effects.

Soluble particles which deposit in conductive airways may dissolve in the mucus, moving with it out of the respiratory tract, or they may be absorbed through the epithelium into the circulatory and/or lymphatic system.

The alveolar region of the lung does not contain cilia or mucus. Particles which reach and deposit within this zone are, however, also subject to clearance, but by different mechanisms, with half-times varying from days to years. Deposited particles may be taken up, via phagocytosis, by specialized cells known as alveolar macrophages, which are present on the alveolar surface. These cells may then follow a number of possible subsequent clearance routes.

The macrophages, and also free particles, may penetrate the alveolar epithelium, entering the interstitial tissue where they may remain (parenchymal sequestration), or become absorbed into the lymphatic system. Macrophages carried in this system may be trapped in lymph nodes, which act as dust stores of the lung, or they may eventually drain with the lymphatic fluid into the blood.

Another possible clearance mechanism is migration, via ameboid movement, of the particle-laden macrophages to the level of the mucociliary blanket, and subsequent clearance via the mucociliary system. Free particles and macrophages may also move to the level of the mucociliary system by mechanisms which may involve respiratory motion of the alveolar walls, capillary action, tensile forces between alveolar fluid and mucus, the influence of ciliary beat in bronchioles, and the shearing effect of layers of mucus. Soluble particles depositing in the alveoli may be removed by direct translocation across the alveolar epithelium into the blood.

Other defense mechanisms of the respiratory system include sneezing, which serves to cleanse the upper respiratory tract, and coughing, which acts to clear the larger bronchial airways. Constriction of bronchi due to the contraction of surrounding smooth muscle following exposure to certain chemicals may act as a defense by decreasing further penetration to the deeper lung areas.

GASES AND VAPORS

The site and extent of absorption of inhaled gases and vapors are, for the most part, determined by their solubility characteristics in water. Highly soluble gases, e.g., SO_2, are largely absorbed in the upper respiratory tract, while those which are less soluble, e.g., NO_2, reach the lower airways. Absorption is also dependent upon the partial pressure of the gas in the inspired air. Gases dissolved in the tracheobronchial tree may be cleared with the mucus.

Gastrointestinal Tract

Chemical contaminants which are waterborne or transported via food chains reach man via the gastrointestinal tract. Ingestion may also contribute to the uptake of chemicals which were initially inhaled, since material deposited on or dissolved in the bronchial mucous blanket is eventually swallowed.

The gastrointestinal tract may be considered as a tube running through the body, but whose contents are actually external to the body (Figure 6–5a). Unless the ingested material affects the tract itself, any systemic response depends upon absorption through the mucosal cells lining the lumen. Although absorption may occur anywhere along the length of the gastrointestinal tract, the main region for effective translocation is the small intestine. The enormous absorptive capacity of this region is due to the presence in the intestinal mucosa of projections, termed villi, each of

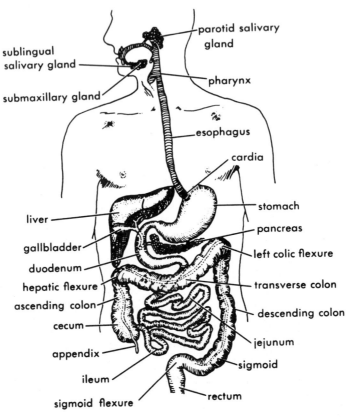

Fig. 6–5 (a) The gastrointestinal tract. The small intestine is comprised of the duodenum, jejunum, and ileum.

From: Digestive System. pp. 150–175 in *McGraw Hill Encyclopedia of Science and Technology*, vol. 4, 1960 Ed. New York: McGraw Hill, © 1960. Used with permission of McGraw-Hill Book Co.

which contains a network of capillaries; the villi result in a large effective total surface area for absorption (Figure 6–5b).

Although passive diffusion is the main absorptive process, certain active transport systems allow essential lipid-insoluble nutrients and inorganic ions to cross the intestinal epithelium. These carrier systems may also be responsible for some contaminant uptake. For example, lead may be absorbed via the system which normally transports calcium ions (4).

Small quantities of particulate material and certain large macromolecules, such as intact proteins, may be absorbed directly by the intestinal epithelium. The mechanism of cellular uptake appears to be pinocytosis; it is, however, more prominent in the newborn than in the adult.

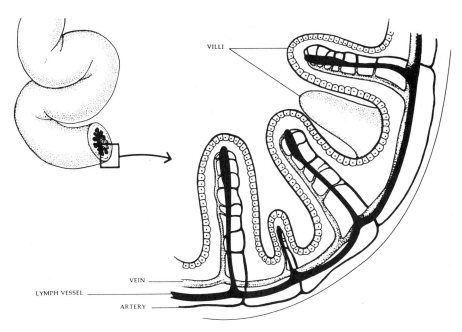

VILLI

VEIN

LYMPH VESSEL

ARTERY

Fig. 6–5 (b) A section of the small intestine. The absorptive area is increased by numerous fingerlike projections, called villi, which cover the entire internal surface of the intestines.

From: Curtis, H. *Biology* New York: Worth, 1968.

Materials absorbed from the gastrointestinal tract enter either the lymphatic system or the portal blood circulation; the latter carries material to the liver, from which it may be actively excreted into the bile, or diffuse into the bile from the blood. The bile is subsequently secreted into the intestines. Thus, a cycle of translocation of a chemical from the intestine to the liver to bile and back to the intestines, known as the enterohepatic circulation, may be established. Enterohepatic circulation usually involves contaminants which undergo metabolic degradation in the liver. For example, DDT undergoes enterohepatic circulation; a product of its metabolism in the liver is excreted into the bile, at least in experimental animals (5).

Various factors serve to modify absorption from the gastrointestinal tract, enhancing or depressing its barrier function. A decrease in gastrointestinal mobility generally favors increased absorption. Specific stomach contents and secretions may react with the contaminant, possibly changing it to a form with different physicochemical properties, e.g., solubility, or they may absorb it, altering the available chemical and changing translocation rates. The size of ingested particulates also affects absorption. Since the rate of dissolution is inversely proportional to particle size, larger particles are absorbed to a lesser degree, especially if they are of a fairly insoluble material in the first place. For example, arsenic trioxide is more hazardous when ingested as a finely divided powder than as a coarse powder (6). Certain chemicals, e.g., EDTA, also cause a nonspecific increase in absorption of many materials.

As a defense, spastic contractions in the stomach and intestine may serve to eliminate noxious agents via vomiting or by acceleration of feces through the gastrointestinal tract.

Skin

The skin is generally a very effective barrier against the entry of environmental chemicals. In order to be absorbed via this route (percutaneous absorption), an agent must traverse a number of cellular layers before gaining access to the general circulation (Figure 6–6). Skin consists of two structural regions, the epidermis and the dermis, which rest upon connective tissue. The epidermis consists of a number of layers of cells, and has varying thickness depending upon the region of the body; the outermost layer is composed of keratinized cells. The dermis contains blood vessels, hair follicles, sebaceous and sweat glands, and nerve endings. The epidermis represents the primary barrier to percutaneous absorption, the dermis being freely permeable to many materials. Passage through the epidermis occurs by passive diffusion.

The main factors which affect percutaneous absorption are degree of lipid solubility of the chemical, site on the body, local blood-flow, and skin temperature. Some environmental chemicals which are readily absorbed through the skin are phenols, carbon tetrachloride, tetraethyl lead, and organophosphate pesticides.

Certain chemicals, e.g., dimethyl sulfoxide (DMSO) and formic acid, alter the integrity of skin and facilitate penetration of other materials by increasing the permeability of the stratum corneum. Moderate changes in permeability may also result following topical applications of acetone, methyl achohol, and ethyl alcohol. In addition, cutaneous injury may enhance percutaneous absorption.

Interspecies differences in percutaneous absorption are responsible for the selective toxicity of many insecticides. For example, DDT is about equally hazardous to both insects and mammals if ingested, but is much

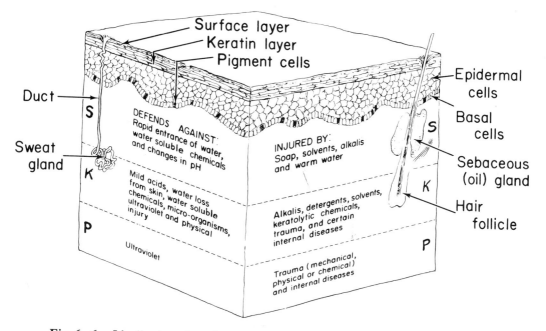

Fig. 6-6   Idealized section of skin. The horny layer is also known as the stratum corneum.

From: Birmingham, D. J. Occupational Dermatoses: Their Recognition and Control. pp. 503–9 in *The Industrial Environment-its Evaluation and Control* U.S. H.E.W., Washington, D.C., 1973.

less hazardous to mammals when applied to the skin. This is due to the poor absorption through mammalian skin, compared to its ready passage through the insect exoskeleton.

Although the main route of percutaneous absorption is through the epidermal cells, some chemicals may follow an appendageal route, i.e., entering through hair follicles, sweat glands, or sebaceous glands.

A minor portal of entry is the eyes. Although they are generally not important for most chemicals, certain materials may be absorbed in sufficient quantities to result in systemic poisoning. An example is fluoroacetate.

FATE OF ENVIRONMENTAL CHEMICALS IN THE BODY

As outlined in Figure 6-2, once absorbed, chemicals may be translocated to storage sites, biotransformation sites, or sites of excretion. We shall

consider each of these pathways before discussing specific biological responses to chemical exposure.

Metabolism of Chemicals: Biotransformation

Chemical contaminants which enter the body may undergo metabolic
conversion in a process known as biotransformation. These chemical reactions are mediated by enzymes and result either in alterations of the parent compound or formation of product(s) via the combination of the parent
with various endogenous substrates.

At one time, it was believed that all contaminants underwent biotransformation into less-toxic forms. Thus, these reactions were often referred
to as detoxification mechanisms. However, it is now clear that certain of
the reactions result in products which are more toxic than the parent
(metabolic toxification).

Contaminants having structures similar to certain normal substrates in
the body may undergo biotransformation by enzymes that normally induce
transformation of these substrates. For example, the enzyme monoamine
oxidase catalyzes the metabolism of endogenous amines such as epineph-
rine; it is also involved in the reactions of foreign, short-chain amines,
such as benzylamine. Most biotransformation reactions are, however, catalyzed by enzymes which have no known endogenous substrates.

The primary site of biotransformation is the liver. Liver cells, like most
other cells, contain a system of internal, smooth- and rough-surfaced
membranes, known as the endoplasmic reticulum. If these cells are ho-
mogenized and then subjected to ultracentrifugation, the smooth fraction
of the endoplasmic reticulum separates into fragments termed micro-
somes. The enzymes involved in biotransformation reactions reside primarily in the microsomal fraction of the cells; most of the information on
these enzymes has been obtained from in vitro studies of this microsomal
fraction. These enzymes are generally nonspecific, i.e., they act upon
various substrates, and are capable of catalyzing a variety of biotransformation reactions. Some chemicals do, however, require specific enzymes for
metabolic transformation. In addition, the enzymes generally do not act
upon lipid insoluble material.

Most biotransformation mechanisms fall into two main classes of reactions: synthetic and nonsynthetic. Nonsynthetic reactions include hydrolysis, oxidation, and reduction. Synthetic reactions, or conjugations, involve the production of a new product from the parent compound or a me-

tabolite, and an endogenous substrate, generally a polar or ionic moiety. Conjugation includes numerous classes of reactions, e.g., acetylation, alkylation, acylation, and esterification.

A number of biotransformation reactions have been found which do not fall into the above basic classes. These miscellaneous mechanisms include aromatic dehydroxylation, aliphatic dehydroxylation, nitrile hydrolysis, dehalogenation, cyclization, and ring scission.

More than one reaction type may be involved in the metabolism of a specific chemical. Sequential reactions involving different mechanisms often occur, and various metabolic products are formed from the parent compound. For example, nonsynthetic reactions often produce metabolites which subsequently undergo conjugation reactions. The exact type of biotransformation reactions which an agent undergoes depends upon the particular structure of the chemical and also upon its route of entry. Route of entry is important because liver cells are not the only sites for biotransformation mechanisms. Enzymes in blood plasma, kidneys, respiratory tract, gastrointestinal tract, and skin are capable of catalyzing metabolic transformation reactions. Certain components of the detoxification systems and the extent of biotransformation activity do, however, differ in the different tissues.

Various host and environmental factors serve to modify biotransformation mechanisms. The host factors include age (e.g., infants have incompletely developed enzyme systems), nutritional status, and genetic deficiencies (e.g., lack of certain enzymes). The environmental factors involve exposure to certain chemicals which result in induction or inhibition of microsomal enzymes.

A variety of chemicals are able to increase the amount and activity of microsomal enzymes in laboratory animals and/or man. These include drugs, such as phenobarbital; organochlorine insecticides, e.g., DDT; certain herbicides, e.g., Herban and Diuron; the food additive BHT; and various polycyclic aromatic hydrocarbons, e.g., benzo(a)pyrene. Liver-enzyme-inducing chemicals may or may not induce enzymes in other sites of biotransformation. Experimentally, enzyme induction follows repeated exposures to the inducing chemical, and is usually temporary, lasting from two to four weeks following termination of exposure.

The metabolism of the chemical inducers is increased as a result of their action in inducing enzymes in the liver. If detoxification is the normal result of metabolism, induction would tend to be protective; if, however, metabolic toxification is the result, induction may increase harmful effects in the host. An example of the former case is the protection against the acute toxicity of certain organophosphate insecticides at least partly afforded by organochlorine insecticides (7, 8); an example of the latter case is the increased toxicity of carbon tetrachloride observed in rats following DDT-mediated enzyme induction (9).

Certain therapeutic drugs are gradually inactivated via metabolism; increased microsomal enzymatic activity would result in a decrease in their duration of action. On the other hand, if metabolism results in activation of the drug, enzyme induction could intensify drug potency.

Environmental chemicals may nonspecifically inhibit microsomal enzymes, at least in laboratory animals. These inhibitors include organophosphate insecticides, carbon tetrachloride, O_3, and CO. However, most studies of inhibition involve drugs. Enzyme inhibition would result in the persistence of a chemical which would normally be metabolized.

Induction and inhibition of biotransformation enzymes is important when one realizes that people are generally exposed to more than one chemical contaminant; thus, exposure to some may increase susceptibility to what alone would be relatively nontoxic levels of other contaminants due to these effects upon the microsomal enzymes.

Contaminants may also be altered by undergoing bacterially mediated and nonenzymatic reactions. For example, gastrointestinal flora may reduce aromatic nitro groups into potentially carcinogenic aromatic amines. Carcinogenic nitrosamines may be formed in the low pH environment of the stomach from secondary or tertiary amines, which naturally occur in a variety of foods, and nitrite, which is used as an additive in smoked meats. Another example is the development of a disease called methemoglobinemia in infants who ingest water and foods having high-nitrate contents. The condition, which results in a decrease in the oxygen-carrying capacity of blood, occurs due to differences in the pH and bacterial content of the gastrointestinal tract of infants compared to adults, resulting in greater conversion of nitrate to nitrite, the actual causative agent, in the former. Bacterially mediated reactions in the gastrointestinal tract may also result in the prolongation of action of certain chemicals via deconjugation, causing reabsorption of a chemical which had been previously conjugated in the liver and secreted with bile into the intestine.

Storage of Chemicals

Some contaminants tend to concentrate in specific tissues due to physicochemical properties, such as selective solubility, or to selective absorption

Table 6.1   Main Storage Sites of Contaminants

Storage site	Example of chemical selectively stored	Mechanism of storage
Fat	DDT, PCB	Dissolution
Bone	Inorganic Lead ^{90}Sr	Substitute for calcium in bone matrix
Kidney	Cadmium	Binds with intracellular protein

onto, or combination with, macromolecules such as proteins. Storage of a chemical often occurs when the rate of exposure is greater than the rate of metabolism and/or excretion.

Storage or binding sites may not be the sites of toxic action. For example, carbon monoxide produces its effect by binding with hemoglobin in red blood cells; on the other hand, inorganic lead is stored primarily in bone, but acts mainly on the soft tissues of the body.

If the storage site is not the site of toxic action, selective sequestration may be a protective mechanism, since only the freely circulating form of the contaminant produces harmful effects. Until the storage sites are saturated, a buildup of free chemical may be prevented. On the other hand, selective storage limits the amount of contaminant which is excreted. Since bound or stored toxicants are in equilibrium with their free form, as the contaminant is excreted or metabolized, it is released from the storage site.

Contaminants which are stored may remain in the body for years without effect, e.g., DDT. On the other hand, accumulation may produce illnesses which develop slowly, as occurs in chronic cadmium poisoning. Table 6–1 lists the main storage sites, with examples of chemicals selectively stored there.

Excretion of Chemicals

The major role in elimination of most toxicants is played by the kidneys. Extrarenal routes may be important in elimination of certain specific substances; these routes are the lungs, gastrointestinal tract, skin, and elimination via body secretions, such as saliva and milk.

When the blood is filtered by the kidneys, contaminants are subject to the same removal mechanisms as are normal metabolic end products. As the filtrate travels through the renal tubules, an agent may be reabsorbed into the blood, or remain in the filtrate and be excreted in the urine. In general, lipid-insoluble substances are poorly reabsorbed, although reabsorption via active transport may occur regardless of solubility characteristics. Biotransformation is an important mechanism for the conversion of chemicals into metabolites which may then be more readily excreted by the kidneys than the parent compound. They may occur, for example, by conversion of a highly lipid-soluble agent into a less lipid-soluble metabolite. On the other hand, biotransformation reactions may also produce metabolites with increased lipid solubility.

Contaminants appear in the feces if they are not absorbed after ingestion, if they are excreted into bile, or are excreted along the gastrointestinal tract. The gastrointestinal tract is the primary route of excretion for many trace metals, e.g., Cd, Hg, Pb, and certain large molecules, e.g., pesticides.

Since the circulation from the gastrointestinal tract goes to the liver prior to entering the general circulation, the liver may serve to remove materials following gastointestinal absorption, preventing distribution to the systemic circulation. The contaminant or its metabolites may then be excreted into the bile, entering the small intestine. Generally, lipid-insoluble substances are more readily excreted into bile; once in the intestines, the agent may be reconverted into a more lipid-soluble form and be reabsorbed, or be excreted with the feces.

The lungs may serve as an organ of excretion for substances that exist in the gas phase at body temperature. Ingested volatile organic compounds, such as CCl_4, may be partially excreted from the lungs as vapors, via diffusion. Very soluble gases, such as SO_2, which are largely absorbed in the upper respiratory tract may be excreted when the blood which contains the dissolved gas reaches the lungs. In general, gases with low blood/gas solubility at body temperature will be more readily excreted through the lungs than will those with high blood/gas solubility.

Although of minor importance, contaminants may be excreted in sweat, tears, saliva, and milk. Generally, lipid-soluble materials are passively diffused into these secretions. Contaminants in mother's milk may be passed along to the suckling infant. Excretion into saliva may result in reabsorption via the gastrointestinal tract. Contaminants can also be eliminated by incorporation into hair and nails, which are periodically trimmed.

BIOLOGICAL RESPONSES TO CHEMICAL CONTAMINANTS

The action of chemical contaminants covers the entire range of effects, from a mere nuisance to extensive tissue necrosis and death, from generalized systemic effects to highly specific attacks on single tissues and even individual enzyme systems. Multiple sources and multiple insults via different environmental routes may increase the risk of disease from environmental chemicals. Biological effects may be the result of exposure to an individual agent, or be due to interaction between different agents. A specific chemical may produce more than one effect upon its target, while completely different agents may result in similar, or even identical, pathological manifestations.

A number of host and environmental factors serve to modify the effects of chemicals; the ultimate response is the result of the interaction between these factors. The main host factors are: (a) age, e.g., older age groups tend to be more susceptible to morbidity and mortality during periods of increased air contamination by sulfur oxides and particulates. This is usually due to chronically reduced cardiovascular and respiratory function, resulting in the inability to cope with additional stresses. Newborns and infants are more sensitive to some toxicants than are adults, partly because of the inability to synthesize specific detoxification enzymes; (b) state of health, e.g., concurrent disease or dysfunction may result in enhanced toxicity following exposure, or may make an organ more susceptible to damage; (c) nutritional status; (d) immunological status; (e) sex, and other genetic factors, e.g., enzyme-related differences in biotransformation mechanisms, such as deficient metabolic pathways, and inability to synthesize certain detoxification enzymes; (f) psychological state, e.g., stress, anxiety; (g) cultural factors, e.g., cigarette smoking, which may affect normal defenses, or may potentiate the effect of other chemicals.

The environmental factors which affect biological response include the concentration, stability, and physicochemical properties of the agent in the exposure environment and the duration, frequency, and route of exposure. Acute and chronic exposures to a chemical may result in different pathological manifestations.

Any organ can only respond in a limited number of ways, although there are numerous diagnostic labels for the resultant diseases. Thus, similar responses may occur in different systems. The following sections discuss the broad types of responses which may occur following exposure to environmental chemicals.

Irritant Response

A pattern of generalized, nonspecific tissue inflammation and destruction may result at the area of contaminant contact. This type of reaction is caused by the class of chemicals termed irritants. It is a portal of entry response, which occurs whether or not the agent is absorbed through the portal and enters the general circulation. A list of some common chemical irritants is presented in Table 6–2.

Some irritants produce no systemic effect because the irritant response is much greater than any systemic effect, e.g., mustard gas in the respiratory tract, or the irritant response is significantly greater than any toxicity, e.g., sulfuric acid in the respiratory tract. Some irritants also have significant systemic effects following absorption, e.g., H_2S absorbed via the lungs.

Exposure to irritants may result in death if critical organs are severely damaged. On the other hand, the damage may be reversible or it may result in permanent loss of some degree of function, such as impaired gas exchange capacity in the lungs or malabsorption of nutrients in the gastrointestinal tract.

The respiratory tract is a common site of irritant exposure. Acute exposure results in a common inflammatory response in all affected regions: rhinitis, pharyngitis, and laryngitis in the upper respiratory tract, bronchitis and bronchopneumonia in the bronchial tree, edema and pneumonia in the alveolar region. Longer-duration exposure may result in a chronic inflammation of small bronchi, and hyperplasia (an increase in size)

Table 6.2   Some Chemical Irritants

Sulfur dioxide
Sulfuric acid mists and other strong acids
Ammonia fumes and other strong bases
Nitrogen dioxide
Hydrogen fluoride
Formaldehyde
Acrolein
Chlorine
Ozone
Peroxyacetylnitrates (PAN)
Aromatic hydrocarbon vapors, e.g., benzene, ether
Particulate sulfates, e.g., zinc sulfate, ammonium sulfate
Organic selenides
Numerous pesticides

and/or metaplasia (transformation of one cell type into another) of mucus-secreting glands and goblet cells. Alveolar responses include deposition of connective tissue (fibrosis), alteration in structural proteins, and loss of lung elasticity.

Changes in respiratory function without development of a specific disease state have been seen in response to exposure to many irritants, e.g., SO_2, NO_2, O_3, dusts, in laboratory studies of both experimental animals and man. These changes are often due to constriction of bronchi caused by a reflex response or by the release of endogenous chemicals such as histamine. Another important effect of exposure to many pulmonary irritants is alteration of the normal functioning of the mucociliary and/or alveolar clearance systems. Those chemicals showing this effect include NO_2, SO_2, and H_2SO_4.

The skin is also a portal of entry frequently subject to environmental chemical irritation. However, most environmental skin disorders arise from occupational exposures to irritants. Most reactions to these are eczematous in nature, and include ulceration and granuloma formation.

Fibrotic Response

A number of environmental agents are etiological factors in the development of a group of chronic lung disorders termed pneumoconioses. This general term encompasses many fibrotic conditions of the lung, i.e., diseases characterized by scar formation in the interstitial connective tissue. Pneumoconioses are due to the inhalation and subsequent selective retention of certain dusts in the alveoli, from which they are subject to interstitial sequestration.

Pneumoconioses are characterized by specific fibrotic lesions, which differ in type and pattern according to the dust involved. For example, silicosis, due to the deposition of crystalline-free silica, is characterized by a nodular type of fibrosis, while a diffuse fibrosis is found in asbestosis due to asbestos-fiber exposure. Certain dusts, such as iron oxide, produce only altered radiology (siderosis) with no functional impairment, while the effects of others range from a minimal disability to death. Some important dusts involved in fibrosis of the lung are listed in Table 6-3.

Dusts initially inhaled may be translocated to other parts of the body. For example, fibrotic nodules due to asbestos are also found in the spleen and peritoneum, possibly due to translocation by lymph or blood, or to swallowing of fibers carried on the mucociliary system (10).

Table 6.3 Pulmonary Fibrotic Agents

Material	Disease designation
Inorganic fibers and dusts	
Crystalline silica	Silicosis
Asbestos	Asbestosis
Talc	Talcosis
Coal (pure)	Coal workers' pneumoconiosis
Kaolin	Kaolinosis
Graphite	Graphite lung
Organic fibers and dusts	
Cellulose	Bagassosis
Cotton	Byssinosis
Flax	Byssinosis
Hemp	Byssinosis
Metallic fumes	
Tin oxide	Stannosis
Iron oxide	Siderosis
Beryllium oxide	Berylliosis

Asphyxiant Response

Some chemicals having the respiratory tract as a portal of entry can cause death by their action in preventing an adequate oxygen supply from reaching the tissues of the body. These chemicals are known as asphyxiants. Some "biologically inert" asphyxiants prevent tissues from receiving enough oxygen simply by displacing the oxygen in the air. Examples of these are CO_2 and CH_4. Others prevent oxygen from being carried by the blood, e.g., CO, or from being utilized by the tissues, e.g., hydrogen cyanide. Chronic exposure to asphyxiants may result in systemic disorders due to a continual impairment of the oxygen supply to vital organs.

Allergic Response

Allergic agents, termed allergens, generally affect the respiratory tract and skin, although they may also affect other organs such as the intestines (allergic colitis). Some chemical allergens are listed in Table 6-4.

Allergic responses involve the phenomenon known as sensitization. Initial exposure to an allergen results in the induction of antibody formation; subsequent exposure of the now "sensitized" individual results in an immune response, i.e., an antibody-antigen reaction (the antigen is the

Table 6.4 Some Chemical Allergens

2,4-D (herbicide)	Epoxy resins
2,4,5-T (herbicide)	Nickel
Organophosphate insecticides	Cobalt
Carbamate herbicides	Arsenic
Toluene diisocyanate (TDI)	Chromium
Acrolein	Beryllium salts

allergen in combination with an endogenous protein). This immune reaction may occur immediately following exposure to the allergen, or it may be a delayed response.

The primary respiratory allergic reactions are bronchial asthma, reactions in the upper respiratory tract which involve the release of histamine or histamine-like mediators following immune reactions in the mucosa, and a type of pneumonitis (lung inflammation) known as extrinsic allergic alveolitis. Skin reactions include urticaria (hives), atopic dermatitis, which is due to ingestion or inhalation of an allergen, and contact dermatitis, due to direct skin contact with an allergen. In addition to these local reactions, a systemic allergic reaction (anaphylactic shock) may follow exposure to some chemical allergens.

Mutagenic and Teratogenic Response

Mutagens are agents which produce mutations, or changes in genetic material, i.e., DNA. Most mutations produce deleterious effects. When the changes occur in the sperm and/or egg cells, the effects may be manifest in future generations. A variety of environmental chemicals have been found to be mutagenic, at least in test systems of cell cultures of mammalian and nonmammalian cells; these may, therefore, pose potential hazards to man (Table 6–5).

A related effect having some overlap with mutagenicity is teratogenici-

Table 6.5 Some Known or Suspected Chemical Mutagens

DDT	Sodium arsenate
2,4-D	Cadmium sulfate
2,4,5-T	Lead salts (some)
Dioxin	Nitrite
Ozone	Benzene

Table 6.6 Some Known or Suspected Chemical Teratogens

2,4,5-T	Cadmium sulfate
Dioxin	Sodium arsenate
Organic mercury	Phenylmercuric acetate
Phthalic acid esters	

ty. A teratogen produces a generative change during early embryonic development, resulting in anatomical defects or other functional or bio-chemical developmental errors. Numerous chemicals, especially drugs, have been shown to be teratogenic in experimental animals (Table 6–6); and these, too, may be potential hazards to man.

Carcinogenic Response

Cancer is a general term for a group of related diseases characterized by the uncontrolled growth of certain tissues. Its development is due to a complex process of interacting multiple factors in the host and the environment.

It has been estimated that between 50–90% of all cancer in the human population in the United States is dependent upon either known or unknown environmental factors (11). In some cases, specific-cancer causing chemicals (carcinogens) are involved, especially in occupational environments. In fact, the first documented cases of cancer produced by exposure to environmental chemicals, in this instance coal tar, were scrotal cancers in men and boys employed as chimney sweeps in 18th-century Britain (12). However, there are numerous carcinogens present in the ambient environment.

One of the great difficulties in attempting to relate exposure to a specific chemical to cancer development in man is the long latent period, typically from 15 to 40 years, between onset of exposure and disease manifestation. Research studies of the carcinogenicity of chemicals, therefore, generally involve bioassay techniques in experimental animals and cell cultures. Although the ultimate relation of any positive results to man is not always clear, evidence for carcinogenicity in these model systems should result in classifying the chemical as a potential human carcinogen.

Certain chemicals as they exist in the environment are carcinogenic; these are termed primary carcinogens. On the other hand, other environmental agents (procarcinogens) can become carcinogenic after they undergo conversion to some other form. Most of these procarcinogens are con-

Table 6.7   Some Environmental Chemicals Carcinogenic in Man

Organic chemicals	Inorganic chemicals
Benzidine	Arsenic (trivalent)
4-aminobiphenyl	Chromate
α-naphthylamine	Nickel carbonyl
β-naphthylamine	Beryllium
Benzene	Cadmium oxide
Vinyl chloride monomer	Asbestos
Bischloromethylether	
Soot, tar (probably due to	Radionuclides
polycyclic aromatic hydrocarbons)	Strontium-90
	Thorium dioxide
	Radium-226
	Radon and daughters

verted by biotransformation reactions, although some undergo nonenzymatic hydrolytic reactions to produce carcinogenic intermediates. It is not always clear whether a specific chemical is a primary or procarcinogen.

Some human carcinogens are listed in Table 6-7. Many other environmental chemicals have been shown to produce cancer in laboratory animals, and thus may be potential human carcinogens. These include N-mustards, epoxides, small-ring lactones, urethane, thioamides, nitrosamine, and azo dyes. The list is quite long.

Current evidence suggests that most organic carcinogens belong to a limited number of chemical classes and, within each class, specific structures appear to be associated with carcinogenicity. However, it is not yet possible to predict from chemical structure whether a specific chemical is carcinogenic, since very slight differences in molecular structure and orientation result in wide variability of carcinogenic potential.

The site of development of malignant tumors may be at various areas of the body other than the initial portal of entry. Thus, for example, inhaled asbestos produces cancer at its portal of entry, the lung, but also in the peritoneum; azo dyes produce tumors in the liver, their site of biotransformation; radium produces bone cancer at its storage site; aromatic amines produce tumors at their site of excretion, the bladder.

Systemic Response

Many environmental chemicals produce a generalized systemic disease due to their effects upon a number of target sites. Table 6-8 presents the main target sites for selected chemicals which may be classified as system-

Table 6.8   Main Target Sites of Selected Systemic Chemical Agents

	Respiratory tract	Hemato-poietic system	Skin	Gastro-intestinal tract	
Mercury	X	X		X	
Lead		X		X	
Cadmium	X	X		X	
Arsenic	X	X	X	X	
Fluoride		X		X	
Molybdenum		X			
Selenium			X	X	
Organophosphate pesticides		X	X		
Organochlorine pesticides					
Carbamate pesticides					
Carbon tetrachloride					
Chlorinated biphenyls			X		

ic toxicants. Some of the metals listed are essential to life at low levels, but quite toxic at high concentrations. In addition, the specific form in which a chemical exists in the exposure environment may affect its toxicity. For example, selenium as selenate and methylated forms of mercury are much more toxic than are other forms of these elements.

Finely dispersed particulates (fumes) of several metal oxides are often associated with an acute systemic syndrome known as metal fume fever. This response is most often due to oxides of cadmium, manganese, and zinc.

INTERACTIONS OF CONTAMINANTS: SYNERGISM, ANTAGONISM, AND TOLERANCE

Most people are generally exposed to more than one contaminant at any one time; thus, interactions of chemicals, as these may relate to health effects, are quite important. The nature of most interactions are, however, largely unknown.

Liver	Kidney	Bones, teeth	Endocrine system	Nervous system: (central and/or peripheral)
X	X			X
	X			X
	X	X	X	X
X	X		X	X
		X		
X				
X	X	X		
				X
X				X
				X
X	X			
X				X

The biological response following simultaneous exposure to a combination of two or more chemicals may be much greater than would be expected from the additive action of each individual agent. This potentiation of the combined effect is termed synergism. For example, at high concentrations, NO_x and CO in combination are more than twice as lethal to cats than if each was inhaled separately (13), the toxicity of malathion may be potentiated by simultaneous administration of the organophosphate pesticide EPN (14), PCB enhances the toxic action of DDT and organophosphate pesticides (15), ethanol potentiates the toxicity of CCl_4 (16), and O_3 and SO_2 act synergistically to depress pulmonary function in man (17).

Nowadays, the generally accepted view of synergism extends beyond simple potentiation to include other increases in toxic reactions, such as gas-particle interaction in the lungs. Gases adsorbed onto particulates may reach deeper lung areas than they normally would if inhaled in the gas phase. For example, the irritant potency of SO_2 is increased in guinea pigs undergoing simultaneous exposure to H_2SO_4 mist (18). Another example of gas-particle interaction is the synergistic effect of hydrogen fluoride gas and beryllium sulfate aerosol (19).

An important area of concern is the synergistic effect of certain environmental chemicals in the pathogenesis of cancer. Some agents termed promoters, or cocarcinogens, potentiate the carcinogenic effect of other chemicals, although they are not carcinogenic themselves. The combination of SO_2 with benzo(a)pyrene results in respiratory tract tumors in hamsters and rats, while no tumors were found following inhalation at similar concentrations of either agent alone or to SO_2 at any level (20). Some phenols and aldehydes also seem to be stimulators of the carcinogenic effect of certain polycyclic aromatic hydrocarbons in laboratory animals (21, 22, 23). Asbestos workers and uranium miners who smoke cigarettes have been found to have a significantly greater risk of developing lung cancer than do their nonsmoking counterparts (24, 25).

Physical agents may also interact with environmental chemicals. For example, it has been suggested that the combination of solar UV radiation and certain chemicals may enhance the action of a known carcinogen, whereas either alone may not (26).

An environmental chemical may act in an antagonistic manner with respect to another, i.e., there is a decrease in toxic effect below that expected by sole exposure to the latter. As opposed to synergism, exposure to both agents does not have to be simultaneous to produce antagonism. For example, the inhalation by mice of particulates plus nitrous oxide resulted in decreased lethality of the nitrous oxide (27). The toxicity of parathion is decreased if rats are pretreated with the pesticides aldrin or chlordane (28). Lung damage due to O_3 in animals may be lessened if the animals are given antioxidants such as Vitamin E (29) and PABA (para-aminobenzoic acid) (30). Carcinogenicity may also be decreased by antagonism, as in the antagonistic effect of selenium to arsenic (31).

Another type of interaction of chemical agents is tolerance. Tolerance is a phenomenon demonstrated in acute studies with experimental animals; tolerant individuals are not killed by levels of certain chemicals which are lethal to their nontolerant counterparts. Exposures of rodents to sublethal levels of O_3 afford protection against subsequent exposure to what would be lethal levels of O_3. Treatment with low concentrations of O_3 also confers protection against the acute effects of other chemicals, e.g., NO_2. This phenomenon, termed cross-tolerance, also occurs between many other oxidants. The importance of tolerance in terms of prolonged exposure in man is unclear, although a recent study suggests that some form of adaptation to O_3 may be occurring in populations in Southern California (32).

A further type of interaction involves environmental chemicals and viable agents. A nonspecific response to certain chemicals, primarily irritants, is a depression of the immune response, making the host more susceptible to viable pathogenic agents. Increases in the susceptibility of experimental animals to respiratory tract infections, e.g., bacterial pneumonia, have been observed following exposures to oxidant gases such as O_3 and NO_2. Subclinical exposure to halogenated hydrocarbons and heavy metals appears to decrease the immune responsiveness of laboratory animals, often increasing their susceptibility to infectious agents (33). In man, the incidence of tuberculosis is enhanced in individuals with silicosis (34).

EPIDEMIOLOGICAL EVIDENCE FOR CONTAMINANT HEALTH EFFECTS

Epidemiology is the study of the distribution and frequency of a disease in a specific population. Epidemiologists seek associations between environmental factors, e.g., levels of certain contaminants, and rates of morbidity and mortality in the exposed population. Effects of contaminants are generally expressed in terms of excess morbidity or mortality, with excess measured relative to the expected statistical mean for the population in a certain time period.

Although studies of certain occupationally exposed groups often show clear-cut associations between diseases and exposure to certain contaminants, there is a lack of data concerning the effects of long-term exposure to contaminants at the lower levels generally found in the ambient environment.

Epidemiologic analyses are beset with inherent difficulties, making it quite hard to obtain a quantitative link between long-term exposure to contaminants and health effects. The main problem is separation of the effects of a specific contaminant from the possible mediating effects of other factors which affect health, such as concurrent disease, diet, living conditions, cultural factors, and occupational exposures. The isolation of the effect of only one factor upon health requires data from large populations which ideally differ only with respect to exposure to the contaminant in question. Because of the many unknown factors, assumptions are often made that these factors are identical for all groups, or vary randomly with respect to levels of the contaminant. Other problems in epidemiologic studies involve: (a) possible synergism, antagonism, and other interaction

Table 6.9   Major Air Pollution Episodes

Location	Date	Topography, meteorological conditions	Chemical agents involved
Meuse Valley, Belgium	Dec. 1–5, 1930	River valley; temperature inversion, weak winds, fog	Responsible agents not definitively know; those implicated include SO_2, fluorides, H_2SO_4. Estimated 9.6–38.4 ppm_v SO_2
Donora, Pennsylvania	Oct. 26–31, 1948	River valley; temperature inversion, weak winds, fog	No conclusive proof for health effects from any single agent; combination of SO_2 and particulates implicated. Estimated 0.5–2.0 ppm_v SO_2, 4 mg/m^3 total particulates
Poza Rica, Mexico	Nov. 24, 1950	Flat terrain; temperature inversion, weak winds, fog	Hydrogen sulfide (H_2S)
London, England	Dec. 5–9, 1952	River plain; temperature inversion, weak winds, fog	Health effects due to any single agent not proven; SO_2 H_2SO_4, particulate levels high; SO_2 maximum of 1.34 ppm_v, average 0.7 ppm_v; smoke particles maximum 4.46 mg/m^3
New York City	Nov. 23–26, 1966	Temperature inversion, weak winds	Believed due to SO_2 and particulates. Maximum 24 hr average of hourly SO_2 concentration = 0.51 ppm_v; maximum hourly SO_2 = 1.02 ppm_v

between individual contaminants, especially since most communities are exposed to combinations which often make isolation of any one contaminant as the primary culprit of observed effects quite hard, (b) the accuracy of the classification and reliability of the records of symptoms during the period under study, and (c) reliability and accuracy of the measurements of ambient levels in the area under study.

Thus, care must be taken in interpretation of epidemiological studies; the finding of an association between specific contaminants and a health effect does not necessarily prove a causal relationship. However, these studies do provide some of the only data on long-term, chronic exposures of large populations to ambient levels of contaminants and, as such, are often used as the basis for policy decisions by governmental agencies.

Main sources	Mortality/Morbidity
Industry (steel mills, glass factories, Zn smelters, H_2SO_4 plants)	63 excess deaths; several hundred attributable illnesses; higher mortality in older age groups. Symptoms were respiratory tract irritation, coughing, chest pain, eye irritation
Industry (steel mills, Zn smelter, H_2SO_4 plant)	20 excess deaths; 5000–7000 attributable illnesses; higher mortality in older age groups. Symptoms were respiratory tract irritation, eye irritation, cough, nausea
Single industrial plant accidental discharge	22 excess deaths; 320 attributable illnesses. Persons of all age groups affected. Symptoms: respiratory tract irritation and nervous system disorders
Space heating using coal in homes and factories	3500–4000 excess deaths; excess morbidity, but no estimate of numbers made; higher mortality in older age groups. Symptoms: respiratory tract irritation, heart disease
General urban sources: household and industry	168 excess deaths; unknown number of attributable illnesses; higher mortality in older age groups

Air Contaminants

ACUTE EXPOSURE – AIR POLLUTION EPISODES

The severe effects of air contaminants upon health were brought to general attention following certain acute air pollution crises known as episodes. Concentrations of some contaminants were reached which were clearly hazardous to human health, as indicated by abrupt increases in morbidity and mortality of the exposed population; there was left little doubt that air pollution was the cause of these effects.

Although some episodes were due to accidental industrial release of chemicals, most were natural buildups caused by abnormally poor ventilation conditions due, in large part, to characteristics of local topography.

Most episodes involved a combination of aerosol and gaseous contaminants, since most occurred under conditions in which water droplets (e.g., fog) were present. The specific contaminant(s) which seemed to be responsible and which did reach greater than normal levels were almost always the reducing type, i.e., characterized by SO_2, related sulfur-containing species, and suspended particulates such as smoke. However, no single agent has been irrevocably indicted as the cause for the excess mortality and morbidity, which occurred mainly among the elderly and people with preexisting pulmonary and/or cardiovascular disease. Table 6–9 presents some of the major air pollution episodes. Note that the concentration and often the nature of specific contaminants during the early episodes were unknown, and were generally estimated at a later date.

The acute episodes presented in Table 6–9 are by no means the only ones which have resulted in excess morbidity and mortality. Many minor air pollution episodes occur regularly in the United States and other areas of the world; only when health effects occur in such startling levels do people become aware of the hazards of air contamination.

CHRONIC EXPOSURE

The acute episodes show levels of contaminants which were associated with almost immediate effects upon human health. Quantitative analyses of effects of low levels of air contaminants are much more difficult to establish.

The available epidemiological evidence suggests that ambient air contaminant concentrations are contributors to the pathogenesis and/or exacerbation of certain chronic diseases in urban populations. Estimates of excess mortality due to air contamination range from 0.1 to 10%; the exact excess morbidity is not clear. The main diseases which are associated with general air contamination involve the respiratory tract—chronic bronchitis, pulmonary emphysema and, perhaps, lung cancer.

Chronic bronchitis is an inflammation of the bronchial tree, accompanied by excessive production of mucus and a persistent, productive cough. Numerous epidemiological studies link air contamination with excess morbidity and mortality from chronic bronchitis in many industrial nations; studies in England and Wales, for example, suggest that it may account for a doubling of bronchitis mortality rates for urban as compared to rural areas (35, 36, 37). The commonly measured pollution indices, i.e., levels of ambient SO_2 and levels of total suspended particulates, appear to be associated with chronic bronchitis incidence.

The main etiologic agent in chonic bronchitis is cigarette smoke, a fact

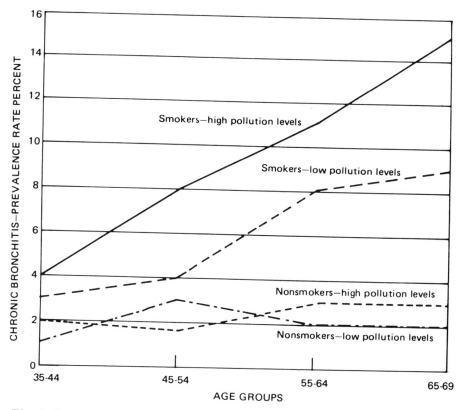

Fig. 6-7   Prevalence of chronic bronchitis in smokers and nonsmokers in areas of high and low air pollution levels.

From: The Burdened Human. in: *Air Pollution*, ed. by V. Brodine.© 1972 by Committee for Environmental Information. Reprinted by permission of Harcourt Brace Jovanovich, Inc.

which complicates analysis of the effects of air contamination. Figure 6-7 presents the results of one study which examined the interaction of smoking and air contamination by comparing the incidence of chronic bronchitis for smokers and nonsmokers in areas of high and low contamination (38). Significantly higher levels of chronic bronchitis were observed in smokers in every age group above 35 who lived in the areas with greater contaminant levels. Thus, air contamination seems to intensify the effects of smoking.

Pulmonary emphysema is a progressive disease characterized by an overdistension and destruction of the alveoli. There is some epidemiologic evidence that air contamination may contribute to emphysema.

For example, in one study (39), autopsy material of lungs from residents of a city with relatively low air contaminant levels (Winnipeg, Manitoba) and a city with greater levels (St. Louis, Missouri) was examined. Emphysema was found to be seven times more common in St. Louis for people in the 20- to 49-year age bracket and twice as common for people over 60 years of age. Although smoking could not be isolated as a causal factor in this study, the incidence of severe emphysema was four times as high among cigarette smokers in St. Louis as among a comparable group of smokers in Winnipeg (Figure 6–8). Although highly suggestive, this study does not prove a causal link between air contaminants and emphysema; there were also climate differences between the two cities. Nevertheless, pulmonary emphysema is increasing in incidence, especially in urban areas.

There is clearly an "urban effect" in the development of lung cancer. People living in urban areas with high contaminant levels are more likely to develop lung cancer than are those living in rural areas, even when the data are corrected for smoking habits. For example, a tenfold difference was found between the cancer mortality rate for English nonsmokers between rural and urban areas in one study (40), while in another (41) the urban mortality rate was twice as high as the rural rate in England and Wales. Mortality rates from lung cancer in California were examined in one study (42). After adjusting for age differences and smoking habits, the rates were found to be 25% greater in the major metropolitan areas than in less-urbanized regions. Among nonsmokers, rates of lung cancer mortality were 2.8–4.4 times as high for the metropolitan areas as for the rural areas. In a review of the evidence relating lung cancer to air pollution, Lave and Seskin (43) found that for smokers the lung cancer mortality rates, adjusted for age and smoking habits, ranged from 26 to 123% greater in urban than in rural areas; for nonsmokers, most differences exceeded 120% (Table 6–10). It is not clear whether the difference in lung cancer rates between urban and rural areas is actually due to higher contaminant levels in urban air. Although arguments have been presented on both sides and other urban factors may be contributory to some degree, a conservative assumption assigns air contamination some role.

Other respiratory ailments may also be associated with ambient air contaminants. Asthma attacks have been related to increased atmospheric levels of, for example, SO_2 and nitrates (44, 45). Acute, nonspecific respiratory disease, i.e., colds, may also be influenced by contaminants. An epidemiologic study in Maryland showed significantly greater prevalence of colds in people living in the more-contaminated area of town than

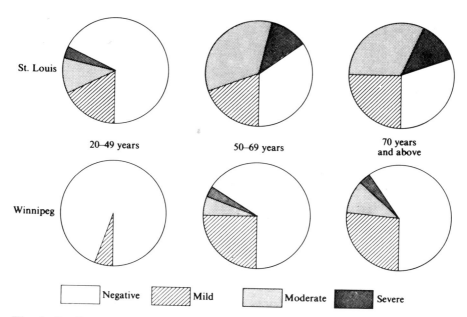

Fig. 6-8   Prevalence and extent of emphysema in St. Louis, Missouri and Winnipeg, Canada.

Redrawn from data of Ref. 40 by Williamson, S. J. *Air Pollution* Reading, Mass.: Addison-Wesley,© 1973.

for those in the less-contaminated area (46). Based upon epidemiological evidence, it has been suggested that the incidence of acute respiratory tract infection may be increased in populations exposed to low ambient levels of NO$_2$. Evaluation of the data indicates that an average annual exposure to 150 μg/m^3, or repeated 2-3 hour exposures to 280 μg/m^3 for 36 days of the year, may be associated with increased susceptibility to respiratory tract infection in children (47). Depression of lung function has also been observed in children living in areas having high levels of NO$_2$ (47).

Children are often studied in epidemiological surveys, since they generally do not smoke cigarettes and are not occupationally exposed to chemical agents. In one such study, New York City children under eight years of age showed a prevalence of respiratory problems which appeared to be related to levels of particulate matter and CO (48). In another study (49), when children from less contaminated (SO$_2$) areas were compared with those from areas with higher levels, the latter were found to have decreased pulmonary function, an increased incidence of lower respiratory tract illness and more irritation of the upper respiratory tract.

Table 6.10 Lung Cancer Mortality Studies (Number of Deaths from Lung Cancer per 100,000 Population*)

Smokers, standardized for age and smoking			Nonsmokers			Region
Urban	Rural	Urban/rural	Urban	Rural	Urban/rural	
101	80	1.26	36	11	3.27	U.S. (males)
52	39	1.33	15	0	—	U.S. (males)
189	85	2.23	50	22	2.27	England & Wales
—	—	—	38	10	3.80	North Ireland
100	50	2.00	16	5	3.20	U.S. (males)

*From: Lave, L. B. and Seskin, E. P. Air pollution and human health, *Science 169*:723–33, 1970. Copyright 1970 by the American Association for the Advancement of Science.

Some studies have attempted to relate various air pollution indices, especially suspended particles and smoke density, with certain nonrespiratory tract diseases. A study in Buffalo and its immediate environs (50) found the mortality rate due to stomach cancer to be over two times as great in areas of high contaminant levels as in areas with low levels. Other studies also suggest a greater mortality rate from nonrespiratory cancers in urban vs. rural areas (43, 51, 52).

Cardiovascular disease has been examined in terms of air contamination. In Nashville the morbidity rate was found to be twice as high in areas of contaminated air vs. clean air (53). A few other studies suggest a correlation between mortality from cardiovascular disease and life in urban areas. In Los Angeles, for example, a correlation was shown between CO

Table 6.11 Categories of Diseases Included in a Study of Mortality and Air Pollution Levels in New York City

Tuberculosis
Cancer of respiratory system
Asthma
Influenza
Pneumonia
Bronchitis
Pneumonia of newborn
Arteriosclerotic heart disease, including coronary disease
Hypertensive heart disease
Rheumatic fever and chronic rheumatic heart disease
Other diseases of the heart, arteries, veins
Certain types of nephritis and nephrosis of the kidneys

Compiled from: Hodgson, T. A., Jr. Short-term effects of air pollution on mortality in New York City. *Environ. Sci. Technol.* 4:589–97, 1970.

levels and mortality rates in patients with preexisting heart problems (54). However, the evidence relating cardiovascular disease to air contamination is much less comprehensive than that which exists for respiratory disorders.

A few attempts have been made to relate daily or weekly mortality and morbidity rates to indices of air pollution obtained during the specific time-interval under study. In one such study (55), the mortality in New York City from respiratory and cardiovascular disease appeared to be significantly related to certain environmental conditions, especially air contamination (the index of which was total concentration of suspended particulates) as well as temperature. Analysis found that about 73% of the variation in daily mortality from the diseases listed in Table 6–11 could be explained by the concurrent variation of levels of air contaminants and temperature, with the former showing a more significant correlation.

In summary, available epidemiological evidence suggests that rather than affecting people only during acute periods, ambient air contamination, especially the reducing type, may continually be increasing morbidity and mortality from various diseases, primarily those of the respiratory tract. The strongest evidence exists for chronic bronchitis; that for emphysema is not as good. Studies of urban and rural populations consistently show greater incidences of lung cancer in the former areas, even when smoking is accounted for. However, because of social and economic differences, this is not definite evidence of an air pollution effect. There is little supporting epidemiological evidence for a link between ambient levels of total oxidant or NO_2 with chronic respiratory tract disease. Eye irritation is common under these circumstances, and some studies do suggest some alteration in respiratory function from smog exposure. Table 6–12 presents selected relations between ambient concentrations and health effects for different air contaminants.

Water Contaminants

Very little work has been done to analyze the effects of drinking water in relation to disease in the general population, even though a vast array of chemicals are found in drinking waters. Recent studies (56, 57) suggest a statistical association between cancer mortality rates and water containing organic chemicals. A few epidemiological studies suggest a relation between certain inorganic contaminants, e.g., asbestos, arsenic, selenium, in drinking water and human cancer (58, 59). However, definitive epidemiological studies have not been conducted, so very little is known of the

Table 6.12   Some Observed Relations Between Contaminant Levels and Health Effects

Contaminant	Concentration level producing adverse health effects	Adverse health effects
Particulate matter and sulfur oxides	1. 80–100 μg/m³ particulates (annual geometric mean)	1. Increased death rates for persons over 50 years of age
	2. 130 μg/m³ (0.046 ppm$_v$) of SO$_2$ (annual mean) accompanied by particulate concentrations of 130 μg/m³	2. Increased frequency and severity of respiratory diseases in schoolchildren
	3. 190 μg/m³ (0.068 ppm$_v$) of SO$_2$ (annual mean) accompanied by particulate concentrations of about 177 μg/m³	3. Increased frequency and severity of respiratory diseases in schoolchildren
	4. 105–265 μg/m³ (0.037–0.092 ppm$_v$) of SO$_2$ (annual mean) accompanied by particulate concentrations of 185 μg/m³	4. Increased frequency of respiratory symptoms and lung disease
	5. 140–260 μg/m³ (0.05–0.09 ppm$_v$) of SO$_2$ (24-hr average)	5. Increased illness rate of older persons with severe bronchitis
	6. 300–500 μg/m³ (0.11–0.19 ppm$_v$) of SO$_2$ (24-hr mean) with low particulate levels	6. Increased hospital admissions for respiratory disease and absenteeism from work of older persons
	7. 300 μg/m³ particulates for 24 hr accompanied by SO$_2$ concentrations of 630 μg/m³ (0.22 ppm$_v$)	7. Chronic bronchitis patients suffering acute worsening of symptoms
Carbon monoxide (CO)	1. 58 mg/m³ (50 ppm$_v$) for 90 min (similar effects) upon exposure to 10 to 17 mg/m³ (10–15 ppm$_v$) for 8 or more hr	1. Impaired time-interval discrimination
	2. Effects upon equivalent exposure to 35 mg/m³ (30 ppm$_v$) for 8 or more hr	2. Impaired performance in psycho-motor tests
	3. Effects upon equivalent exposure to 35 mg/m³ (30 ppm$_v$) for 8 or more hr	3. Increase in visual threshold
Photochemical oxidants (O$_3$ and peroxy-organic nitrates)	1. In excess of 130 μg/m³ (0.07 ppm$_v$)	1. Impairment of performance by student athletes
	2. 490 μg/m³ (0.25 ppm$_v$) maximum daily value. (This value would be expected to be associated with a maximum hourly average concentration as low as 300 μg/m³ ([0.15 ppm])	2. Aggravation of asthma attacks
	3. 200 μg/m³ (0.1 ppm$_v$) maximum daily value	3. Eye irritation

health effects of long-term exposure to ambient levels of most water contaminants.

One exception involves nitrite and nitrate. Ingested nitrate per se has little toxicity. However, nitrate may be reduced to nitrite by bacteria in the gastrointestinal tract, resulting in the development of methemoglobinemia in infants. Nitrite in drinking water may also be involved in the formation of carcinogenic nitrosamines in the environment, or in the gut.

Some epidemiologic studies show an association between cardiovascular disease and soft water (water with a low-mineral content). Higher cardiovascular death rates were found in soft-water regions compared with areas having hard water; however, the specific etiologic agents responsible for the disease/soft-water relationship are not known. In a recent study (60), the factors related to coronary heart disease in two populations in Finland were examined. Lower concentrations of chromium and higher levels of copper in the well water used for domestic purposes correlated most conspicuously with a high death rate from coronary disease.

There have been incidences of chronic poisoning due to high levels of certain chemical contaminants in water. Two of the better-known ones involved cadmium and mercury. In the late 1950's, a large factory on the Jinzu River in Japan discharged cadmium into the water which was subsequently used to irrigate rice fields. A number of people developed cadmium poisoning, characterized by softening of bone and kidney failure. In the second episode, also in Japan, which occurred between 1953–1960, about 111 people were disabled and, of those, 43 died as a result of the industrial discharge of inorganic mercury waste into Minimata Bay. The mercury was methylated in the aquatic sediments and eventually reached the population via fish they ate.

REFERENCES

1. Hammond, P. B. Lead poisoning. An old problem with a new dimension, In: *Essays in Toxicology*, F. R. Blood (ed.). New York: Academic Press, 1969.
2. Klaassen, C. D. and Shoeman, D. W. Biliary excretion of lead. Proceedings of the Fifth International Congress of Pharmacology, (abstract) p. 757, 1972.
3. Kehoe, R. A. Metabolism of lead under abnormal conditions. *Arch. Environ. Hlth.* 8:225–43, 1964.
4. Sobel, A. E., Gawron, O., and Kramer, B. Influence of vitamin D in experimental lead poisoning. *Proc. Soc. Exp. Biol. Med.* 38:433–37, 1938.
5. Hayes, W. J., Jr. Review of metabolism of chlorinated hydrocarbon insecticides especially in mammals. *Ann. Rev. Pharmacol.* 5:27–52, 1965.

6. Schwartze, E. W. The so-called habituation to "arsenic": variation in the toxicity of arsenious oxide. *J. Pharmacol. Exp. Ther.* *20*:181 – 203, 1923.

7. DuBois, K. P. Combined effects of pesticides. *Can. Med. Assoc. J.* *100*: 173 – 79, 1969.

8. Murphy, S. D. Mechanism of pesticide interactions in vertebrates. *Residue Rev.* *25*:201 – 21, 1969.

9. McLean, A. E. M. and McLean, E. K. The effect of diet and 1,1,1-trichloro-2,2-bis-(p-chlorophenyl) ethane (DDT) in microsomal hydroxylating enzymes and on sensitivity of rats to carbon tetrachloride poisoning, *Biochem. J.* *100*:564 – 71, 1966.

10. Meurman, L. Asbestos bodies and pleural plaques in a Finnish series of autopsy cases. *Acta Pathol. Microbiol. Scand. Suppl. 181*:1, 1966.

11. Higginson, J. and Muir, C. S. Epidemiology of Cancer, pp. 241 – 306 in *Cancer Medicine*, J. F. Holland and E. Frei, III (eds.). Philadelphia: Lea and Febiger, 1973.

12. Pott, P. Chirurgical observations relative to the cancer of the scrotum, 1775. Reprinted in *Nat. Cancer Inst. Monogr. 10*:7 – 13, 1963.

13. Stokinger, H. E. Effects of air pollution on animals, pp. 282 – 334 in *Air Pollution*, vol. 1, A. C. Stern (ed.). New York: Academic Press, 1962.

14. Frawley, J. P., Fuyat, H. N., Hagan, E. C. Blake, J. R., and Fitzhugh, O. G. Marked potentiation in mammalian toxicity from simultaneous administration of two anticholinesterase compounds. *J. Pharmacol. Expt. Therap. 121*: 96 – 106, 1957.

15. Waldbott, G. L. *Health Effects of Environmental Pollutants*. St. Louis, C. V. Mosby, 1973.

16. Cornish, H. H. and Adefuin, J. Ethanol potentiation of halogenated aliphatic solvent toxicity. *Amer. J. Ind. Hyg.* *27*:57 – 61, 1966.

17. Bates, D. V. and Hazucha, M. The short-term effects of ozone on the human lung, in: *National Research Council Assembly of Life Science, Proceedings of Conference on Health Effects of Air Pollutants*, Senate Committee on Public Works, 1973. Serial No. 93 – 15. Washington, D.C.: Government Printing Office, 1973.

18. Amdur, M. O. Physiological response of guinea pigs to atmospheric pollutants. *Int. J. Air Pollut. 1*:170 – 83, 1959.

19. Stokinger, H. E., Ashenberg, N. J., Devoldre, J., Scott, J. K., and Smith, F. A. Acute inhalation toxicity of beryllium: II: The enhancing effect of the inhalation of hydrogen fluoride vapor on beryllium sulfate poisoning in animals. *Arch. Ind. Hyg. 1*:398 – 410, 1950.

20. Kuschner, M. The causes of lung cancer. *Amer. Rev. Resp. Dis. 98*:573 – 90, 1968.

21. Bingham, E. and Falk, H. L. The modifying effect of co-carcinogens on the threshold response. *Arch. Environ. Hlth. 19*:779 – 83, 1969.

22. Segal, A., Katz, C., and VanDuuren, B. L. Structure and tumor-promoting activity of anthralin (1-8-dihydroxy-9-anthrene) and related compounds. *J. Med. Chem. 14*:1152 – 54, 1971.

23. Feron, V. J. Respiratory tract tumors in hamsters after intratracheal instilla-

tions of benzo(a)pyrene alone and with Furfural. *Cancer Res.* 32:28–36, 1972.

24. Selikoff, I. J., Hammond, E. C., and Chung, J. Asbestos exposure, smoking and neoplasia. *J. Amer. Med. Assoc.* 204:106–12, 1968.

25. Archer, V. E. and Wagoner, J. K. Lung cancer among uranium miners in the United States. *Health Physics* 25:351–71, 1973.

26. Suskind, R. R. Ultraviolet radiation carcinogenesis: UVR and atmospheric contaminants, pp. 285–98 in *Sunlight and Man: Normal and Abnormal Photobiologic Responses*, M. A. Pathak, (ed.). Tokyo: Univ. of Tokyo Press, 1974.

27. LaBelle, C. W., Long, J. E., and Christofano, E. E. Synergistic effects of aerosols. *Arch. Industr. Hlth.* 11:297–304, 1955.

28. Ball, W. J., Sinclair, J. W., Crevier, M., and Kay, K. Modification of Parathion's toxicity for rats by pretreatment with chlorinated hydrocarbon insecticides. *Can. J. Biochem. Physiol.* 32:440–45, 1954.

29. Roehm, J. N., Hadley, J. G., and Menzel, D. B. The influence of vitamin E on the lung fatty acids of rats exposed to ozone. *Arch. Environ. Hlth.* 24:237–42, 1972.

30. Goldstein, B. D., Levine, M. R., Cuzzi-Spada, R., Cardenas, R., Buckley, R. D., and Balchum, O. J. p-Aminobenzoic acid as a protective agent in ozone toxicity. *Arch. Environ. Hlth.* 24:243–47, 1972.

31. Ferm, V. H. and Carpenter, S. J. Malformations induced by sodium arsenate. *J. Reprod. Fertil.* 17:199–201, 1968.

32. Hackney, J. D., Linn, W. S., Karuza, S. K., Buckley, R. D., Law, D. C., Bates, D. V., Hazucha, M., Pengelly, L. D., and Silverman, F. Effects of ozone exposure in Canadians and Southern Californians: Evidence for adaptation? *Arch. Environ. Hlth.* 32:110–16, 1977.

33. Moore, J. A. and Faith, R. E. Immunologic response and factors affecting its assessment. *Environ. Health Perspective* 18:125–131, 1976.

34. Vorwald, A. M., Droorski, M., Pratt, P. C., and Delehant, A. B. BCG vaccination in silicosis. *Natl. Tuberc. A. Tr.* 46:188–207, 1950.

35. Stocks, P. Cancer and bronchitis mortality in relation to atmospheric deposit and smoke. *Brit. Med. J.* 1:74–79, 1959.

36. Stocks, P. On the relation between atmospheric pollution in urban and rural localities and mortality from cancer, bronchitis and pneumonia, with particular reference to 3,4 benzopyrene, beryllium, molybdenum, vanadium and arsenic. *Brit. J. Cancer* 14:397–418, 1960.

37. Ashley, D. J. B. The distribution of lung cancer and bronchitis in England and Wales. *Brit. J. Cancer* 21:243–59, 1967.

38. Brodine, V. *Air Pollution.* New York: Harcourt, Brace Jovanovich, 1973.

39. Ishikawa, S., Bowden, D. H., Fisher, V., and Wyatt, J. P. The emphysema profile in two midwestern cities in North America. *Arch. Environ. Hlth.* 18:660–66, 1969.

40. Stocks, P. and Campbell, J. M. Lung cancer death rates among non-smokers and pipe and cigarette smokers. Evaluation in relation to air pollution by benzpyrene and other substances. *Brit. Med. J.* 2:923–29, 1955.

41. Daly, C. Air pollution and causes of death. *Brit. J. Prev. Soc. Med.* 13:14–27, 1959.

42. Buell, P., Dunn, J. E. Jr., and Breslow, L. Cancer of the lung and Los Angeles-type air pollution. Prospective Study. *Cancer 20*:2139–47, 1967.

43. Lave, L. B. and Seskin, E. P. Air pollution and human health. *Science 169*: 723–33, 1970.

44. Knelson, J. H. and Lee, R. E. Oxides of nitrogen in the atmosphere: Origin, fate and public health implications. *Ambio 6*:126–30, 1977.

45. Zeidberg, L. D., Prindle, R. A., and Landau, E. The Nashville air pollution study. I. Sulfur dioxide and bronchial asthma. A preliminary report. *Amer. Rev. Resp. Dis. 84*:489–503, 1961.

46. Heimann, H., Reindollor, W. F., Brinton, H. P., and Sitgreaves, R. Health and air pollution. A study on a limited budget. *Arch. Ind. Hyg. Occ. Med. 3*: 399–407, 1951.

47. Bolin, B. and Arrhenius, E. (eds.) Nitrogen—An essential life factor and a growing environmental hazard. Report from Nobel Symposium No. 38. *Ambio 6*:96–105, 1977.

48. Mountain, I. M., Cassell, E. J., Wolter, D. W., Mountain, J. D., Diamond, J. R., and McCarroll, J. R. Health and the urban environment. VII. Air pollution and disease symptoms in a "normal" population. *Arch. Environ. Hlth. 17*:343–52, 1968.

49. Lunn, J. E., Knowelden, J., and Roe, J. W. Patterns of illness in Sheffield Junior School children. *Brit. J. Prev. Soc. Med.* 24:223, 1970.

50. Winkelstein, W. Jr. and Kantor, S. Stomach cancer. Positive association with suspended particulate air pollution. *Arch. Environ. Hlth. 18*:544–47, 1969.

51. Hagstrom, R. M., Sprague, H. A., and Landau, E. The Nashville air pollution study. VII. Mortality from cancer in relation to air pollution. *Arch. Environ. Hlth. 15*:237–48, 1967.

52. Levin, M. L., Haenszel, W., Carroll, B. E., Gerhardt, P. R., Handy, V. H., and Ingraham, S. C., II. Cancer incidence in urban and rural areas of New York state. *J. Natl. Cancer Inst. 24*:1243–57, 1960.

53. Zeidberg, L. D., Horton, R. J. M., and Landau, E. The Nashville air pollution study. VI. Cardiovascular disease mortality in relation to air pollution. *Arch. Environ. Hlth. 15*:225–36, 1967.

54. Bodkin, L. D. Carbon monoxide and smog. *Environment 16*:35–41, 1974.

55. Hodgson, T. A., Jr. Short-term effects of air pollution on mortality in New York City. *Environ. Sci. Technol. 4*:589–97, 1970.

56. U. S. Environmental Protection Agency. *A Report: Assessment of Health Risks from Organics in Drinking Water.* Ad Hoc Study Group to the Hazardous Materials Advisory Committee, Science Advisory Board, Washington, D.C., 1975.

57. Page, T., Harris, R. H., and Epstein, S. S. Drinking water and cancer mortality in Louisiana. *Science 193*:55–57, 1976.

58. Granata, A., DeAngelis, L., Piscaglia, M., and Drago, G. Relationship between cancer mortality and urban drinking water metal ion content. *Minerva Medica 16*:36, 1970.

59. Tramp, S. W. Possible effects of geographical and geochemical factors in development and geographic distribution of cancer. *Schweiz Z. Pathol. 18*:929, 1955.

60. Punsar, S., Erämetsä, O., Karvonen, M. J., Ryhänen, A., Hilska, P., and Vornamo, H. Coronary heart disease and drinking water: A search in two Finnish male cohorts for epidemiologic evidence of a water factor. *J. Chron. Dis. 28*: 259–87, 1975.

7

Effects of Contaminants on Environmental Quality

INTRODUCTION

Man is not independent of his natural environment. He uses its resources, and puts into it the waste products of his biological and industrial activities. However, the wastes are often put into systems that cannot handle them effectively, often upsetting those very systems which are essential to human survival.

Chemical contaminants may affect the environment in various ways; the result is often expressed in terms of changes in "environmental quality." This term is hard to define, for environmental quality means different things to different people and is related to cultural and social attitudes. To some, a quality environment is a virgin forest, while to others it may be a thriving metropolis. Yet all would probably agree that no matter what specific environment is "ideal," certain factors can reduce its quality. A quality environment may thus be defined as one which does not adversely affect the health or welfare of its inhabitants.

AIR CONTAMINANTS

Effects on Animals

Like man, both domestic and wild animals are subject to the acute and chronic effects of air contaminants. During the various acute air pollution episodes, many animal pets and livestock became ill, and some died from cardiopulmonary disorders. In December, 1952, during the "Killer Fog," numerous animals in the London zoo showed increased incidences of

Table 7.1 Effects of Air Contaminants on Animals

Chemical	Symptoms
Arsenic	Colic, ulcers, hair loss, scleroderma (increase in thickness of upper layer of skin), bone malformation
Lead	Nervous system disorders, swollen joints, emaciation
Fluoride	Loss of appetite, weight loss, gastrointestinal disturbances, decreased milk production, lameness, worn teeth
Organic mercury	Nervous system disorders, listlessness, vomiting
Selenium	Loss of hair, abnormal hoof growth, systemic effects
Molybdenum	Gastrointestinal disturbances, limb stiffness, hematological disorders

bronchitis and pneumonia as well as other chest disorders. Incidences of severe pollution damage to animals are also common in the vicinity of some primary industries, such as smelters and fertilizer plants.

Chronic poisoning of animals is more common than acute, and arises from inhalation exposure or feeding on forage upon which air contaminants have accumulated. Chronic poisoning of livestock due to arsenic from smelters, as well as arsenic poisoning of game animals and bees, were reported as long ago as the early part of this century (1, 2). Airborne fluorides cause more damage to domestic animals, on a worldwide basis, than any other air contaminant. One of the earliest descriptions of fluoride poisoning was provided in 1937 and involved livestock near an aluminum smelter in Italy (3). Numerous insects have been victims of fluoride poisoning due to contaminant deposition on plant surfaces (4).

Another air contaminant of importance in terms of damage to animals is lead. For example, in Germany in 1955, cattle and horses near a foundry developed symptoms of lead poisoning (5). More recently, cattle and horses grazing within 5 km of a smelter in Canada were found to have lead-induced damage (6). Captive animals in city zoos are also subject to lead poisoning; as they lick their fur, these animals ingest lead accumulated primarily from automobile exhaust emissions (7).

Table 7–1 describes the effects of selected air contaminants upon animals.

Effects on Vegetation

In the early history of air pollution, there were well-documented cases of the destruction of vegetation around certain industries. Today, the scope of the problem has changed, with more widespread, but generally less severe, effects predominating. Although industrial sources do account for a certain amount of injury to vegetation, the larger problem is associated with air contaminants, such as O_3 and SO_2, which are common to urban centers. For example, large natural forests in Southern California are being severely depleted by oxidant damage to pine trees. Pollution damage to vegetation is an increasing problem in the corridor extending from Washington, D. C., to Boston.

Phytotoxic air contaminants may cause damage to natural plant communities, as well as agricultural crops and ornamental vegetation. Different species of plants, different varieties of one species, and even different parts of one plant may vary widely in sensitivity to a specific chemical. In addition, some contaminants are associated with specific manifestations in particular plants; this often allows identification of the type and range of a specific agent. The sensitivity of any plant is affected by a number of factors, such as the age of the tissue, amount of moisture in the soil, air temperature, plant nutritional status, and intensity of light.

Contaminants may enter plants following wet or dry deposition on aerial structures (leaves, stems), or from the soil via the roots. Most gases enter via leaves, in the course of normal respiration, through the stomata, small openings between cells on the lower leaf surface. Particulates are generally not injurious unless they are corrosive, or deposit in very heavy layers.

Injury to vegetation may be acute or chronic. Short-term exposure to relatively high concentrations of phytotoxicants produce necrotic patterns on leaves due to cell collapse, and perhaps eventually plant death. More commonly, however, plants are subject to long-term exposure to lower contaminant levels. Numerous types of chronic injury may occur. The most common are stunting of growth, destruction of leaf tissue, and chlorosis, a reduction and loss of chlorophyll. Other responses are premature aging, leaf abscission, small fruit, and flowering or fruiting abnormalities. Genetic and biochemical changes, such as alterations in activity of certain enzymes, may occur without any visible injury. Chronic exposures may also increase the susceptibility of plants to other environmental stresses, such as pests and disease.

Table 7-2 lists some of the more important phytotoxic air contaminants, presents typical effects and some examples of the more sensitive plants. The list is not all-inclusive. Other contaminants, such as mercury vapors, ammonia, hydrogen sulfide, sulfuric acid and, of course, herbicides, may affect plant life. Furthermore, combinations of contaminants may produce synergistic effects in some plants. Plant activity may also be

Table 7.2   Effects of Air Contaminants on Vegetation

Chemical	Symptom	Sensitive plants*	Examples of concentration for sensitivity
Chlorine	Bleaching, leaf-tip and margin necrosis, leaf abscission, spotting chlorosis	Radish, alfalfa, peach, buckwheat, corn, tobacco, oak, white pine	Radish, 1.3 ppm$_v$
Fluorides	Leaf-tip and margin necrosis, chlorosis, dwarfing, leaf abscission, decreased yield	Gladiolus, tulip, apricot, blueberry, corn, grape, blue spruce, white pine	Gladiolus, apricot, 0.1 ppb$_v$
Nitrogen oxides	Brown spots on leaf, suppression of growth	Azalea, sunflower, mustard, tobacco, pinto bean	Pinto beans, 3 ppm$_v$
Sulfur dioxide	Bleached spots on leaf, chlorosis, suppression of growth, early abscission, reduced yield	Barley, pumpkin, alfalfa, cotton, wheat, lettuce, apple, oats, aster, zinnia, birch, elm, white pine, ponderosa pine	Alfalfa, barley, cotton, 0.3 ppm$_v$
Ozone	Reddish brown flecks on upper surface of leaf, bleaching, suppression of growth, early abscission, premature aging	Alfalfa, barley, bean, oat, onion, corn, apple, grape, tobacco, tomato, spinach, aspen, maple, privet, white pine, ponderosa pine	Tomato, tobacco, 0.05 ppm$_v$
Oxidant gases, e.g., peroxyacetyl nitrate (PAN)	Glazing, silvering or bronzing of lower surface of leaf	Pinto bean, mustard, oat, tomato, lettuce, petunia, blue grass	Petunia, lettuce, 0.2 ppm$_v$
Unsaturated hydrocarbons, e.g., ethylene	Leaf abscission, dropping of flowers, loss of flower buds, epinasty, chlorosis, suppression of growth	Orchid blossom, carnation blossom, azalea, tomato, cotton, cucumber, peach	Orchids, 0.005 ppm$_v$ Tomatoes, 0.1 ppm$_v$

*Certain varieties of these plants are sensitive.

indirectly affected by specific classes of contaminants, such as the restriction of growth due to alterations in soil chemistry caused by acidic rainwater.

Some plants may accumulate chemicals. For example, high levels of lead have been found in vegetation near highways (8, 9). In some cases a contaminant may be beneficial to vegetation. It has been suggested that large increases in atmospheric CO_2 may enhance photosynthetic activity in terrestrial forests, resulting in increased biomass (10).

Effects on Materials

There is significant damage to nonliving material by air contaminants in many areas. Material corrosion rates, for example, are significantly higher in contaminated urban and industrial atmospheres than in rural atmospheres (11). Damage may be the result of various mechanisms. These include abrasion destruction and the deposition of particulates on surfaces, resulting in the soiling of their appearance. The most important mechanism, however, is chemical attack. This may be direct, i.e., the contaminant reacts directly with the material, as in the tarnishing of silver and the blackening of lead-based paints by H_2S, or indirect, i.e., damage is due to a product of chemical conversion following absorption of the contaminant, as in the absorption of SO_2 by leather and conversion to sulfurous and sulfuric acid.

A number of factors affect the extent of deterioration. Degree of moisture is one of the most important. Oxides of sulfur, carbon, and nitrogen in moist air may be converted into sulfuric, carbonic, and nitric acids, respectively, leading to corrosion of numerous materials. The increasingly acidic rainfall in certain regions of the world has resulted in enhanced corrosion. On the other hand, rain may decrease corrosion rates to the extent that corrosive contaminants are washed away.

Another mediating factor in material deterioration is temperature, which affects the rate of those chemical reactions responsible for deterioration and also has an influence on the degree of moisture condensation on surfaces.

Other factors affecting deterioration include sunlight, which may promote photochemical reactions; air movement, which carries contaminants to surfaces; and position of the surface in space with respect to deposition or accessibility to a contaminant. Table 7-3 presents a survey of the types of damage to material caused by the major air contaminants.

Air contaminants are also responsible for damage to art treasures in many areas of the world. For example, Cleopatra's Needle has undergone more deterioration since it was moved to New York City than in previous thousands of years in Egypt. In Athens, the Parthenon has been subject to cracking due to the growth of nitrate, sulfate, and chloride salts penetrating into cracks in the building structure.

Effects on Climate

One of the most potentially devastating ways in which air contaminants may affect the environment is alteration of climate. As discussed in Chap-

Table 7.3   Effects of Air Contaminants on Materials

Chemical	Primary materials attacked	Typical damage
Carbon dioxide	Building stones, e.g., limestone	Deterioration
Sulfur oxides	Metals	
	Ferrous metals	Corrosion
	Copper	Corrosion to copper sulfate (green)
	Aluminum	Corrosion to aluminum sulfate (white)
	Building materials (limestone, marble, slate, mortar)	Leaching, weakening
	Leather	Embrittlement, disintegration
	Paper	Embrittlement
	Textiles (natural and synthetic fabrics)	Reduced tensile strength, deterioration
Hydrogen sulfide	Metals	
	Silver	Tarnish
	Copper	Tarnish
	Paint	Leaded paint blackened due to formation of lead sulfide
Ozone	Rubber and elastomers	Cracking, weakening
	Textiles (natural and synthetic fabrics)	Weakening
	Dyes	Fading
Nitrogen oxides	Dyes	Fading
Hydrogen fluoride	Glass	Etches, opaques
Solid particulates (soot, tars)	Building materials	Soiling
	Painted surfaces	Soiling
	Textiles	Soiling

ter 2, the earth-atmosphere energy balance is maintained by minor atmospheric constituents, e.g., CO_2; this suggests that small changes in their levels due to man's activities could have major effects on this balance.

There are essentially three ways in which the balance may be subject to perturbation. These are (a) increasing temperature, (b) increasing levels of CO_2, and (c) increasing particulate load. The potential effects of each of these are discussed in the following sections.

ELEVATION OF AMBIENT TEMPERATURE

Waste heat released into the atmosphere has the potential to alter climate, primarily by causing an increase in the average surface temperature of the earth. The primary sources of waste heat are electric-power-generating plants and space heating.

On a global scale, the rate of energy used and released as heat is small compared with the solar input to the earth's surface; net solar radiation is over 6,000 times greater than the anthropogenic source strength (12). However, energy use is usually concentrated on a local or regional scale in highly populated areas; in these regions, the natural and anthropogenic source inputs may be on the same order of magnitude. For example, in 1970, the Los Angeles basin generated heat equal to about 6% of the solar energy absorbed at the ground (13). In fact, in extremely densely populated areas, anthropogenic heat input may even exceed the solar input in colder winter months. According to one projection, by the year 2000, the northeast United States may approach a thermal waste of about 5% of the absorbed solar energy (13).

Urban heat release has affected the character of the local climate, producing what is known as the "urban heat island." Average urban temperatures in many areas often exceed those in surrounding rural areas by 1–2° C, with nighttime differences as great as 5° C. The thermal capacitance for solar input of buildings and streets in built-up areas is also a factor in formation of this heat island.

Local or regional heat islands could conceivably result in changes in atmospheric motions that would be global in scope; however, it has been estimated (14) that anthropogenic heat output would have to increase by approximately 50-fold before there would be climate changes comparable to the natural year-to-year variations on a global scale in general atmospheric circulation patterns.

It is more likely that local heat islands would disturb the character of natural regional climates. One possibility is that very large concentrations of surface heating could, under appropriate environmental conditions, trigger convective instabilities that would lead to convective storms, e.g., thunderstorms, hailstorms, tornadoes (15). Thermal loading may also contribute to an increase in general cloudiness.

PARTICLES

Particulates suspended in the troposphere may affect the radiative energy balance and, therefore, climate—either directly or indirectly. The direct effect involves the absorption and scattering of both incoming solar radiation and long-wave terrestrial radiation. The indirect effect involves influences in the formation and structure of clouds.

Tropospheric concentrations of suspended particulates appear to be increasing in a belt which girds the Northern Hemisphere between 30° and 70° N latitude. A decrease in surface-incident radiation has been noted in many areas in this belt, with the most pronounced effect near urban areas. The possible climatic effects of the increasing tropospheric particulate load are quite conflicting. By absorbing and scattering the incoming solar radiation, particles could tend to reduce the amount of this radiation reaching the ground, resulting in a reduction in heating both within and below the particulate layer. By absorption of the long-wave terrestrial radiation, particles would tend to increase the heating in the absorption layer, enhancing the greenhouse effect. The net effect is not clear, and depends upon a number of factors. These include optical characteristics of the aerosol, such as refractive indices at various wavelengths, concentration, aerosol-size distribution, and reflectivity of the underlying earth surface. For example, the size of the aerosol affects the relative proportion of radiation which is scattered in the forward versus the backward (towards space) direction, as well as the interaction with specific wavelengths of radiation (the greater the size, the greater is the potential to interact with longer wavelengths).

Models of the atmosphere, based upon physical principles, are often used to predict the effects of an increase in particulate loading upon climate. Early models indicated that a global increase in aerosols would result in a net cooling of the surface of the earth. However, the situation is not entirely clear, since the early models underestimated the radiant energy absorption of these aerosols. In addition, increases in particulate load are mainly regional, rather than global, in extent.

A recent model attempted to predict the rate of increase of global background opacity, and its effect on temperature, due to increasing the injection of particles into the troposphere. A fourfold increase in global background opacity (possible, according to this study, in 50-100 years) would decrease global temperature by up to 3.5° C (16). Some models predict a mean surface temperature decrease of 1-3° C if present ambient particulate levels double (16, 17, 18), while still others predict a net warming effect (13).

In addition to any direct effect, particulates may alter climate by influencing the type, structure, formation, location, or optical properties of clouds. This could affect the earth-atmospheric energy balance, since clouds are a contributory factor both to the amount of solar radiation which is reflected back to space, and to the greenhouse effect. Many anthropogenic aerosols can act as condensation nuclei for water vapor, increasing cloud droplet number and reflectivity to incoming solar radiation.

On the other hand, increased cloud cover would also absorb more outgoing terrestrial radiation. The net effect of increasing cloud cover is not clear. However, the increased particulate levels and effects on cloud formation may be a factor in the increased precipitation which occurs in, and downwind of, urban areas compared to upwind and remote rural areas.

CARBON DIOXIDE

The global background level of atmospheric CO_2 prior to the Industrial Revolution is estimated to have been 290–300 ppm$_v$, a level believed to have prevailed for about 10,000 years (19). Currently, the level is approximately 330 ppm$_v$ (Figure 7–1). This increase in atmospheric CO_2 is due primarily to the combustion of fossil fuels. The annual rate of increase in CO_2 concentration jumped from 0.7 ppm$_v$/year in the late 1950's (when the first reliable measurements were made), to 1.3 ppm$_v$/year in the early 1970's.[*]

Numerous predictions of future trends of atmospheric CO_2 levels have been made; however, accuracy is difficult, since future levels depend upon whether the current rate of increase in fossil-fuel consumption will continue, decline, or level off. Nevertheless, most predictions call for an increase in atmospheric CO_2 concentration to at least 379 ppm$_v$ – 400 ppm$_v$ by the year 2000 (13, 20).

Even though it is a trace constituent of the atmosphere, CO_2 plays a major role in the energy balance. Because it absorbs terrestrial long-wave radiation and reradiates most of it back to earth, increases in atmospheric CO_2 could result in increases in the earth-surface temperature, the so-called "runaway greenhouse effect."

Numerous atmospheric models have been employed in attempts to ascertain the magnitude of any temperature change. They predict that an increase in CO_2 to 400 ppm$_v$ could raise the average global surface temperature of the earth by about 0.5 – 1.0° C (15, 21), while a doubling of current atmospheric levels could increase the average temperature by 2–3 ° C (22,23).

However, one model (16) predicts that there will not be a "runaway greenhouse effect" but, rather, as more and more CO_2 is released into the atmosphere, the rate of temperature increase would become proportionately less. The explanation for this is that the main absorption band of CO_2 would become "saturated," so additional CO_2 would not substantially in-

[*]A slowing to 0.6 ppm$_v$/year occurred in 1973–74, possibly due to an oil embargo.

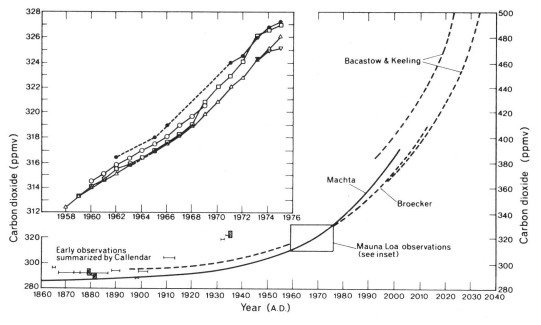

Fig. 7–1 The record of carbon dioxide concentration from 1860 to 1975, measured at several locations, and some estimates of future trends. The early data were critically reviewed by Callender (*Tellus*, *10*:243–248, 1958). The observations for Mauna Loa, Hawaii (□-□) are those reported by Keeling et al. (*Tellus*, *28*: 538–551, 1976); for the South Pole (△-△) by Keeling et al. (*Tellus*, *28*:552–564, 1976); for American Samoa (▽-▽) and Point Barrow, Alaska (●-●) by National Oceanic and Atmospheric Administration (1975); and for Swedish aircraft observations (○-○) by Bolin and Bischof (*Tellus*, *22*:431–432, 1970). The model calculations predicting future carbon dioxide levels are those of Machta (in *Carbon and the Biosphere*, US AEC, 1973), Bacastow and Keeling (in *Carbon and the Biosphere*, US AEC, 1973), and Broecker (*Science*, *189*:460–463, 1975).

From: Kellogg, W. W. Global Influences of Mankind on the climate. pp. 205–227 in *Climatic Change*, J. Gribbin, ed., Cambridge: Cambridge University Press, 1978.

crease the infrared opacity of the atmosphere, and the earth-surface temperature would, therefore, tend to become stabilized, but at some higher than present level.

Certain other gaseous contaminants, e.g., fluorocarbons used as propellants in aerosol-spray products and as refrigerants, N_2O and NH_3, have strong infrared absorption bands, and often absorb in the spectral regions where H_2O vapor and CO_2 are poor absorbers, thus effectively interfering with the atmospheric window which transmits most of the reradiated infrared radiation from the earth's surface and lower atmosphere to space. Thus, anthropogenic emissions of these gases may also affect climate by increasing the surface temperature of the earth (24).

OVERVIEW OF CLIMATIC EFFECTS

Over the course of the earth's history, the climate has changed numerous times. These changes have all been attributed to various natural processes, such as alterations in incident solar radiation or orbital changes in the earth. However, since the Industrial Revolution and with increasing growth in energy use, man has the potential to produce climatic changes which may be global in extent.

Between about 1880 and 1940, the mean earth-surface temperature of the Northern Hemisphere increased about 0.6° C. From 1940 to 1970, a cooling of 0.2–0.3 ° C has occurred (25, 26) (Figure 7–2). However, it is not clear to what extent, if any, these changes are due to man's actions. Furthermore, it has been suggested that the current cooling at high northern latitudes is actually accompanied by a warming at high southern latitudes since about 1955.

Any theory of climatic change is difficult to prove, especially when fluctuations are within the range of normal variability. In addition, many factors control the energy systems which determine climate, and much uncertainty exists in our knowledge of climate cause-and-effect links.

The effects upon overall climate of any isolated contaminant is hard to predict, since possible synergism and numerous feedback mechanisms in the real climatic system are largely unknown or hard to model. Some of the possible mitigating feedback mechanisms are outlined below:

a. An increase in earth-surface temperature due to high CO_2 levels would probably be accompanied by an increase in cloudiness. This increased cloudiness, by acting to decrease incoming solar radiation, could counteract any warming trend. On the other hand, CO_2 is less soluble in water at higher temperatures, so that increases in sea-surface temperature due to increasing atmospheric CO_2 levels could result in a decrease in ocean uptake, enhancing the warming trend by upsetting the ocean-atmospheric CO_2 balance.

b. If earth-surface temperatures increase, snow and ice cover may melt to some extent. This would decrease surface albedo, increase absorption of solar radiation, and thus enhance any warming effect. On the other hand, cooling of the surface could increase ice cover, increasing the albedo, decreasing absorption and enhancing the cooling effect.

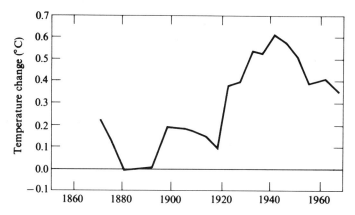

Fig. 7–2 Change in the mean temperature of the Northern Hemisphere (from 0° to 80° N. Latitude) since 1880.

From: Mitchell, J. M., Jr. A preliminary evaluation of atmospheric pollution as a cause of the global temperature fluctuation of the past century. In *Global Effects of Environmental Pollution*, ed. S. F. Singer, p. 139. New York: Springer-Verlag, 1970.

c. Differential heating of the earth sets up heat gradients, which drive wind and ocean currents. Altered gradients due to temperature changes could feed back on winds, etc., to change many other aspects of climate.

Different mixing rates of CO_2 and particles may affect climate differently in the Northern and Southern Hemispheres. It has been suggested that because of the fast latitudinal mixing of CO_2 in the two hemispheres, differences in CO_2 levels between them would be small (27, 28); however, particles are unlikely to be completely mixed globally. Thus, any effects of particles on climate would be intensified in the Nothern Hemisphere, whereas effects due to CO_2 would be more equal in both regions.

Over the last decade, there has been significant progress in the development of mathematical models to predict climatic changes due to air contaminants. Nevertheless, inadequacies of the models still result in inconclusive results, which may merely be order-of-magnitude guesses. Our current knowledge of climate theory is insufficient to eliminate fully even the most pessimistic of the present estimates of climatic change.

The generally accepted view is that continued release of heat, CO_2, and particulates into the atmosphere has the potential to produce significant changes in climate, although the magnitude of the change and its direction are not accurately known. However, in dense urban areas, local climatic changes have already occurred, and as urban areas are extended and energy production increases, the scale of influences upon climate may increase from this local urban scale to larger regional areas.

The most potentially devastating climatic effect of air contaminants is alteration of the earth-surface temperature. In terms of relative significance, a difference of 6°C distinguished the glacial from the interglacial conditions during the Pleistocene Ice Age (25). Smaller surface temperature changes, perhaps as low as 2°C (29), could result in an advance or retreat of the polar ice caps, depending upon the direction in which the temperature was altered. Changes in levels of the oceans would result from changes in the size of the ice caps. Temperature changes may also affect the evaporation-condensation pathway of the hydrologic cycle, with possible significant alterations in the amount and global distribution of precipitation.

A slight but persistent change in temperature could have serious ecological consequences. It could tip the balance in favor of certain organisms which may have been held in check by previous climatic conditions. Changes in patterns of winds and currents may also disrupt previously adapted organisms. Finally, any increase in the instability of global weather patterns is a definite threat to world food production.

EFFECTS ON THE STRATOSPHERE

Most of the potential effects of air contaminants upon climate involve the troposphere. However, certain alterations of the stratosphere may also affect climate.

Jet plane flight in the stratosphere, on a large-scale basis, may increase the cloudiness of the upper atmosphere, via formation of contrails, i.e., trails of water vapor. By thus increasing stratospheric albedo, this could intensify the atmospheric greenhouse effect (24).

Solid particles injected into the stratosphere, because of their longer residence time than those in the troposphere, may also have an effect upon climate. To date, however, the main stratospheric aerosols are volcanic dust. Although these dusts warm the stratosphere as a result of absorption, surface air may be cooled due to a reduction in radiation reaching the earth's surface (13).

By far, the stratospheric effect which has generated the greatest interest is the potential depletion of the ozone layer due to release of certain air contaminants. The center of the stratospheric ozone mass occurs at a height of approximately 20 km from the surface of the earth. In addition to its role in the radiative energy balance of the stratosphere, the ozone layer shields the earth from solar ultraviolet (UV) radiation. On the average, a decrease in ozone content of 1% would result in a 2% increase in UV radiation striking the surface of the earth; the exact relation between ozone loss and UV penetration depends on time of year, latitude, and other factors.

The biological consequences of a substantial global decrease in the stratospheric ozone content and resultant increase in the intensity of UV radiation at ground level are hard to assess. One possible effect is an increase in the incidence of skin cancer; it has been estimated that a 1% decrease in the total ozone content would result in a 2% increase in the incidence of skin cancer (30). Ultraviolet radiation may act synergistically with chemical contaminants to adversely affect other aspects of health. The growth and photosynthetic activity of plants and the behavior of some insects are also affected by UV radiation.

The sensitivity of the ozone layer to gaseous contaminants is a question which is widely debated. The stability of this layer can be decreased by catalytic chemical reactions which destroy ozone. Because many of these reactions involve regenerating chains, a small change in contaminant level may produce large secondary effects. Although numerous models have been used to compute the possible effects of these contaminants, no clear and definite picture of the impact upon the stratosphere has emerged. Any prediction is complicated by the fact that ozone levels vary in natural cycles.

The normal, primary limiting factor of the stratospheric ozone content is the catalytic destruction by nitrogen oxides emitted from natural sources. Nitric oxide and nitrogen dioxide released at ground level are almost completely removed from the troposphere by chemical transformations and precipitation scavenging. However, nitrous oxide emitted at ground level is not removed in the lower atmosphere to any significant extent, and it thus reaches the stratosphere, where it undergoes photochemical decomposition, producing nitric oxide.

Increasing amounts of fixed nitrogen are being introduced into the environment each year, especially in the form of inorganic fertilizers. Normal denitrification processes in soils and waters would lead to increased production and release to the atmosphere of both N_2 and N_2O. One prediction is that 1–6% of the nitrogen in agricultural use will be released as N_2O within the next ten years (31). Thus, the increasing use of fertilizers has been implicated in diminution of the stratospheric ozone content. One estimate places the decrease at 2% by the year 2000 (32), although predictions are quite difficult to make based upon our present knowledge of N_2O cycles and sinks. Aside from fertilizers, possible increases in atmospheric release rates of N_2O may also occur due to an increase in the combustion of coal (32).

The complexity of the problem is made evident by hypotheses that any loss of stratospheric ozone will actually be recovered, due to feedback mechanisms. For example, decreased ozone will result in increased UV penetration into the troposphere, a resultant increase in photolysis of N_2O in the lower atmosphere, and a decline of nitrogen oxides in the stratosphere (33).

Another potential problem is the direct emission of NO and NO_2 within the stratosphere by high-flying aircraft, e.g., supersonic transports. A large fleet of aircraft flying at altitudes above 17–20 km could have an effect upon the ozone layer (34); however, the current planes fly below 17 km, and based upon the type and number of aircraft currently planned, there should be little danger (34).

One other group of chemicals which has been implicated in ozone depletion is the fluorocarbons (35). Since there is no known tropospheric sink for these, they enter and accumulate in the stratosphere, undergoing photolysis at altitudes above 25 km. This reaction releases chlorine atoms, which may then undergo chain reactions with ozone. Whether or not any reduction due to this mechanism has yet occurred is a controversial issue; one estimate places the loss at 0.5–1.0% (36). However, even after ground-based fluorocarbon release ends, ozone destruction would continue due to the long residence times for the fluorocarbons already in the atmosphere.

Aside from the ecologic and biologic consequences of increased UV penetration, changes in the height distribution and amounts of ozone may also result in climate changes. Although a decrease in ozone content may lower the temperature of the upper stratosphere, the resultant climatic consequence of this change cannot be reliably predicted because of the complexities of stratospheric-tropospheric interactions.

Acidification of Surface Waters and Terrestrial Soils

Ordinary rain should have an average pH of about 5.7, based upon atmospheric CO_2 in equilibrium with the precipitation. However, large regions of eastern North America (northeastern United States and southeastern Canada) and southern Scandinavia receive highly acidic precipitation, having a weighted average pH of 4.0–4.4 (37), and a range extending as low as 2.8 (38). The acidity is due primarily to atmospheric formation of H_2SO_4 and, to some extent, HNO_3 from SO_2 and NO_x, respectively. Acid precipitation occurs especially in areas downwind (up to thousands of ki-

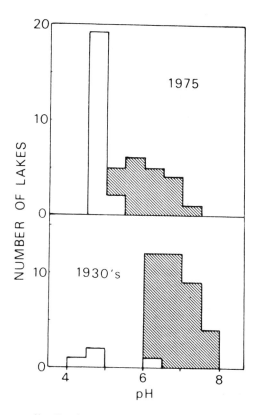

Fig. 7–3 Frequency distribution of pH and fish population in 40 Adirondack (NY) lakes greater than 610 m elevation, surveyed during 1929–1937 and again in 1975.

From: Schofield, C. L. Acid Precipitation: Effects on Fish. *Ambio* 5:228–30, 1976.

lometers) of dense urban and industrial complexes. Figure 7–3 shows the change in pH in some lakes of the Adirondack Mountains of New York State.

The pH of an aquatic environment reflects both the acidity of the precipitation and the ability to neutralize incoming acid via chemical weathering and ion exchange processes in the watershed. Numerous lakes and rivers, especially those with naturally poorly buffered waters, i.e., having low mineral content, have shown the most dramatic increase in acidity, often having a pH less than 5. With continued acid precipitation, however, even those waters with better buffering capacity may soon show changes.

Acidic waters are also characterized by high levels of certain trace metals because of enhanced solubility due to acidification. Many of these metals are common air contaminants.

Acidification of waters affects all aspects of the aquatic environment. Extensive depletion of fish populations in acidified lakes and streams has been seen in many of the areas receiving acid rain (Figure 7-3). The causes are usually direct toxic action of acid waters on eggs and/or larval fish stages, reduction in the number of eggs due to interference with reproductive physiology, and changes in food supply due to the effects of acidity upon other aquatic biota. If the waters are very acidic, a direct toxic action to adult fishes results via interference with ion exchange across the gill membranes .

Like waters, soils having high buffering capacity are not as susceptible to increased acidification. However, many areas do show increased soil acidity. This results in the alteration of some properties of the soil, with an increase in the leaching of certain ions. The ecological effects include a decrease in the growth rate of forests and changes in nutrient cycling.

Aesthetics

It is often the changes in aesthetics caused by the clearly visible effects of air contaminants which spur the public to demand the control of the sources of contamination. Some aesthetic effects of air contaminants have already been mentioned, such as the soiling of buildings and damage to ornamental plants. Two others of importance, decreased visibility and odors, are discussed in the next sections.

VISIBILITY

One of the most obvious effects of air contamination is the deterioration of visibility, which is not only an aesthetic problem, but also a safety hazard. Airline pilots have reported increases in ground haze and air pollution domes over cities since the 1950's and 1960's; cities previously visible from aerial distance of about 30-60 km are now often not visible for more than 16 km or even less during pollution episodes.

Visibility may be defined in meteorologic terms as the greatest distance in a given direction at which it is just possible to see and identify, with the unaided eye, (a) in the daytime, a prominent dark object against the sky at the horizon, and (b) at night, a known, preferably unfocused, moderately intense light source (39). After visibilities are determined around the entire horizon circle, they are resolved for reporting purposes into a single value of prevailing visibility.

The reduction of visibility is due both to the absorption and scattering of visible light by gas molecules and particles in the atmosphere, although scattering is the primary mechanism responsible.

Scattering is the deflection of the direction of travel of light by airborne matter. Particulates in the size range of 0.4–0.7 μm are most effective in scattering, because their size is close to the wavelength of visible light. In addition to size, the degree of scattering is also influenced by particle shape, surface roughness, and refractive index.

The coloration of some contaminated atmospheres is due to the selective absorption of specific wavelengths of light by various atmospheric contaminants.

ODORS

Malodors due to air contaminants are mainly a problem of aesthetics, although property values may be adversely affected in areas where odors are common. Some odorants are merely a nuisance, while others will also produce systemic effects, such as nausea and appetite loss. In addition, odors may interact synergistically or antagonistically.

The olfactory system is quite sensitive. Even when gases are present in the ppb$_v$ range, odors may still be so annoying as to affect personal comfort. However, people do differ widely in their sensitivity to odorants. In addition, the sense of smell is often rapidly fatigued, i.e., continued exposure results in loss of smell of the odorant.

Table 7.4   Selected Odorants

Chemical	Character of odor	Odor threshold*
Acetic acid	Sour	1.0
Ammonia	Sharp, pungent	46.8
Chlorine	Pungent, irritating, bleach	0.314
Ethyl mercaptan	Decayed cabbage, sulfidelike	0.001
Hydrogen sulfide	Rotten eggs, nauseating	0.00047
Isocyanides	Sweet, repulsive	†
Ozone	Slightly pungent, electric sparks	†
Selenium compounds	Putrid, garliclike	†
Sulfur dioxide	Pungent, mustiness	0.47
Tars	Rancid, skunklike	†

*This represents odor recognition thresholds for a panel of four trained observers. The concentrations are in ppm$_v$ and are those at which all panelists recognized the odor.
†Not tested.

Compiled from: Leonardos, G., Kendall, D. and Barnard, N. Odor Threshold Determinations of 53 Odorant Chemicals. *Journ. of Air Poll. Control Assoc.* 19:91–95, 1969.

Many natural sources of odor exist; these are all primarily due to sulfur compounds formed during microbial decomposition of plants and animals, largely in stagnant and insufficiently aerated waters, such as swamps and sewers.

Anthropogenic odorants are the result primarily of industrial operations, although moving combustion sources, such as gasoline and diesel engines, are also contributors in certain areas. The principal industries responsible for odorant release are petroleum refineries, natural gas plants, chemical plants, food processing plants, smelters, paper mills, and tanneries. Table 7 – 4 lists some common chemical odorants.

WATER CONTAMINANTS

The hydrosphere has long been considered an infinite sink for the disposal of all kinds of wastes. It was assumed that running waters were always able to "purify" themselves, and the vastness of the oceans was such that they were able to assimilate any and all of the chemicals we cared to dump into them. Although water does have certain self-purification capabilities, very often these are simply overwhelmed by contaminant discharge. Today, all aquatic environments are threatened, some to a greater degree than others. These include coastal regions near cities, as well as remote lakes and ocean waters thousands of miles from land areas.

Water quality is a value judgment, since water which is unfit for one use may be fine for another, or water unfit for one species may be quite suitable for another. Even natural waters are never 100 % pure. Therefore, an aquatic environment may be considered as polluted if the water becomes unsuitable for its intended use, or changes occur due to man's activities which disrupt the natural ecological balance.

Contaminants may alter aquatic environments in various ways. The type and degree of effect depends upon the type and amount of contaminant and characteristics of the receiving waters, such as temperature, pH, flow rate, degree of dilution and mixing, and the presence of other contaminants which may interact with each other.

The effects of water contaminants upon aquatic life may be divided into two broad types: direct and indirect. An agent is said to exert direct action if it affects characteristics of the organism itself, e.g., growth, reproduction, physiology. Indirect effects involve alterations of the organism's environment, either abiotic or biotic, so as to make the habitat unsuitable for continued existence or reproduction, e.g., affecting food resources, changing turbidity, DO content. Indirect effects may be just as critical to survival as are the direct effects.

Contaminants may be quite selective in their action, or they may affect many types of organisms. Often, only certain tolerant species survive, resulting in a change in the population balance. As food sources are destroyed, animals on higher trophic levels die or move out. In addition, stressed individuals are often more susceptible to disease and parasites, further reducing the reproductive potential of the population. Certain chemicals act as repellents to aquatic life, so that, for example, fish are driven out of an area, making the entire region biologically unproductive. Other types of contaminants may result in extreme overproduction and overenrichment of a body of water. The end result of any disturbance is an alteration, often slow and subtle, of the character of the waterway over a certain period of time.

All contaminants do not adversely affect aquatic life. Some may affect the aesthetics of the water, such as color, or the ability to use it for specific purposes, such as drinking or recreation, without affecting the natural biotic communities.

In the following sections, water contaminants will be discussed by broad classes. Effects upon the physicochemical properties of the aquatic system are presented together with discussions of the effects upon aquatic life.

As with air contaminants, the aesthetic effects of water contaminants contain components of the entire range of environmental problems. People enjoy waterways of high quality which may be used for recreational purposes, or which may add a degree of natural beauty to some area. No one likes a beach blackened by floating oil, or a lake which smells due to rotting sewage.

Waste Heat

Waste heat is released into receiving waters by electric-power-generating stations, especially those using nuclear fuel, and by the many industrial processes which require cooling waters. Although cooling towers are often used in attempts to dissipate the heat before the water is discharged into the environment, the water is still released warmer than when it was taken up.

Certain other of man's actions may affect water temperatures. Removal of forest canopies, removal of brush and shade trees along streams, impounding of river waters, reduction of stream flows, and return of irrigation waters often lead to increases in temperature in the nearby waterways.

The temperature of water naturally varies from season to season, year to year, and even between night and day. Man's activities have often served to extend this normal range of variability, sometimes beyond the limits to which the native biota are adapted.

Except for birds and mammals, other aquatic organisms are poikilotherms ("cold blooded"); their body temperature is at or near the temperature of their environment, and is subject to change as the environmental temperature changes. Within a certain range of tolerance, however, aquatic organisms can adjust and survive. Problems may arise when the temperature exceeds this limited range.

The chemical processes of life are, like all chemical reactions, very sensitive to the temperature of the environment in which they occur. As environmental temperature changes, so does the rate of metabolic reactions, which increase with rising temperature. As the metabolic rate increases, so does the demand for oxygen. However, with rising water temperature, the solubility of oxygen in the water decreases; thus, just as increased demand occurs, less oxygen is available, and levels of DO may fall below the critical value. Of course, if temperatures are high enough, complete enzyme inactivation and cellular disruption may occur.

Aside from affecting metabolic reactions, temperature influences all facets of life, such as hormonal and nervous control, digestion, respiration, osmoregulation, and behavior. Behavioral changes may be just as detrimental as physiologic changes, e.g., attempts to spawn too early in the season or premature migration triggered by changes in temperature.

Many nuisance species of plants and animals thrive at higher temperatures than do more desirable species. For example, blue-green algae survive quite well at around 24–34°C, while certain more desirable plankton have their optimal range at about 14–24°C. As the water temperature increases, the algae have a selective competitive advantage, and their numbers increase, fouling the stream, and further reducing DO upon their death and decomposition. In extreme conditions, i.e., above 60° C, only a few bacterial species are able to survive.

The effects of waste heat also depend, to some extent, upon the season and weather conditions, tending to be more harmful in hot climates than in cold areas. Rapid changes in temperature and intermittent or nondependable fluctuations, such as those due to intermittent discharges, are often more harmful to biota which may otherwise have been able to adapt to slower, more constant changes of temperature greater than their optimal level. Waste heat is also a greater problem in small streams and rivers than in larger waterways having better circulation and mixing.

In addition to direct biological effects, waste heat may affect the action of other contaminants in the water. Temperature affects the susceptibility to some chemicals, resulting in increased hazard from these agents. Some examples are the synergistic effect of heat and methyl mercury (40), PCB (40), and a number of insecticides (41).

Suspended Solids

Solids may produce effects on aquatic environments while carried in suspension or after settling to the bottom of the waterway.

Suspended solids increase the turbidity of the water, thus decreasing the penetration of light by absorption and scattering of the sun's rays. This directly affects the photosynthetic activity in the waterway, which ultimately appears as a reduction in biological productivity. Extreme turbidity can result in almost complete cessation of photosynthesis. Increased turbidity may also interfere with the vision of animals, possibly preventing them from sighting their prey and thus reducing the efficiency of food use.

If the concentration of suspended particulates is high enough, they may interfere with the feeding of filter-feeding organisms; in addition, physical injury to delicate eye and gill membranes may occur by abrasion.

Once solids settle to the bottom, the benthic environment may be disturbed, e.g., by actually burying bottom habitats. Destruction of bottom dwellers and fish larvae, essential parts of certain food chains, could disrupt entire biological communities.

Plant Nutrients: Organic

As described in Chapter 5, organic material introduced into receiving waters undergoes a normal process of decomposition by microbial action, which depends primarily on the oxygen dissolved in the water. If sufficient amounts of organic matter are introduced, the rate of oxidative decomposition could exceed the rate of oxygen replenishment, and the DO concentration in the receiving waters will decline. The exact rate and extent of this decline depends upon the differential between the rates of oxidation and replenishment.

Most healthy streams and rivers have DO levels of 5-7 mg/l. Values consistently less than 4 mg/l. tend to indicate organic overloading. Except for the purest of natural waters, all waters have a measurable BOD, which may be as high as 5 mg/l. Many domestic and industrial wastes have BODs of several hundred mg/l. which, if inadequately diluted in receiving waters, will produce severe oxygen depletion.

Organic wastes released into waters are, of course, sources of nutrients and food for the microbial decomposer organisms. These wastes may encourage their growth, and the resultant increased production may result in other problems, such as health risks and malodors. Suspended microorganisms may increase turbidity and result in discoloration of the water. Organic waste discharges often lead to the growth of bacteria and other microorganisms in the stream bottom, producing what is known as sewage fungus. Not only is this an aesthetic problem, but it may clog water supply inlets or the nets of fishermen, and may change the benthic environment to make it unsuitable for its natural inhabitants.

Organic waste discharge and resultant oxygen depletion may have both direct and indirect effects upon all aspects of aquatic life. It can result in fish kills due to oxygen depletion, or make an area unsuitable for desirable species. In lakes, as DO levels decrease in deep waters, certain species may disappear, while those better able to tolerate reduced oxygen tension may increase in number. In rivers and streams, the typical example of ef-

Table 7.5   The Zonation of Contaminated Streams

Zone	Level of contamination	DO	Predominant Biota	Bacterial count (per ml H_2O)
I (polysaprobic)	Heavy	Zero to very low	Bacteria, sludge worms, maggots, some algae	$>10^6$
II* (mesosaprobic)	Strong, but diminishing	Low to fully saturated	Bacteria, protozoa, worms, rotifers, midge larvae, diatoms, algae, carp	10^4-10^5
III (oligosaprobic)	Low or none	Fully saturated	Diverse plants and animals including game fish	$<10^3$

*This zone is actually separated into a heavily contaminated subdivision which adjoins Zone I, and a less contaminated region which adjoins Zone III.

Compiled from: Kolkwitz, R. and Marsson, M. Ökologie der pflanzlichen Saprobien. *Berichte der Deutschen Botanischen Gesellschaft 26a*:505–519, 1908. In *Biology of Water Pollution*, eds. L. E. Keup, W. M. Ingram, and K. M. Mackenthun, pp. 47–52. Federal Water Pollution Control Administration, U.S. Dept. of the Interior, 1967.
and
Kolkwitz, R. and Marsson, M. Ökologie der tierischen Saprobien. Beiträge zur Lehre von der biologischen Gewässerbeuteilung. *Inter. Revue der Gesamten Hydrobiologie und Hydrographie.* 2:126–52, 1909. Translated: Ecology of animal saprobia. In *Biology of Water Pollution Control*, eds. L. E. Keup, W. M. Ingram, and K. M. Mackenthun, pp. 85–95. Federal Water Pollution Control Administration, U.S. Department of the Interior, 1967.

fects on aquatic life is the succession of communities downstream from a sewage outfall, a process discussed in Chapter 5.

The classic scheme for describing the biological effects of organic waste discharge was introduced by Kolkwitz and Marsson (42, 43). This is known as the saprobic system of zones of organic enrichment, which classifies aquatic biota according to areas in which these species are found. Later systems have been patterned after this one, with differences in nomenclature and in delineation of specific zones. Although there are difficulties in attempting to present a rigid and arbitrary classification of an environment as dynamic as a river, the zonation system is of value in ordering, to some extent, interrelationships in rivers receiving organic wastes. The zonal regions proposed by Kolkwitz and Marsson are presented in Table 7–5 in simplified form. It should be kept in mind that in nature there are no clear boundaries between classes, and both the extent and location of any zone within river reaches may vary with time, with flow conditions, and with the seasons of the year.

Plant Nutrients: Inorganic

The primary inorganic nutrients are nitrates and phosphates, one or both of which are generally the limiting factors to plant growth. The main effect of either of these is the "fertilization" of water, resulting in accelerated eutrophication.

Eutrophication is the term generally applied to the process of natural evolution of a lake. It involves the slow change with time from a biologically unproductive, nutrient-deficient lake having a sparse biotic community (oligotrophic lake), to one which is highly productive, nutrient-rich, and supports diverse biota (eutrophic lake). The gradual addition and buildup of nutrients in the lake from natural sources in the surrounding watershed and internal mineral fixation leads to this enrichment and biological development. The rate of normal eutrophication in a lake depends upon such factors as soil type, type of vegetation in the drainage basin, and local geology.

At some point in this process, a stable condition results where the rate of biological productivity approximately equals the rate of decomposition. This results in a fairly constant, homeostatic chemical and biologic aquatic environment where small disturbances may cause rates of productivity and decomposition to vary, but in balance. By introducing excessive amounts of inorganic plant nutrients, man is responsible for accelerated eutrophication in many aquatic environments, as these homeostatic mechanisms are overwhelmed.

There are predilective sites for accelerated eutrophication. The prime ones are impounded (naturally or artificially) bodies of water, such as small lakes and dammed reservoirs. In these, added nutrients may recycle for many years, since there is little removal via water exchange of any introduced chemicals. Semienclosed bodies of water such as estuaries and coastal regions of large lakes, especially in those areas downstream to waste discharge sites of large urban areas, are also prone to accelerated eutrophication. The rate of accelerated eutrophication is dependent upon the size of the body of water, the rate of nutrient input, and the homeostatic condition which existed prior to any disturbance.

Nutrient input initially results in the increased growth of some plankton species, primarily blue-green algae, and certain rooted aquatic weeds. The previously dominant phytoplankton disappear, decreasing the variety of food for herbivores and reducing total plant species diversity.

Upon algal death, oxidative decomposition in the sediments increases the BOD of the water, decreasing the level of DO. Deep stratified lakes that are eutrophic have little or no DO in the hypolimnion. As the deeper waters are robbed of oxygen, those fish species which require deep, well-oxygenated waters, e.g., trout and herring, disappear, being replaced by other species, e.g., sunfish and carp, which require less oxygen and are supported on shorter food chains. In addition, upon algal death and decomposition, the nitrogen and phosphorus bound in the algae are released back into the water. The spring overturn carries these nutrients to surface layers, where they may promote new blooms. Thus, as mentioned, added nutrients may continually recycle in lakes.

With increasing algal growth, the water becomes more turbid, a condition which restricts the penetration of light to deeper water layers, further depressing photosynthesis below the surface layer. At its worst, the lake becomes stagnant, supporting only anaerobic organisms involved in decay food chains; the decay results in the gradual filling of what is now an odorous (NH_3, H_2S, amines), discolored, and generally unsightly body of water. Thus, the biotic and chemical composition of the lake has changed, making it unsuitable for domestic, industrial, or recreational uses, or even for other wildlife.

The most important example of accelerated eutrophication in the United States is Lake Erie. Effects in this lake occurred mostly since 1910, with the large increase in population within its drainage basin. The main source of water into Lake Erie is Lake Huron, via a river system which services industrial Detroit. Lake Erie also receives additional water from rivers that drain agricultural lands and flow through industrial cities such as Toledo and Cleveland. Over the past 100 years there have been major changes in the chemical composition of the lake and severe oxygen depletion in the central basin. The accelerating deterioration of the lake led to a joint United States – Canadian effort to reverse the trend, and significant progress has been made since the mid-1960's in controlling the discharge of nutrients into Lake Erie; this has resulted in improved water quality.

That control and elimination of nutrient sources to a lake can sometimes prevent algal blooms is also evidenced by the reversal of accelerated eutrophication in Lake Washington due to the elimination of phosphate input by the diverting of sewage effluents (44).

Although the term eutrophication is generally not directly applied to rapidly running waters such as rivers and streams, a response similar to those in lakes does occur following enrichment, with increased algal and bottom growth (Chapter 5). This response usually abates if effluent discharge is ended.

Dissolved Solids and Minerals

The total content of dissolved minerals is referred to as salinity. Excessive mineral content may affect the taste of drinking water and be harmful to people who have specific illnesses. "Hard water," i.e., with excess minerals, primarily calcium and magnesium ions, affects use of the water for domestic, irrigation, and many industrial purposes.

Freshwater fauna are adapted to low salt levels, and discharges which produce excess salt in freshwater environments may be lethal.

Excess minerals may affect certain aspects of water chemistry. The addition of chloride and sulfate, primarily as sodium salts, affects the solubility equilibrium of waters, primarily via absorption of the sodium onto clay particles. This produces a decrease in the pH, especially if the sodium is exchanged with a less alkaline cation. These changes will affect carbonate dissociation and buffering capacity of the water.

The discharge of inert brines of, for example, Na_2SO_4 or $NaCl$, may promote eutrophication. This occurs because phosphate tends to be released from bottom sediments formed under a region of low total dissolved salt but bathed by waters having higher ionic strength.

Some dissolved chemicals may color the water, with a possible decrease in light penetration and, therefore, photosynthetic activity.

Industrial Chemicals

Multitudes of industrial wastes are discharged into aquatic environments. These may be derived from specific factory point sources via direct discharge or accidental release, or be general watershed runoff.

Many waters contain certain levels of natural toxic chemicals leached from surrounding rocks and soils, e.g., mercury, or produced during the decomposition of plants and animals, e.g., hydrogen sulfide. Depending upon the specific mechanism which has evolved to handle it, an aquatic organism will most likely tolerate a certain amount of these natural toxins. However, man's activities have often resulted in increasing the levels of many natural toxic chemicals above the tolerable levels, and have also resulted in the introduction of other chemicals against which the organisms have no defense.

The effects due to discharge of large amounts of toxic chemicals are quite different from those due to sewage discharge described above. While certain species may become severely depressed by sewage release, others become quite abundant. On the other hand, most toxic chemicals do not benefit aquatic organisms, and are probably deleterious to most, except for a few types of microorganisms which may be able to use specific chemicals as a source of energy. Thus, large increases in populations of particular aquatic species are usually not found after toxic chemical release.

High concentrations of toxic materials have obvious deleterious effects. The most dramatic of these are fish kills, which are due primarily to acute effects of industrial poisons. In 1973, the EPA (45) reported that 57.8 million fish were killed due to accidental or intentional discharges of contaminants into waterways.

More often, levels of chemicals which are released are not high enough to produce fish kills but, rather, result in other effects which may or may not be clear cut. These may be, for example, tumors, fungal diseases, systemic pathology, abnormal larval growth, and decreased reproductive potential. Behavioral changes involving migratory, territorial, feeding, and social behavior may also be induced. All of these effects may ultimately result in decreases of population size and biologic productivity of the waterway.

Certain chemicals may accumulate in bottom sediments and in food chains. In 1971, the FDA temporarily banned the sale of swordfish because a large number of samples tested had excessive levels of mercury. High levels of other chemicals often result in the temporary banning of commercial fishing in many areas of the United States.

Table 7.6 Effects of Selected Industrial Chemical Contaminants in Aquatic Environments

Contaminant	Effect upon aquatic environment
Petroleum	Contains many water-soluble compounds which are toxic to plant and animal species; coats benthic environment when it sinks; oil films decrease light penetration and oxygen absorption; coats body surfaces of aquatic birds and destroys waterproofing; some products may impart unpleasant flavors to fish and shellfish; may act to concentrate other fat-soluble toxic chemicals in the water; interferes with recreational use and is aesthetically unpleasing; oil-laden sediments may move with bottom currents to other areas
Organochlorine pesticides (e.g., DDT)	Reduction in thickness of eggshells in birds which feed upon fish containing residues; decreased survival of fish fry; inhibits phytoplankton photosynthesis; acutely toxic to some aquatic fauna via food chain accumulation
Polychlorinated biphenyls	Similar to DDT; eggshell thinning in aquatic birds
Heavy metals	Toxic to many forms of aquatic life; some may be carcinogenic to fish and shellfish

No aquatic environment has an unlimited ability to accomodate contaminants. Even open ocean areas thousands of miles from any population center show evidence of chemical contaminants. For example, waste oil and associated debris have been found in the central Atlantic Ocean (46). A 20% decline in plankton primary production since the 1950's has been reported in the North Atlantic (47). Table 7–6 presents some important industrial chemical contaminants found in aquatic environments, and describes their biological effects. The list is by no means inclusive; it would take a whole volume to present the industrial chemicals found in aquatic systems. Bear in mind that environmental conditions, such as water pH and temperature, and the presence of other chemicals, may modify the hazard from each of these contaminants.

Numerous chemicals, when present in low, sublethal levels may impart unpleasant tastes to fish and shellfish used as food by man. Only trace amounts are sometimes enough to produce a noticeable effect, without changing the distribution or abundance of biota. For example, unpleasant tastes in fish will occur when chlorophenol is present at levels of 0.0001 mg/l. (48). Other chemicals which impart tastes at low levels are benzene, oil, and 2,4-D. This action affects recreational and commercial fishing, and may also make the water undrinkable.

Acidity

Aside from acid rainfall, a major source of acidity, especially in streams, is acid mine drainage, primarily from coal mines. When exposed to air during mining operations, iron sulfide ores (pyrite) become oxidized, resulting in numerous products, including sulfates. Water draining through the mines dissolves these products, and the resulting acidic solution (H_2SO_4) runs off into surface waters and may percolate to groundwater. These acidic waters often contain several kinds of metals, whose solubility in water is increased at the lower pH. The effects of acidity upon aquatic life have been described previously, in the discussion of acid rain.

CHEMICAL CONTAMINANTS AND THE DISRUPTION OF NATURAL ECOSYSTEMS

Natural ecosystems, upon which man ultimately depends, are complete functional units, which create and maintain patterns of energy flow, nutrient cycles, growth, sanitation and, within limits, self-regulation and self-restoration. If these limits are exceeded, however, disruption and degradation may occur. It may only require imbalance in a few components to upset the entire ecosystem.

Unbalancing of Biogeochemical Cycles

Biogeochemical cycles are integral components of the ecosystem, serving to maintain steady-state levels of various chemical elements via equilibrium in a number of continuing processes.

Disruption of biogeochemical cycles is one of the greatest problems caused by man's action as a contaminator of his environment. Many contaminants enter natural cycles, becoming redistributed and/or concentrated at some stage as the cycle strives to regain an equilibrium. What results is, as Hutchinson has aptly stated, a case of "too little here, and too much there" (49); this redistribution of cycle intermediates may result in an effect on human health and welfare. Local, and in some cases, even global disruptions of cycles are already in evidence.

The carbon cycle shows evidence of a global disruption. Exchange of CO_2 between the hydrosphere and atmosphere had resulted in a balanced, relatively constant CO_2 concentration in the latter until about the time of the Industrial Revolution. Since then, the ever-increasing combustion of fossil fuels is returning to circulation carbon which had been fixed into biota, and then subsequently trapped as coal and petroleum millions of years ago. This has resulted in an upsetting of the dynamic equilibrium between the atmosphere and hydrosphere.

Although the total amount of CO_2 injected into the atmosphere is small when compared to the total amount of CO_2 in global circulation, the atmospheric reservoir is also small. Carbon dioxide cannot be absorbed into the much larger hydrospheric reservoir as fast as it is being generated. The result of this, as discussed previously, is a gradual increase in atmospheric levels of CO_2.

The nitrogen cycle shows evidence of imbalances on a more local scale. For example, runoff of nitrates from fertilized land areas into lakes and streams may result in overenrichment of these waters. Balanced nutrient cycles would have maintained optimal levels of plant nutrients in these waterways.

Organic nitrogen from farm wastes and untreated or inadequately treated domestic sewage may ultimately be oxidized to nitrate, again resulting in accelerated eutrophication.

Plants which are grown in soils containing high levels of nitrogen fertilizers may assimilate nitrate faster than it is fixed into organic molecules, resulting in an accumulation of nitrate within the plant. These high nitrate levels may adversely affect livestock and people feeding on these plants.

Imbalances, especially excesses, at certain stages of the cycle may have other implications in terms of human health. High levels of nitrate in potable water may cause methemoglobinemia in infants. Nitrate is converted into nitrite naturally via microbial action; reaction of nitrite with secondary or tertiary amines (these occur in certain pesticides and animal wastes) may result in formation of potentially carcinogenic nitrosamines in the environment. Bacteria have already been implicated in the formation of some of these compounds.

Prior to the Industrial Revolution, the amount of N_2 removed from the atmosphere via biological fixation was essentially balanced by the amounts returned by denitrification processes. However, the rate of nitrogen fixation in the biosphere is now increasing very rapidly due to the large amount of fixation by various industrial processes, such as fertilizer manufacture and various combustion processes, and the increased agricultural cultivation of legumes which fix nitrogen through symbiosis with certain bacteria. This accelerated rate of fixation may overwhelm the ability of natural denitrification processes to keep pace, or result in increases in denitrification products, producing an imbalance in the nitrogen cycle. If sustained for continued periods of time, this imbalance could result in overall global consequences of which we are not as yet aware.

As with nitrogen, steady-state levels of certain sulfur-cycle intermediates normally maintained in undisturbed ecosystems may be upset by man's activities. For example, bacterial reduction of sulfate to H_2S under anaerobic conditions is part of the normal sulfur cycle. However, release of sulfate-rich effluents from, for example, paper mills may be responsible for fish kills in waterways, as secondarily derived H_2S reaches lethal levels.

Various naturally occurring toxic chemicals have, due to ecological processes controlling their synthesis and degradation, tended to maintain a global distribution which remains fairly constant. As man produces many of these materials and releases them into the environment, an imbalance between degradation and synthesis may occur. A classic example is mercury.

Mercury occurs naturally in the environment, but generally only in trace amounts. The biologic mercury cycle is shown in Figure 7–4. Under partial to total aerobic conditions in aquatic sediments, microorganisms convert inorganic and phenyl compounds to methyl and, to some extent, dimethyl mercury. These alkyl compounds are the most toxic forms. They are quite stable and, in addition, if taken up by biota, are only very slowly excreted; thus, they tend to become concentrated in biological organisms and may be long-term environmental hazards.

At concentrations of less than 0.1 ppb$_w$, methyl mercury causes a decrease in the growth rate of phytoplankton (50); toxicity to fish has been noted at levels of 3 ppb$_w$ (51). These toxic levels compare with normal methyl mercury levels in surface waters of less than 0.001 ppb$_w$ (52).

As with other cycles, interconversions of various mercury compounds result in a homeostatic system and, under undisturbed conditions, a steady-state concentration of highly toxic methyl mercury in the aquatic sediments. Man has often upset this balance.

A number of other elements besides mercury may be methylated by microbes in aquatic sediments and terrestrial soils. These include gold, palladium, platinum, tin, thallium, selenium, and tellurium. Like methyl mercury, most of these alkyl metals are quite stable in the aquatic environment. The use of palladium and platinum catalysts in automobile emission control systems and heavy metal alkyls in gasoline additives instead of lead alkyls may thus pose a potential environmental problem.

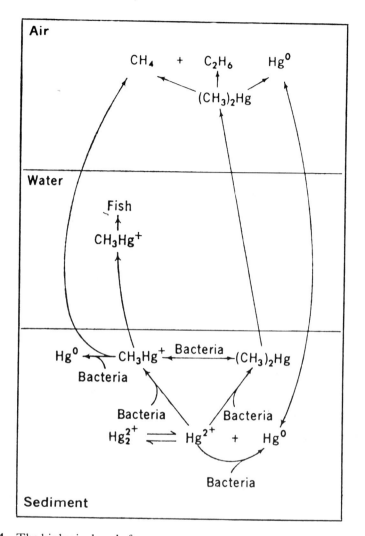

Fig. 7-4 The biological cycle for mercury.

From: Wood, J. M. Biological Cycles for Toxic Elements in the Environment. *Science* 183:1049–52, 1974. Copyright 1974 by the American Association for the Advancement of Science.

Disruption of Ecosystem Structure

The introduction of chemical contaminants into the environment often leads to complex changes in overall ecosystem structure and function. Details of the pattern of alteration have been demonstrated clearly in only a few instances. Although the effects of some contaminants may be more conspicuous than others, the effects of all were found to be parallel; the resultant changes followed similar and broadly predictable patterns for both terrestrial and aquatic ecosystems (53).

A mature, undisturbed ecosystem is a dynamic unit which maintains some degree of stability. An ecosystem reaches this stage via an orderly development with time, if not disturbed by man or by natural catastrophies; this evolutionary process is termed ecological succession. Succession involves the interplay between the biotic and abiotic components of the environment, leading towards development of stable communities, with maximum biomass and diversity of organisms in relation to the physical constraints of the environment. Succession proceeds through a sequence of biological community types until a climax community is reached; this is the final and most stable stage. The climax is self-perpetuating if not disturbed, and its complexity results in resistance to perturbances of the physical environment. The biogeochemical cycles are "tight," i.e., only very small amounts of nutrients are lost, as, for example, via drainage water from terrestrial ecosystems.

A broad outline of the pattern of ecosystem alteration due to environmental contaminants has been described by Woodwell (53), and is presented below.

The disturbance of any ecosystem is accompanied by a reduction in biotic community structure, with a resultant trend toward simplification and monotony of total ecosystem structure. A diversity of specialized species of plants and animals is shifted towards a monotony of a larger number of only a few species of plants or animals which are smaller and have a more rapid reproduction rate. For example, hardy shrubs replace trees in a forest, and algae replace the diverse phytoplankton of a lake. Food chains become shortened. Any disruption of community structure poses the greatest risk to those consumers at the apex of the trophic pyramid, e.g., man.

As a result of the reduction in community structure, a change in cycling of nutrients occurs. Terrestrial communities tend to become "leaky," with nutrient depletion, while nearby aquatic communities become overburdened due to accelerated nutrient input.

The ultimate result of disturbance is that a balanced, stable ecosystem becomes unbalanced and unstable. Often, short-lived disturbances may be "repaired" by the natural successional processes of the ecosystem. However, cumulative and chronic effects may produce more severe changes, with long time intervals necessary for recovery, if corrective measures are taken in time.

Much is still unknown about physicochemical and biological processes in the environment. Thus, accurate predictions of effects are not always possible, and long-term changes are difficult to assess. This is especially true since individual ecosystems are not discrete entities but are all interconnected with each other in some fashion, especially via food webs. The nutrient loss from one ecosystem may be the input to another, and with passage of time each system responds to a net loss or gain by changes in structure and function as each tends towards a new balance. However, the new, balanced system which develops following a disturbance may not prove desirable to man.

CONTAMINATION AND EVOLUTION

The diverse forms of life in the biosphere are the products of evolution, the mechanism of which is natural selection as proposed by Charles Darwin in 1838. Natural selection is the power of the environment to select characteristics of the next generation of a population, leading to organisms which are better adapted to their surroundings.

Chemical contamination is an environmental change which produces new selection pressures. Given the presence of heritable variation in the population, these new pressures may cause a change in gene frequency and, therefore, a change in the genetic makeup of the population involved. Thus, environmental contamination may be expected to cause evolutionary changes in exposed populations of plants and animals. The evolution of populations resistant to the destructive effects of specific chemicals, or otherwise adapted to life in contaminated areas, enables continued existence of the species under conditions which might otherwise have resulted in their extinction.

Contaminant-related evolution is widespread among many plant and animal species, and in response to many different chemicals. For example, there is evidence of populations of plants which are specifically adapted to most known chemical contaminants (54).

Most evolutionary processes are slow. The rapidity, often in one or two generations, of some evolution is evidenced by the presence of plants adapted to life under conditions of contamination which have been in existence for only a short time, e.g., lead from car exhaust (55), triazine herbicides, (56) and heavy metals from refinery industries (57). Examples of rapid natural selection in the animal kingdom are the development of strains of mosquitoes resistant to DDT, and the case of melanism in the peppered moth of the British Isles induced by industrial pollution (58).

Thus, changes in population characteristics may be one of the effects of exposure to chemical contaminants, and some of these changes have direct implications for human health and welfare, such as pesticide-resistant insects.

REFERENCES

1. Harkins, W. D. and Swain, R. E. The chronic arsenical poisoning of herbivorous animals *J. Amer. Chem. Soc. 30*:928–46, 1908.
2. Prell, H. Die Schädigung der Tierwelt durch die Fernwirkungen von Industrieabgasen. *Arch. Gewerbepathol.–Gewerbehyg. 7*:656–70, 1937.
3. Bardelli, P. and Menzani, C. La fluorosi (fluorosis), part 2. *Atti Ist. Veneto Sci. 97*:623–74, 1937.
4. Lillie, R. J. *Air Pollutants Affecting the Performance of Domestic Animals: A Literature Review.* Agriculture Handbook No. 380. Washington, D.C.: U.S. Dept. of Agriculture, 1970.
5. Hupka, E. Über Flugstaubvergiftungen in der Umgebung von Metallhütten. *Wiener. Tierärztl. 42*:763, 1955.
6. Schmitt, N., Brown, G., Devlin, E. L., Larsen, A. A., McCausland, E. D., and Savillo, J. M. Lead poisoning in horses. *Arch. Environ. Health 23*:185–97, 1971.
7. Bazell, R. J. Lead poisoning: zoo animals may be the first victims. *Science 173*:130–31, 1971.
8. Cannon, H. L. and Bowles, J. M. Contamination of vegetation by tetraethyl lead. *Science 137*:765–66, 1962.
9. Chow, T. J. Lead accumulation in roadside soil and grass. *Nature 225*:295–96, 1970.
10. Botkin, D. B. Forest, lakes and the anthropogenic production of carbon dioxide. *Bioscience 27*:325–31, 1977.
11. Kucera, V. Effects of sulfur dioxide and acid precipitation on metals and anti-rust painted steel *Ambio 5*:243–48, 1976.
12. *Inadvertent Climate Modification: Study of Man's Impact on Climate (SMIC).* Cambridge, Mass.: MIT, 1971.
13. *Man's Impact on the Global Environment: Report of the Study of Critical Environmental Problems (SCEP).* Cambridge, Mass.: MIT, 1970.

14. Sawyer, J. S. Can man's waste heat affect the regional climate. First Special Assembly, IAMAP, Melbourne, Australia, 1974.
15. Schneider, S. H. and Dennett, R. G. Climatic barriers to long-term energy growth. *Ambio 4*:65–74, 1975.
16. Rasool, S. I. and Schneider, S. H. Atmospheric carbon dioxide and aerosols: Effect of large increases on global climate. *Science 173*:138–41, 1971.
17. Yamamoto, G., and Tanaka, M. Increase of global albedo due to air pollution. *J. Atmos. Sci. 29*:1405–12, 1972.
18. Weare, B. C. and Snell, F. M. A diffuse thin cloud atmospheric structure as a feedback mechanism in global climatic modeling. *J. of Atmos. Sci. 31*: 1725–34, 1974.
19. Barrie, L. A., Whelpdale, D. M. and Munn, R. E. Effects of anthropogenic emissions on climate: A review of selected topics. *Ambio 5*:209–212, 1976.
20. Seinfeld, J. H. *Air Pollution: Physical and Chemical Fundamentals.* New York: McGraw Hill, 1975.
21. Manabe, S. Dependence of atmospheric temperature on the concentration of carbon dioxide, pp. 25–29 in *Global Effects of Environmental Pollution* S. F. Singer (ed.). New York: Springer-Verlag, 1970.
22. Schneider, S. H. Carbon dioxide. Climate confusion. *J Atmos. Sci. 32*:206–66, 1975.
23. Manabe S. and Wetherald, R. Effects of doubling the carbon dioxide concentration on the climate of a general circulation model. *J. Atmos. Sci. 32*:3–15, 1975.
24. Wang, W. C., Yung, Y. L., Lacis, A. A., Mo, T. and Hansen, J. E. Greenhouse effects due to man-made perturbations of trace gases *Science 194*:685–90, 1976.
25. Mitchell, J. M., Jr. A preliminary evaluation of atmospheric pollution as a cause of the global temperature fluctuation of the past century, p. 139 in *Global Effects of Environmental Pollution* S. F. Singer (ed.). New York: Springer-Verlag, 1970.
26. Mitchell, J. M., Jr. Natural breakdown of the present interglacial and its possible intervention by human activities. *Quaternary Res. (NY) 2*:436–45, 1972.
27. Damon, P. E. and Kunen, S. M. Global cooling? *Science 193*:447–53, 1976.
28. Hoffert, M. I. Global distributions of atmospheric carbon dioxide in the fossil-fuel era: A projection *Atmos. Environ. 8*:1225–49, 1974.
29. Sellers, W. D. A global climatic model based on the energy balance of the earth-atmosphere system. *J. Appl. Meteorology 8*:392–400, 1969.
30. Urbach, F., Berger, D. and Davies, R. E. pp. 523–35 in *Proceedings of the Third Conference on the Climatic Impact Assessment Program.* A. J. Brodeerick and T. M. Hard (eds.). Washington, D.C.: U.S. Department of Transportation, 1974.
31. Nitrogen-An essential life factor and a growing environmental hazard. Report from Nobel Symposium No. 38. B. Bolin and E. Arrhenius (eds.). *Ambio 6*:96–105, 1977.
32. Crutzen, P. J. and Ehhalt, D. H. Effects of nitrogen fertilizers and combustion on the stratospheric ozone layer. *Ambio 6*:112–117, 1977.
33. Chameides, W. L. and Walker, J. C. G. Stratospheric ozone: The possible

effects of tropospheric-stratospheric feedback. *Science 190*:1294–95, 1975.

34. WMO Statement on modification of ozone layer due to human activities and some possible geophysical consequences. World Meteorological Org., WMO/R/STW/2 Annex.

35. Molina, M. J. and Rowland, F. S. Stratospheric sink for chlorofluoromethanes: Chlorine atom-catalyzed destruction of ozone. *Nature 249*:810–12, 1974.

36. Ahmed, A. K. Unshielding the sun: human effects. *Environment 17*:6–14, 1975.

37. Wright, R. F. and Gjessing, E. T. Acid precipitation: Changes in the chemical composition of lakes. *Ambio 5*:219–23, 1976.

38. Likens, G. E., Bormann, F. H. and Johnson, N. M. Acid rain. *Environment 14*: 33–40, 1972.

39. R. E. Huschke, (ed.) *Glossary of Meteorology*. Boston; American Meteorological Society, 1959.

40. Arrhenius, E. Health Effects of multipurpose use of water. *Ambio 6*:59–62, 1977.

41. Macek, K. J., Hutchinson, C. and Cope, O. B. The effects of temperature on the susceptibility of bluegills and rainbow trout to selected pesticides. *Bull. Environ. Contam. Toxicol.* 4:174–83, 1969.

42. Kolkwitz, R. and Marsson, M. Ökologie der pflanzlichen Saprobien. *Berichte der Deutschen Botanischen Gesellschaft 26a*:505–19, 1908. Translated: Ecology of plant saprobia, pp. 47–52 in *Biology of Water Pollution* L. E. Keup, W. M. Ingram and K. M. Mackenthun, (eds.). Federal Water Pollution Control Administration, U. S. Dept. of the Interior, 1967.

43. Kolkwitz, R. and Marsson, M. Ökologie der tierischen Saprobien. Beiträge zur Lehre von der biologischen Gewässerbeurteilung. *Inter. Revue der Gesamten Hydrobiologie und Hydrogeographie*. 2:126–152, 1909. Translated: Ecology of animal saprobia, pp. 85–95 in *Biology of Water Pollution Control* L. E., Keup, W. M. Ingram, and K. M. Mackenthun (eds.). Federal Water Pollution Control Administration, U.S. Department of the Interior, 1967.

44. Edmondson, W. T. Phosphorus, nitrogen and algae in Lake Washington after diversion of sewage. *Science 169*:690–91, 1970.

45. *Environmental Quality: Sixth Annual Report of the Council of Environmental Quality*. Washington, D.C.: Govt. Printing Office, 1975.

46. Heyerdahl, T. The voyage of Ra II. *National Geographic* Jan. 1971, 44–71, 1971.

47. Sinderman, C. J. Some biological indicators of marine environmental degradation. *Washington Acad. of Sci. Journal 62*:184–89, 1972.

48. Boetius, J. Foul taste of fish and oysters caused by chlorophenol. *Medd Denmarks Fishlog Havundersdg. N. S. 1*:1, 1954.

49. Hutchinson, G. E. On living in the biosphere. *Scient. Monthly 67*:393–98, 1948.

50. Harriss, R. C. White, D. and MacFarlane, R. Mercury compounds reduce photosynthesis in plankton. *Science 170*:736–37, 1970.

51. Matida, Y. Toxicity of mercury compounds to aquatic organisms and accumu-

lation of the compounds by the organisms. *Bull. Freshwater Fish Res. Lab. 21*: 197, 1971.

52. Andren, A. and Harriss, R. C. Methyl mercury in estuarine sediments. *Nature 245*:256–57, 1973.

53. Woodwell, G. M. Effects of pollution on the structure and physiology of ecosystems. *Science 168*:429–33, 1970.

54. Bradshaw, A. D. Pollution and evolution, pp. 135–159 in *Effects of Air Pollutants on Plants*. T. A. Mansfield (ed.). Cambridge: Cambridge University, 1976.

55. Briggs, D. Population differentiation in *Marchantia polymorpha* in various lead pollution levels. *Nature 238*:166–67, 1972.

56. Ryan, G. F. Resistance of common ground soil to simazine and atrazine. *Weed Science 18*:614–16, 1970.

57. Edroma, E. L. Copper pollution in Rwenzori National Park, Uganda. *J. of Ecol. 11*:1043–56, 1974.

58. Bishop, J. A. and Cook. L. M. Moths, melanism and clean air. *Scientific American 232*:90–99, 1975.

8

Contamination Criteria and Exposure Limits

INTRODUCTION

Excessive levels of chemical contaminants can clearly produce adverse health effects and degradation of environmental quality. Since the extent and frequency of the effects vary with the magnitude and duration of the exposure, it follows that the effects can be eliminated, or at least reduced in frequency and severity, by reducing the exposures.

Eliminating exposures entirely can sometimes by accomplished by banning the use or consumption of a material, as was done by the FDA ban on the artificial sweetener, cyclamate. Since cyclamates are not found in nature, human exposures gradually fell toward zero as inventories were depleted. The ban was readily accepted by the public, in part because of the availability of a "safe" alternative artificial sweetener, i.e., saccharin. When a ban on saccharin was proposed by FDA in 1977, there was a much greater reluctance on the part of the public to accept this action, and implementation was delayed.

The alternative to a ban on the use of a toxic material is a permissible or acceptable content or level of exposure. There are permissible levels or "tolerances" for the carcinogen aflatoxin in peanuts and peanut products, for coliform organisms in dairy products, for rodent hair and feces in bakery products, for mercury in fish, etc. A low level of contamination is acceptable for such food products, because a societal judgment has been made that a total ban on their consumption is unacceptable. In practical terms, a contamination level is acceptable as long as it produces no observable adverse effects. Acceptable levels are changed periodically to reflect a wider range of observations of effects and to reflect shifting societal values on what actually are acceptable effects.

Statements of the degree of quality for air or water which would ensure protection of the public health and/or prevent environmental degradation are known as contamination criteria (e.g., air-quality criteria, water-quality criteria, etc.). These are technical statements which may form the basis for the setting of specific exposure limits.

The process by which contamination criteria and exposure limits are established and subsequently modified is inherently difficult, slow, and contentious. There are generally conflicting forces at play, with major economic interests, public health and environmental quality concerns, and professional reputations at stake. The available effects data are always inadequate, sometimes pathetically so. The standards-setting process, once begun, can seldom be delayed sufficiently to await the availability of better data. Studies of useful laboratory biological effects generally are expensive and time consuming. Similarly, large scale surveys of environmental quality or public-health parameters are frequently very expensive, even when the measurement technology is adequate. Often, however, successful field evaluations require the prior development of better measurement techniques, and such developments often take many years. Criteria and limits are, therefore, heavily dependent upon "informed judgment"; they are educated guesses by well-informed professionals who have considered and weighed the available data and the views of the interested advocacy groups. While tolerance limits for contaminants in food have been established by the federal government for more than 40 years, it is only within the past ten years that the federal government has become heavily involved in standards setting in other aspects of the environment. Occupational exposure limits were initially proposed by professional society committees and/or by consensus-type standards-setting organizations. These recommendations usually were incorporated into code limits by local governments. Air and water contaminant limits were also established by state and municipal governments. As a result, there were different standards in different jurisdictions and relatively little real incentive to enforce those that existed. Essentially all standards setting in the United States is now being done by the federal government, and the procedures have become much more formalized and elaborate.

BASES FOR ESTABLISHING CONTAMINATION CRITERIA AND EXPOSURE LIMITS

There are many possible bases for the establishment of contamination criteria and the setting of exposure limits. These include: (a) epidemiological studies of populations exposed in their occupations, or through contamination of food, drinking water or community air. Such studies can provide statistical associations between contaminant levels and reported effects; (b) toxicologic studies, i.e., studies of groups of animals exposed intentionally in controlled laboratory experiments where it is possible to define the doses, their frequency of application, and their route of administration. Such studies can provide more complete information on metabolic pathways, storage depots, and the types and degrees of biological damage that the agents produce, including information on whole body effects; (c) extrapolation of available epidemiological and toxicological data on other related materials to the material in question, based on their similarities in chemical structure and metabolism, and perhaps their effects in simplified biological test systems, such as bacteria and cell or organ tissue cultures.

Existing contaminant limits are not all designed to protect human health. Lower limits may be needed to prevent health effects in domestic and wild animals than for protection of the public health. Some materials, such as ethylene gas, which are essentially nontoxic to mammalian species, can produce severe damage to rooted plants and trees. Controls on the levels of aerosol mass have been based on the soiling properties of particles and the attendant costs of cleaning and earlier replacement of clothing, building materials, etc. The effects of submicrometer particles on atmospheric visibility and the effects of carbon dioxide on atmospheric temperature may eventually lead to controls on their airborne concentrations.

Epidemiological Studies

The primary advantage of epidemiological data for establishing safe limits for human exposure is that they are not complicated by the uncertainties of interspecies variations inherent in the interpretation of animal data. On the other hand, the number of materials whose standards can be set on the basis of measured adverse health effects in human populations is, fortunately, quite limited. Intentional exposures of humans to known or potentially toxic materials are severely constrained by ethical and legal considerations, and, therefore, controlled studies in animals are generally needed in place of, or to supplement, available human data.

The application of available epidemiological data in the standards-setting process is also limited by their generally poor quality and other inherent failings. This is less a reflection on the epidemiologists than on the materials that are generally available to them. As discussed in Chapter 6, the populations at risk may be hard to define in terms of their number; their exact ages; ethnic and educational backgrounds; prior smoking habits; occupational and residential histories; the number, extent, and severity of prior diseases and medical treatments; their past and current consumption of alcohol or drugs; their past and current dietary practices; etc. These factors can affect the incidence of many of the reportable diseases of nonspecific etiology and may affect the development of the disease of interest with respect to the environmental factor in question. One of the most important and difficult tasks in designing an epidemiologic study is the selection of a suitable control population. Ideally, it should share all of the pertinent characteristics of the study population, with the single exception of the exposure of interest.

Another major problem is the characterization of the health endpoint. While death is a generally reliable and easily definable statistic, its cause(s) may be more difficult to obtain. It is necessary to locate the place of death and have access to the death certificate. The reporting physician may not have entered the secondary causes, and these may be the most informative for the study in question. A sandblaster with advanced silicosis may die of a heart attack; but it is important to know that the lung disease was present, since it put a great strain on the heart. The influence of contributory factors, such as cigarette smoking, on mortality are still more difficult to obtain than the data on the death certificate, and few investigations have the resources to do so.

Morbidity statistics have additional complications. There is an appalling lack of consistency in the criteria for defining cases of disease and/or dysfunction. Data collected for other purposes are generally not usable. While physicians must report deaths of patients in their care, they are under no obligation to report cases of most diseases. Furthermore, they tend to differ on the severity of symptoms needed to classify an individual as diseased. In addition to the individual variability in the reporting of disease in a given geographic area, there are also regional and national differences in reporting criteria.

The utility of both mortality and morbidity studies is often limited by the absence or poor quality of the data on the exposure of the population of interest and/or the control population to the agent under study. In some cases, no environmental measurement data are available at all. In others, measurement data are available for the recent past, but not for the exposures in earlier times. In any case, the environmental parameters measured may not be the most appropriate ones. In many community pollution problems, the evidence for health effects may be an increase in the incidence of a chronic nonspecific disease, and the specific causative agent, if any, may not be known. Finally, even if the measurements are made on the appropriate chemical species, and are accurate, they still may not be indicative of the population exposure. For example, the relation between the concentration of an air contaminant as measured on the roof of a public school building, and the actual concentration in the breathing zones of various individuals in the community, is uncertain and variable. There are spatial variations in the outdoor environment, and there may be large differences between indoor and outdoor concentrations.

Epidemiologic studies on community air and water contaminants are generally not definitive with respect to cause and effect, and only rarely establish reliable dose-response relationships. This should not be surprising considering the extraordinary difficulty of characterizing population exposures, and the fact that most diseases associated with exposure to chemical contaminants can also be caused or exacerbated by smoking, dieting, and other factors. Epidemiologic studies of working populations can more frequently establish cause and effect and dose-response, because industrial exposures to chemical agents may be much higher than community exposures. They are, therefore, more likely to produce clear-cut symptoms and/or lesions, including many not commonly seen in the general population. Also, the populations exposed are more easily defined, and their exposure levels and intervals can be more easily determined. Furthermore, supplementary indications of exposure and biological effects can frequently be obtained by physiological testing and collection and analysis of samples of blood, urine, hair, exhaled air, etc. Information may also be available on other factors which affect their vital statistics, e.g., smoking histories, preexisting diseases, residential locations, ethnic backgrounds, etc.

The major problems with epidemiologic data on occupational groups are generally the relatively small population sizes at risk for any given exposure, and the reluctance of some managements to publish data which might increase their legal and financial liabilities. The problem of limited population size can sometimes be overcome by establishing industry-wide studies, although these generally can only be performed by the National Institute for Occupational Safety and Health (NIOSH) or by organizations acting for NIOSH under the authority of the Occupational Safety and Health Act of 1970.

The limited amount of reliable human health effects data available on occupationally exposed populations are of considerable value in establishing exposure limits for workers. Fortunately, there aren't very many materials which have adversely affected enough workers and, therefore, most occupational exposure limits have been based on studies in experimental animals.

In any case, the human health effects data from occupational exposures are of limited value in establishing safe limits for general community exposures. Community exposure limits must be much lower than occupational exposure levels for a variety of reasons. Among the more important of these are:

a. *Intermittent vs. Continuous Exposure:* Employee groups are exposed during working hours and generally have at least 16 hours per day of essentially negligible exposure between their workday exposures. During this time, some of the accumulated dose can be eliminated and other fractions can be neutralized or immobilized. Community exposures to some contaminants are, on the other hand, essentially continuous.

b. *Selected vs. Total Population:* Working populations are inherently more resistant to environmental stresses since they do not include the more vulnerable segments of the overall population, such as young children, the aged and infirm, and those people with chronic diseases or disabilities too severe to permit them to work.

c. *Voluntary vs. Involuntary Exposure:* While working populations are entitled by right and by law to safe and healthy working environments, there will always be some risk of accidents and excessive chemical exposures due to unknown or unanticipated events and toxicities of materials. In accepting employment, each employee implicitly accepts the risks. The people in the surrounding community, on the other hand, obtain no direct benefit from contaminants emitted into their air and water and should, therefore, not have to accept any significant risk. At most, the risk should be proportionate to the indirect benefits of the activities generating the contaminant releases, e.g., a healthy local economy and minimal rises in the cost of living.

To the above, one may add an additional practical, if somewhat cynical, reason for different standards for the same chemicals between occupational and community exposures. This is the probability of a detectable incidence of measurable effects. As indicated earlier, industrial populations exposed to a particular chemical under defined conditions are generally relatively small and seldom exceed a few thousand. Thus, considering population variabilities, conditions which elevate the number of cases of, for example, emphysema, lung cancer, liver cirrhosis, and heart disease by

1 in 1,000 will never be detected. On the other hand, an increase in the incidence of a significant disease by 1 in 1,000 in the total U.S. population would result in more than 200,000 excess cases and would be considered a major public-health catastrophe.

Toxicological Studies

The primary advantage of toxicological data for establishing safe exposure limits is that the experimental design and protocols are set by the investigator and the number of uncontrolled variables can be minimized.

Some specific advantages include:

A. Statistical uncertainties can be reduced by using large numbers of animals in each test group.
B. Inherent interindividual variability can be held down by (a) using inbred strains of test animals, (b) using animals of one sex, and (c) using animals of the same age.
C. The effects of extraneous environmental factors can be limited by (a) providing a diet compatible with the study objective and one constant over time, (b) maintaining a uniform temperature and humidity in the housing and dosing facilities, (c) using well trained and highly motivated animal handlers who will keep the facility clean and free of infectious agents and other contaminants.
D. The effects of the exposure protocols themselves can be compensated for by giving sham exposures to the control animals.
E. The exposures can be made quite uniform in each test group with respect to the manner and amount of administered dose to each individual animal and with respect to the number of doses and the intervals, if any, between them.
F. The observations, test, and lab evaluations on each animal can be made in a systematic manner to avoid biased results.
G. The kinds of observations and tests to be performed on each animal, and the periods of exposure, testing, and follow-up can be selected to optimize the prospects of achieving the study objectives.

Unfortunately, many of the potential advantages are not realized. There are never enough funds, facilities, and trained personnel to investigate (a) all of the materials of interest, (b) all the appropriate routes of administration, (c) effects on all of the species of interest, (d) effects in all the tissues of interest, (e) all of the potential effects manifested long after the expo-

Table 8.1   Animal Species Commonly used for Toxicological Testing

Species	Approx. adult weight (kg)	Approx. life-span (years)	Major usage in toxicological evaluations				Physiological effects	Carcino-genesis
			Toxicological effects					
			Feeding	Skin	Eye	Inhalation		
Mouse	0.03	2	X	X		X		X
Hamster (Syrian)	0.1	2	X			X		X
Rat	0.3	4	X	X		X		X
Guinea pig	0.6	7				X	X	
Rabbit	3	6		X	X			
Monkey	5	15	X	X		X	X	
Dog (Beagle)	10	15				X	X	X

sures and, especially, (f) effects at minimal exposures where the incidence of response may be very low and the magnitude of the effects very small.

To the extent that the animal test system is idealized, it becomes somewhat unrealistic as a model for human exposure. Many of the effects which environmental agents produce in humans are seen primarily in people who have preexisting diseases and/or physiological dysfunctions. There are animal models available which mimic some human diseases, but the extent of the correspondence may be debatable. In any case, purebred strains can have very different responses or sensitivities to some chemicals than other strains in the same species, and there can be substantial further variations among the common animal test species.

There are well-established reasons for much of the interspecies variability. Many materials are not very toxic themselves, but undergo biotransformation into much more toxic chemicals. The enzymes which catalyze these transformations and, hence, the products of metabolism, may vary substantially among species.

Thus, even with animal test data that are consistent and reliable, the degree of confidence with which they can be extrapolated to indicate the potential effects in humans is always fairly limited. It is generally desirable to have data on several different species, and to have some other information to indicate which species most closely resemble man in terms of metabolism of the chemicals in question.

Table 8–1 lists the animal species most commonly used for toxicological evaluations, some of their more important characteristics, and the kinds of tests most frequently performed. These species are readily avail-

Table 8.2   An Outline of Animal Toxicologic Tests

I. Acute Tests (single dose)
   A. LD_{50} determination (24-hour test and survivors followed for 7 days)*
      1. Two species (one that is not a rodent)
      2. Two routes of administration (one by intended route of use if other than by topical contact)
   B. Topical effects on rabbit skin (if intended route of use is topical; evaluated at 24 hours and at 7 days)

II. Prolonged Tests (daily doses)
   A. Duration – three months
   B. Two species (usually rats and dogs)
   C. Three dose levels
   D. Route of administration according to intended route of use
   E. Evaluation of state of health
      1. All animals weighed weekly
      2. Complete physical examination weekly
      3. Blood chemistry, urinalysis, hematology, and function tests performed on all ill animals
   F. All animals subjected to complete autopsy including histology of all organ systems

III. Chronic Tests (daily doses)
   A. Duration – one to two years
   B. Species – Selected from results of prior prolonged tests, pharmacodynamic studies on several species of animals, possible single dose human trial studies. Otherwise use two species, one of which is not a rodent.
   C. Two dose levels
   D. Route of administration according to intended route of use
   E. Evaluation of state of health
      1. All animals weighed weekly
      2. Complete physical examination weekly
      3. Blood chemistry, urinalysis, hematologic examination and function tests on all animals at six-month intervals and on all ill or abnormal animals
   F. All animals subjected to complete autopsy including histologic examination of all organ systems

IV. Special Tests
   A. For potentiation with other chemicals
   B. For effects on fertility
   C. For teratogenicity
   D. For carcinogenicity

*The LD_{50} (median lethal dose) is the dose at which 50% of the exposed population will die within a specified time.

From: Loomis, T. A. *Essentials of Toxicology*, 2nd Ed. Philadelphia: Lea and Febiger, 1974.

Table 8.3   Signs and Symptoms Obtainable by Observation and Physical Examination of Animals Undergoing Toxicologic Tests

OBSERVATION	PHYSICAL EXAMINATION
Bizarre physical positions	Altered muscle tone
Bizarre tail positions	Catatonia
Exploratory behavior	Muscle tremors
Aggressiveness toward some species	Aggressiveness toward experimenter
Inactivity	Coma
Convulsions, spontaneous	Convulsions to touch
Dyspnea	Alterations in cardiac rate and rhythm
Sedation	Paralysis
Nystagmus	Change in pupillary size
Cyanosis	Sensitivity to pain
Abnormal excreta	Skin lesions
Salivation	Corneal opacities
Nasal discharge	Pacing reflexes
Piloerection	Righting reflexes
Phonation	Grasping reflexes
	Pinnal reflexes
	Death

From: Loomis, T. A. *Essentials of Toxicology*, 2nd Ed. Philadelphia: Lea and Febiger, 1974.

able, at reasonable cost, from well-established breeding colonies. Because these animals are so widely used, their anatomy, physiology, pathology, and pharmacology are well-characterized. This knowledge is important in experimental studies, where the effects produced by the toxic materials under investigation may be minor or subtle, and would be difficult to distinguish against an unknown or variable pattern in the control animals.

For most studies, the more commonly used laboratory animals listed in Table 8–1 provide adequate models for human toxicity or physiological effects studies. When they do not, studies can be performed on other animal species which have the desired similarities in anatomic, physiologic, metabolic, histologic, biochemical, and disease progression patterns. Frequently, such patterns can be found in large domestic animals, e.g., swine, sheep, cattle, donkeys, horses, etc. Swine are similar to humans in skin characteristics and airway branching patterns. Equines are good models for studies of pulmonary function and pathology. Comprehensive reviews of anatomic, physiologic, and metabolic similarities and dissimilarities between humans and experimental animals, the uses of some common animals in research studies, and of the more uncommon, uniquely useful animal models are provided by Mitruka et al. (1).

In evaluating the toxicity of a newly synthesized chemical, a standard-

Table 8.4 Procedures Commonly Employed in Animal Toxicologic Tests

Blood chemistry studies:	Hematology:
Sodium	Hematocrit
Potassium	Total red blood cell counts
Blood urea nitrogen	Total and differential
Glucose	white blood cell counts
	Thrombocyte counts
Urinalysis:	Organ function tests:
pH and specific gravity	Bromsulphalein retention
Protein	(liver function)
Glucose	Thymol turbidity (liver function)
Ketones	Serum alkaline phosphatase
Crystals	(liver function)
Blood cells	Blood urea nitrogen (kidney function)
Bacteria	

From: Loomis, T. A. *Essentials of Toxicology*, 2nd Ed. *Philadelphia*: Lea and Febiger, 1974.

ized pattern of testing is generally employed. Table 8–2 lists the kinds of evaluations that may be needed. Carcinogenic testing usually involves an exposure interval of at least 18 months, and observations throughout the natural life of the animals. The standard kinds of observations and examinations of the animals are listed in Table 8–3 and some common laboratory procedures used on biological samples are listed in Table 8–4.

The tests, observations, and procedures listed in these tables are not all performed on all chemicals, and the degree of postexposure follow-up is quite variable, depending on the objectives and persistence of the investigator. Thus, the absence of observed effects does not necessarily mean that there was no toxicity, and animal tests can never provide absolute certainty that a chemical is safe for its intended uses. They are, however, essential screening steps for any chemical for which industrial workers and the general public will receive significant exposures, and exposure limits based on these tests may be expected to be protective to the extent that they are observed. The adverse effects that do occur from exposures below these limits should be low in frequency. They should also be limited in severity, except for materials that are found by subsequent evaluations to be carcinogens.

Extrapolation of Epidemiological and Toxicological Data

Chemical contaminants in the environment can be found in almost all possible molecular and ionic forms, and many form isomers with different geometric configurations. They can vary from monomers, to dimers, to polymers, and can also have varying degrees of hydration. These differences in structure and charge levels can have major effects on the expression of toxicity. It is obviously impossible to perform toxicity tests on all of the forms of a chemical in the environment. Thus, frequent use is made of the available toxicity data, in conjunction with available information on the effects of physical form and structure as modifiers of biological effects, to develop estimates of acceptable levels of those forms not tested.

Many concentration limits for air, water, and food may appear to be specific, but actually are generic. This is especially true for the metals. The toxicity may vary considerably among the various compounds, but the only distinction made, if any, will be between "soluble" and "insoluble" forms, or between organic and inorganic compounds. In such cases, experimental data on one or several compounds have, in effect, been extrapolated to cover related materials.

In one particular field, the development of standards for internally deposited radionuclides, essentially all of the exposure limits are based upon extrapolations. This was the case because it was necessary to estimate potential health effects at doses which were orders of magnitude below the lower limit of actual human experience, and which were delivered over long periods of time.

The one exception involved radium, where there were data from low-level, long-term exposures of radium dial painters during World War I and the following two decades. For radium, the critical organ, i.e., the organ receiving the highest radiation dose, is bone; the biological effects were bone cancer and bone degeneration. Using knowledge of the measured body burdens in the survivors and in tissues of the deceased, and a knowledge of radium metabolism, it was possible to relate the internally deposited radiation dose to the biological effects. It was concluded that no effects would be observed among workers for a critical body burden of 0.1 μg of ^{226}Ra.

Except for standards for bone seekers, which are all based upon radium, current standards are based upon extrapolation from experience with many different sources of ionizing radiation, obtained largely since World War II.

The first standards were solely based upon X-ray exposure. The extension of standards to new sources and types of radiation was made possible by the use of units which allowed comparison of radiation doses from different sources.

The unit of absorbed radiation dose is the rad (D), and is equivalent to

100 ergs per gram of tissue. To compare the biological effects of different types of radiation having differing energies, a unit of dose equivalent, the rem, is used. The rem is equal to the product of a number of factors. These are (a) the absorbed dose (rads); (b) a quality factor, determined by the type of radiation and the pattern of its energy deposition in tissue, and (c) other modifying factors, e.g., the distribution of the absorbed dose in space and time.

The deposition of energy in tissue from the decay of internally deposited radionuclides depends on many factors. One is the type of radiation. Some heavy elements, such as radium, uranium, and plutonium, decay by the emission of alpha (α) particles, which are helium nuclei. These heavy decay particles leave a dense track of ionization in tissue, but have a very limited linear range. Most lighter radionuclides decay by emitting beta (β) particles (electrons) and gamma (γ) rays (high-energy electromagnetic radiation). The beta particles produce a less-dense ionization track and penetrate further than the α-particles, and γ-rays penetrate further still. Thus, the β's and γ's affect more cells, but deposit less energy in each. Also the damage caused by α-irradiation is more concentrated and may be more irreversible than that produced by β and γ irradiation. These differences are recognized by the use of a correction, called a quality factor, in dose calculations.

The basic criterion for the upper limit of permissible occupational exposure is that accumulated whole-body exposure should not be greater than 5(N-18) rem, where N is the employee's age. This reflects the assumption that there should be no exposure to people under 18 years of age. Considering specific organ radiosensitivities, the standards recognize that some organs may receive a greater dose than permissible for whole-body exposure. Occupational limits are 5 rems (per year) for whole body, 75 rems for the hands, 30 rems for the forearms, and 15 rems for the skin and most other organs. The exception is for pregnant women, whose body dose may not exceed 0.5 rem (2).

Much greater safety factors are built into exposure limits for the general population, on the basis that this is a nondiscretionary risk, that a small increase in mutation of genetic material in a large population can be of great significance, and upon the assumption of no threshold for genetic effects. The maximum average permissible dose to the general population is set at 0.17 rem per year above natural background.

A major factor affecting the pattern of energy deposition is metabolism. Radionuclides follow the same metabolic pathways as do nonradioactive nuclides of the same element. Thus, a knowledge of the element's metabolism can be used, in conjunction with a knowledge of the rate of intake into the body, to predict the concentration of the nuclide of interest in each particular organ of the body. With knowledge of the organ burden, and mode and rate of radioactive decay, one can calculate the energy deposition in that organ.

If certain simplifying assumptions are made, the basic dose considerations outlined in the preceeding discussion can be used to generate specific permissible concentrations for all of the radionuclides. For occupational exposure limits the basic premises are:

a. That exposure begins at age 18 or greater, and continues at a constant rate for 50 years.
b. That one consumes 2×10^7 cm³ of air and 2,200 ml of water per day, half of that occurring during an eight-hour workday.

The equation for Maximum Permissible Concentrations (MPCs) for air (MPC$_a$), and water (MPC$_w$) in microcuries/unit volume are:

$$(MPC_a) = \frac{10^{-7} qf}{T f_a (1 - e^{-0.693\ t/T})} \ \mu C_i/cm^3 \qquad (8.1)$$

$$(MPC_w) = \frac{9.2 \times 10^{-4} qf}{T f_w (1 - e^{-0.693\ t/T})} \ \mu C_i/ml \qquad (8.2)$$

Where:

q = maximum permissible body burden, in μC_i
f = fraction of body burden in the critical organ
f_a = fraction of inhaled nuclide which reaches the critical organ
f_w = fraction of ingested nuclide which reaches the critical organ
t = 50 years × 365 days/year
T = effective half-life in critical organ, in days.

The National Committee on Radiation Protection (NCRP) has tabulated values of q, f_a, and f_w for most of the nuclides of interest (3). They have also calculated and tabulated MPC$_a$ and MPC$_w$ values for both 8 hr/day and 24 hr/day exposures.*

*Within the past year, the International Commission on Radiological Protection (ICRP) has published proposed new guidelines for radiation protection standards (ICRP Publication No. 26, Annuals of the ICRP v. 1, #3, 1977, Pergamon Press, NY). As of this writing, these proposals are being reviewed by NCRP for possible adoption.

Contamination Limits for Environmental Protection

Many concentration limits have been established to control environmental contamination for purposes not directly related to human health, but rather are based upon recognized adverse effects on animals, vegetation, or building materials, economic losses, or evidence of aesthetic degradation of surface air or water; the latter are considered to be effects upon "public welfare." The U.S. Environmental Protection Agency (EPA) has promulgated dual standards for some air contaminants, calling those intended to protect the public health "primary standards," and those intended to protect the public welfare "secondary standards."

In some cases, the evidence used to establish secondary standards is better defined and more susceptible to realistic cost accounting than that used to establish standards based on human-health effects. It is not very difficult or time consuming to estimate the impact of air contaminants on the soiling of clothing, home furnishings, and building materials; on the reduction in their useful lives; and on their maintenance and replacement costs. Similarly, direct economic impacts due to damage to crops and livestock can readily be calculated. On the other hand, there are some effects whose costs are more intangible. These include, for example, the limitation of tree and ornamental plant species which can be grown in contaminated airsheds and the limitation or alteration of aquatic species which thrive in contaminated surface waters. An example of intangible costs which, however poorly defined, are deemed unacceptable by many are effects on the survival of natural species. The successful pressures on EPA to ban general U.S. usage of DDT were based largely on its alleged effects on the viability of the eggs of wild birds. Its potential human-health effects were, and still are, quite speculative.

BASES FOR ESTABLISHING EMISSION LIMITS AND SOURCE CONTROLS

Contaminant concentration limits provide a standard of comparison for measured ambient levels. When ambient levels exceed the limit, the focus of concern shifts to the reduction in source strengths or elimination of major sources of the contaminant of interest. When one source is dominant, e.g., fluoride contamination downwind of a fertilizer manufacturing plant, that source may be held accountable for whatever problems result from the contamination. Typically, the ambient contaminant level is due to many sources, and excessive levels can only be effectively reduced by (a) an across-the-board reduction in emissions by all sources, or (b) by eliminating or greatly reducing the source strengths of a limited number of major sources.

The attainment of reduced exposure levels through effective control of emissions requires the establishment of rational and enforceable emission limits. The first step is generally an inventory of sources and their strengths. It is also important to know the location of the discharge point and the temporal pattern of emissions, since the latter affects contaminant dispersion. Other important factors are the technological and economic feasibility of the application of controls. Finally, it is important that there be some positive incentives for the installation and maintenance of controls and/or penalties for not installing them or maintaining their effectiveness. For example, the effectiveness of the manufacturer-supplied motor-vehicle emission controls has been considerably less than their potential because of improper maintenance and some deliberate disconnections and alterations of the system components. In most states, there are no requirements for mandatory inspection or maintenance of auto-emission controls.

Other types of emission controls have been, predictably, more effective. These include those for lead, achieved through the reduction of the lead content of gasoline, and sulfur dioxide, achieved through reductions in the sulfur content of fossil fuels used in boilers. In practice, it is much easier to obtain compliance of standards from the relatively small number of suppliers than from the relatively large number of consumers.

ESTABLISHMENT OF CONTAMINATION CRITERIA AND EXPOSURE LIMITS

Exposure levels have considerable temporal variability. Some toxic effects are determined primarily by cumulative exposure, while others are influenced more by peak levels of exposure. As a result, several different kinds of exposure limits are needed, some based on overall intake or accumulation, and some based on rate of intake. Another basic factor affecting the choice of a numerical standard is the degree of protection that is selected. As discussed previously, limits need to be set lower for general population exposures as compared to occupational exposures. Within the occupational environment, it may be desirable to have relatively high emergency limits as well as routine or regular exposure limits. Such emergency limits are usually based on brief single exposures at levels which may be expected to produce clearly demonstrable, but largely reversible, adverse effects. The rationale for accepting such effects is that it permits action which can prevent more serious consequences. These actions include rescue of injured or unconscious workers, fire fighting, and gaining access to equipment or controls in order to halt releases of toxic or explosive materials.

Occupational Health

In the first half of the twentieth century the greatest need for exposure limits for chemical contaminants was in occupational health protection, where clear-cut cases of chemical intoxication and chronic diseases were very common. In response to this need, the first organized and continuing mechanisms for establishing hygienic limits were begun by groups of concerned professionals.

In 1940, the American Standards Association (ASA) formed its Z37 Committee to formulate allowable air concentrations of carbon monoxide, lead dust, and various solvent vapors. The ASA is a voluntary, consensus-type of organization, consisting of representatives from industry, labor, insurance companies, and governmental agencies. Since October, 1969, the organization has been known as the American National Standards Institute, Inc. (ANSI). Its standards currently are known as "maximum acceptable concentrations" (MACs) and are interpreted as upper bounds on the excursions of air concentrations at industrial operations. Since ANSI is a consensus-type organization, standards are adopted by unanimous approval. While this leads to careful consideration, it also ensures that the process can only advance slowly. Between 1941 and 1969, MACs were established for 24 chemicals.

An alternate source for occupational exposure limits has been the American Conference of Governmental Industrial Hygienists (ACGIH). For more than 30 years, its Threshold Limits Committee has annually updated and expanded its list of recommended Threshold Limit Values (TLVs); these values refer to airborne concentrations of substances. The 1947 list contained about 140 listings, while the 1977 list has more than 600. The lists in recent years also contain guidance on the interpretation of occupational exposures to carcinogens, substances of variable composition, mixtures, nuisance particulates, and asphyxiants.

Full membership in the ACGIH is limited to professionals in industrial hygiene and coordinate disciplines who are employees of governmental agencies. University employees can be associate members and participate in the organization's activities, but cannot be officers. The members of the TLV Committee have been people close to the problems who are not directly concerned with the economic consequences to industry. They have made their recommendations as "best estimates." By and large, their work has been well received and has stood the test of time. The basic modus operandi of the TLV Committee has been described by Stokinger (4).

Until 1963, all Threshold Limit Values were defined as time-weighted average (TWA) concentrations over the course of the workday, representing conditions under which it is believed that nearly all workers may be repeatedly exposed daily without any adverse effect. Since 1964, some listings have been given a "C", or ceiling, notation which, in effect, makes them equivalent to a MAC, i.e., an absolute upper limiting value below which all concentrations should fluctuate. Ceiling values apply to chemicals having a prompt response. For most compounds, where the response develops relatively slowly, the TWA concentrations are more appropriate to hazard evaluations.

In 1976, the TLV Committee of ACGIH introduced the concept of a short-term exposure limit (STEL) to supplement its time-weighted average TLVs. These STEL values were defined as maximal concentrations to which workers can be exposed for a period up to 15 minutes continuously without suffering from (a) irritation, (b) chronic or irreversible tissue damage, or (c) narcosis of sufficient degree to increase accident proneness, impair self-rescue, or materially reduce work efficiency. The provisos are that no more than four excursions per day are permitted, with at least 60 minutes between excursion periods, and that the daily TLV-TWA is not exceeded. The STEL should be considered a maximum allowable concentration, or absolute ceiling, not to be exceeded at any time during the 15 minute excursions. ACGIH also specifies that a STEL value should not be used as an engineering design criterion or considered as an emergency exposure level (EEL).

Some specific EEL values have been recommended by the Toxicology Committee of the American Industrial Hygiene Association (5), and by Zielhuis (6). The concept of EEL intended applications are described by the Toxicology Committee as follows:

> Emergency Exposure Limits are concentrations of contaminants that can be tolerated without adversely affecting health but not necessarily without acute discomfort or other evidence of irritation or intoxication. They are intended to give guidance in the management of single, brief exposures to airborne contaminants in the working environment.
>
> In recommending these Limits, the Committee makes several assumptions: (1) Exposures at these levels will be accidental, not the result of

engineering controls designed to hold exposures at these levels. (2) Normally prevailing values of airborne contamination will be below Threshold Limit Values or Maximal Acceptable Concentrations. (3) These accidental exposures will be single events, i.e., if a man is exposed at these levels, further exposure will be prevented (e.g., by removing the man from the work area) until he regains his normal resistance. (4) Men who could be exposed under these conditions are not idosyncratic, hypersensitive, or otherwise predisposed to disease from the specific contaminant. (5) Men who could be exposed under such conditions are under medical surveillance. (6) The probable severity of injury due to secondary accidents, including those resulting from impairment of vision, judgment and coordination, must be considered by those applying these values. In connection with this last point, a degree of intoxication, even though reversible that prevents self-rescue is not considered acceptable; these Limits are believed sufficiently low so that exposure at these levels will not prevent such self-rescue operations as walking out of the area or shutting off valves.

In contrast to Threshold Limit Values recommended by the American Conference of Governmental Industrial Hygienists, Maximal Acceptable Concentrations recommended by the American Standards Association, or Hygienic Guides of the American Industrial Hygiene Association, these Emergency Exposure Limits are not intended to be used as guides in the maintenance of hygienic working environments, but rather as guidance in advanced planning for the management of emergencies. They are peak values which should not be exceeded except in circumstances where impairment to health is justifiable in order to prevent a still more serious event. The "safety factor," often used in deriving air-quality guides, is not applied in these Limits except in cases of low confidence in the extrapolation to man of data derived from animals.

The MACs and the TLVs were conceived of as guidelines for use by professionals who understood their nature and limitations. This is well stated in the preface to the TLV list (7):

> Threshold limit values refer to airborne concentrations of substances and represent conditions under which it is believed that nearly all workers may be repeatedly exposed day after day without adverse effect. Because of wide variation in individual susceptibility, however, a small percentage of workers may experience discomfort from some substances at concentrations at or below the threshold limit; a smaller percentage may be affected more seriously by aggravation of a pre-existing condition or by development of an occupational illness.
>
> Tests are now available (J. Occup. Med. 15: 564, 1973; Ann. N.Y. Acad. Sci., 151, Art. 2: 968, 1968) that may be used to detect those individuals hypersusceptible to a variety of industrial chemicals (respiratory irritants, hemolytic chemicals, organic isocyanates, carbon disulfide).
>
> Threshold limits are based on the best available information from industrial experience, from experimental human and animal studies, and when possible, from a combination of the three. The basis on which the values are established may differ from substance to substance; protection against impairment of health may be a guiding factor for some, whereas

reasonable freedom from irritation, narcosis, nuisance or other forms of stress may form the basis for others.

The amount and nature of the information available for establishing a TLV varies from substance to substance; consequently the precision of the estimated TLV is also subject to variation and the latest Documentation should be consulted in order to assess the extent of data available for a given substance.

The committee holds to the opinion that limits based on physical irritation should be considered no less binding than those based on physical impairment. There is increasing evidence that physical irritation may initiate, promote or accelerate physical impairment through interaction with other chemical or biologic agents.

In spite of the fact that serious injury is not believed likely as a result of exposure to the threshold limit concentrations, the best practice is to maintain concentrations of all atmospheric contaminants as low as is practical.

These limits are intended for use in the practice of industrial hygiene and should be interpreted and applied only by a person trained in this discipline. They are not intended for use or for modification for use, (1) as a relative index of hazard or toxicity, (2) in the evaluation or control of community air pollution nuisances, (3) in estimating the toxic potential of continuous, uninterrupted exposures or other extended work periods, (4) as proof or disproof of an existing disease or physical condition, or (5) for adoption by countries whose working conditions differ from those in the United States of America and where substances and processes differ.

The application of TLVs to novel work schedules has been discussed in several journal articles in recent years. The various authors (8–11) have developed guidelines for modified TLVs for work schedules which depart significantly from the eight hours per day, five days per week schedule assumed as normal by the TLV Committee.

Special consideration is given to some TLVs by the added notation "skin." The substances involved are those for which there is a substantial potential contribution to the overall exposure by the cutaneous route, including mucous membranes and eye, either via air, or more particularly, by direct contact with the substance. Many solvents can alter skin absorption. This attention-calling designation is intended to suggest appropriate measures for the prevention of cutaneous absorption so that the threshold limit is not invalidated.

While enforcement of airborne concentration limits is generally the best approach to control of occupational overexposures, there are many situations where it is difficult to characterize airborne exposure, and where routes other than inhalation are important to the overall exposure. This is frequently the case for maintenance workers, whose irregular schedules and work locations may be difficult to characterize or keep track of. Among the other means which may be useful in determining the extent of exposure are Biologic Limit Values (BLVs). These values represent limiting amounts of substances (or their effects) to which the worker may be exposed without hazard to health or well-being as determined by analysis of his tissues and fluids or exhaled breath. The biologic measurements on which the BLVs are based can furnish two kinds of information useful in the control of worker exposure: (a) measure of the individual worker's overall exposure; (b) measure of the worker's individual and characteristic response. Measurements of response furnish a superior estimate of the physiologic status of the worker, and may be made of (a) changes in amount of some critical biochemical constituent, (b) changes in activity of a critical enzyme, (c) changes in some physiologic function. Measurement of exposure may be made by (a) determining the amount of a substance to which the worker was exposed, by analysis of blood, urine, hair, nails, and other body tissues and fluids; (b) determination of the amount of the metabolite(s) of the substance in tissues and fluids; (c) determination of the amount of the substance in the exhaled breath. The biologic limits may be used as an adjunct to the TLVs for air, or in place of them. The BLVs, and their associated procedures for determining compliance with them, should thus be regarded as an effective means of providing health surveillance of the worker.

There are no comprehensive compilations of BLVs currently endorsed by any of the professional societies. Values for specific suggested BLVs have appeared in the literature (12, 13) and some have been included in a number of the recommended standards proposed by the National Institute for Occupational Safety and Health (NIOSH) in recent years. The recommendations of NIOSH and the official standards of the Occupational Safety and Health Administration (OSHA) will be described in the discussion which follows.

FEDERAL OCCUPATIONAL HEALTH STANDARDS

The TLVs, MACs and EELs are stated as numbers with one or, at most, one and one-half significant figures. The epidemiological and/or toxicological data base is generally inadequate, especially in view of inherent intersubject variability, to establish more precise limits. Thus, they were not intended to provide fine lines separating safety and hazard or to provide a basis for regulatory control. For many years, the TLV preface included a disclaimer to the effect that the list was not intended nor suitable for inclusion in legislative codes. However, by the late 1960's most states had established codes for permissible occupational exposure levels based on TLVs, primarily because numerical standards were needed for code enforcement, and no other reasonably comprehensive and authoritative source was available.

While almost all states based their codes on TLV lists, states still varied considerably in permissible limits, because their codes were based on the limits recommended in the years preceding the adoption of the code and the frequency of periodic updating varied considerably among the states. As the list was continually updated and expanded, many states were left with outdated exposure limits.

Some of the confusion of varying state exposure limits was eliminated by the Occupational Safety and Health Act of 1970. Under this Act, the Secretary of Labor was directed to promulgate uniform federal standards. Since Congress recognized that permanent standards-setting would be time-consuming, provision was made for the adoption of interim standards.

Twenty-two MACs are being used as interim standards in preference to the TLVs for those materials. Where consensus-type standards exist, it is the policy of the federal government to favor them over all other external recommendations. Where TLVs existed, but not MACs, as was the case for approximately 280 other materials having interim standards, the TLVs were adopted. However, the 1968 TLVs were adopted instead of the most recent values. The rationale was that the U.S. Department of Labor had already incorporated the 1968 values in regulations pertaining to occupational health and safety for federal government contractors under provisions of the earlier Walsh-Healey Act.

The interim standards were to be replaced, in due time, by permanent standards. The procedure for establishing permanent federal occupational health standards begins with NIOSH, an agency within the U.S. Department of Health, Education, and Welfare (DHEW). NIOSH, or a NIOSH contractor, prepares a draft criteria document. After internal and external reviews by qualified personnel and representatives of professional societies, there are public hearings. The responses are weighed by NIOSH, and a modified criteria document is prepared and, with the endorsement of the Secretary of the U.S. Department of Health, Education, and Welfare, is transmitted to the Secretary of the U.S. Department of Labor (DOL). After further consideration within OSHA, and further public hearings, the Secretary of Labor may promulgate a permanent standard.

In practice, the process has advanced quite slowly. Between the passage of the OSHA legislation and the end of 1977, only four permanent standards have been promulgated by the Secretary of Labor: (a) for

Table 8.5 NIOSH Criteria Documents on Air Contaminants

Substance	Year issued	Revision issued	Substance	Year issued	Revision issued
Acetylene	1976		Hydrogen cyanide and	1976	
Acrylamide	1976		cyanide salts	1976	
Alkanes	1977		Hydrogen fluoride	1976	
Allyl chloride	1976		Hydrogen sulfide	1977	
Ammonia	1974		Isopropyl alcohol	1976	
Arsenic, inorganic	1974	1975	Kepone	1976	
Asbestos	1972	1976	Lead, inorganic	1973	1975
Benzene	1974		Mercury, inorganic	1973	
Benzoyl peroxide	1977		Methyl alcohol	1976	
Beryllium	1972	1975	Methyl parathion	1976	
Boron trifluoride	1976		Methylene chloride	1976	
Cadmium	1976		Nitric acid	1976	
Carbaryl	1976		Nitrogen oxides	1976	
Carbon dioxide	1976		Organotin compounds	1976	
Carbon disulfide	1977		Parathion	1976	
Carbon monoxide	1972		Phenol	1976	
Carbon tetrachloride	1975		Phosgene	1976	
Chlorine	1976		Polychlorinated		
Chloroform	1974		biphenyls (PCBs)	1977	
Chloroprene	1977		Refined petroleum solvents	1977	
Chromic Acid	1973		Silica, crystalline	1974	
Chromium (VI)	1975		Sodium hydroxide	1975	
Coke oven emissions	1973		Sulfur dioxide	1974	
Cotton dust	1974		Sulfuric acid	1974	
Decomposition products of			1,1,2,2-Tetrachloroethane	1976	
fluorocarbon polymers	1977		Toluene	1973	
Dioxane	1977		Toluene diisocyanate	1973	
Epichlorohydrin	1976		Trichloroethylene	1973	
Ethylene dibromide	1977		Tungsten and cemented		
Ethylene dichloride	1976		tungsten carbide	1977	
Fibrous glass	1977		Vanadium	1977	
Fluorides, inorganic	1975		Vinyl chloride	1974	
Formaldehyde	1976		Xylene	1975	
			Zinc oxide	1975	

asbestos, (b) for 14 carcinogens, (c) for vinyl chloride, and (d) for coke-oven effluents. The asbestos standard of 1972 specified that the TWA air concentration would initially be 5 fibers/cm^3, and that it should drop to 2 fibers/cm^3 on July 1, 1976. In December, 1976, NIOSH recommended that the permissible TWA air concentration limit for asbestos be lowered to 0.5 fibers/cm^3. This proposal is under consideration by OSHA. As of July, 1977, NIOSH had also completed the preparation of other criteria documents and transmitted them to the Secretary of Labor; these are listed in Table 8–5.

It is apparent from Table 8–5, that the rate of issuance of NIOSH criteria documents has accelerated in recent years; this is due to increased efforts by NIOSH in this area. It has become clear that the process was too time-consuming and expensive to permit the development within the foreseeable future of permanent standards for all of the more than 600 substances on the TLV list. In fact, there are generally more new materials being developed or more widely used than there are standards being approved for known toxicants, so that it would be difficult to keep from falling further behind.

NIOSH recognized the problem and, in 1974, launched its Standards Completion Program, acknowledging that it would not be possible to develop enough complete permanent standards in the near future and embarking on a so-called mini-standards program to fill the interim needs. Under this program, guidelines were prepared to cover approximately 380 substances, and these were made available in 1976 and 1977.

OCCUPATIONAL HEALTH STANDARDS IN OTHER NATIONS

The ACGIH's TLVs have been used as a pattern for standards-setting by many other countries throughout the world. However, exposure standards in the Soviet Union and in the countries which have followed their lead are quite different, and frequently much lower, than the corresponding TLVs. The Soviet standards, which are all ceiling values, are based primarily on neurophysiological responses in experimental animals and on behavioral responses in humans. The standards are set at the no-response level. On the other hand, the Soviet standards are not, in practice, treated as limits to be enforced, but rather as goals and as design criteria for new facilities.

In general, occupational health professionals in the United States have rejected the Soviet approach. Their basis for doing so has been well summarized by Hatch (14), as follows:

Recognizing the many unknowns in the whole dose-response relation, the U.S.S.R. approach provides a greater factor of safety, but is it necessary to go so far to insure adequate health protection? To assume that any significant departure from the normal state of a healthy and previously non-exposed animal is evidence of too much stress seems to deny the basic purpose of the homeostatic processes and to be in conflict with a common definition of a state of health which requires constant use of the adaptive processes to maintain health tone.

Others who have recently discussed the occupational health standards used in various countries include Zielhuis (15) and Holmberg and Winell (16).

Air Contamination

The air pollution episode which occurred in Donora, Pennsylvania, in 1948 was found, in retrospect, to have resulted in an estimated 20 excess deaths and in other health effects among 43% of a total population of 13, 800. The team of engineers and physicians dispatched by the U.S. Public Health Service to investigate this disaster was drawn from the staff of its Division of Occupational Health. At that time there was no air pollution group in the Public Health Service, indicative of the fact that air pollution was not considered a significant health problem at the national level. The occupational health team did a good job of investigation. After all, they were quite familiar with the health consequences of excessive exposure to dusts and gases.

In the early 1950's, the Public Health Service formed the nucleus of an air pollution research program. The federal government's involvement has grown substantially over the years and is now carried forward by the U.S. Environmental Protection Agency.

In 1948, air contamination was still a local concern, and was considered more of a nuisance problem than a health hazard. In eastern and midwestern cities it was primarily a soot problem, involving soiling and light extinction. In Los Angeles, people were beginning to wonder why their eyes smarted at midday on 10 to 20 days a year in the summer and fall months. In both cases, contaminants in the air were causing the effects, but it was not at all clear what the specific offending chemicals were, or what the threshold levels were for their effects.

The only community air contamination standard which was in widespread use in 1948 was an emission standard for black smoke from point sources. This was the Ringelmann Chart, named after the French engineer who developed it in 1895. It consisted of a row of squares having 20%, 40%, 60%, and 80% in coverage with black ink (Figure 8-1).

The Ringelmann chart was simple, inexpensive, easy to use, and unambiguous in interpretation. A smoke inspector would hold it up as he looked at a smokestack. All he had to do was decide which square was closest in blackness to the plume from the stack. Ringelmann charts are still used as enforcement tools in the late 1970's.

One other index of air pollution had widespread use before 1948; this was dustfall. Dustfall is the weight of the particles which fall into an open pot in one month, normalized for the cross-sectional area of the entrance to the pot. The results were reported in terms of tons/sq. mi/month, and were useful in gauging trends, provided that there were no transient sources, like construction activities, and no significant changes in airflow patterns around the sampling site. Dustfall is an index of contaminant particles, but is useless as an index of health risk. Particles small enough to be inhaled do not fall into the pot.

Since air contamination was a local problem, standards for contaminant levels that did exist were established for local jurisdictions such as cities, counties, and states. Different jurisdictions regulated different air contaminants and often had different standards for the same contaminants. The most frequently regulated contaminants were smoke, carbon monoxide, and sulfur dioxide. The extent of the confusion is illustrated in Table 8-6, which lists a variety of ambient air-quality standards for one compound, carbon monoxide.

A complete listing of all of the various ambient air-quality standards occupies many pages containing fine print, and additional pages of footnotes which amplify and/or clarify the numerical listings. The most comprehensive of these compilations are those in the three editions of Stern's multivolume reference work, *Air Pollution* (17, 18, 19).

With the passage of the Air Quality Act of 1967, air pollution control became a federal responsibility to be shared with states which developed control programs meeting federal standards. Since air contaminant problems were not confined to single political jurisdictions, provision was made to establish Air Quality Control Regions based in part on airshed configurations. Other provisions of the Act directed the Secretary of DHEW to issue Air Quality Criteria and Control Technology documents covering specific contaminants. Following official notice of the release of these documents by announcement in the Federal Register, the governors of states with designated air-quality control regions within their bounda-

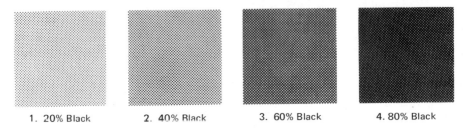

1. 20% Black   2. 40% Black   3. 60% Black   4. 80% Black

Fig. 8–1   Modified Ringelmann's scale for grading smoke density.

ries had 90 days to notify the Secretary of DHEW that they intended to adopt air-quality standards for the contaminants. They then had 180 days to hold public hearings and adopt standards, and another 180 days to adopt plans for enforcement. The standards adopted by the states are subject to review and approval. In the event no local standard was approved, DHEW had the right to impose one.

Other provisions of the Air Quality Act of 1967 included (a) the establishment of the President's Air Quality Advisory Board, the Advisory Committee on Criteria, and the Advisory Committees for the various contaminants, comprised of experts from industry, academia, and other agencies outside the federal government, to assist in developing Criteria and Control documents; (b) the conducting of comprehensive cost studies to assess the economic impact of air standards on industry; (c) the expansion

Table 8.6 Pre-1969 Ambient Air Quality Standards for Carbon Monoxide

Jurisdiction	Concentration mg/m³	ppm$_v$	Averaging time
California	33	30	8 hrs
	132	120	1 hr
Czechoslovakia	1	0.9	24 hrs
	6	5.4	30 min
New York	16.5	15	8 hrs
	66	60	1 hr
Ontario	5.5	5	30 min
Pennsylvania	27.5	25	24 hrs
Poland	0.5	0.45	24 hrs
	3	2.7	20 min
U.S.S.R.	1	0.9	24 hrs
	3	2.7	20 min

Compiled from: Stern, A. C. (ed.). *Air Pollution* – 2nd Ed. New York: Academic Press, 1968.

of research and development programs for air pollution control; (d) the study of emissions from aircraft and national emission standards; and (e) the registration of fuel additives.

Criteria and control documents were issued in 1969 and 1970 for particulates, sulfur dioxide, carbon monoxide, photochemical oxidants, hydrocarbons, and nitrogen oxides. The primary and secondary air-quality standards for these materials are listed in Table 8–7. As of 1977, these are the only federal ambient air standards which have been promulgated.

Since 1970, the EPA has been responsible for the enforcement of federal air pollution legislation; most of the standards setting activity related to air pollution has been focused on emissions standards. The 1970 Clean Air Act amendments were concerned with (a) controlling existing mobile or stationary sources of contaminants to bring air quality to levels defined by the national ambient air quality standards; (b) setting national emission standards for new or existing hazardous air contaminants for which ambient air quality standards were not applicable, e.g., asbestos, beryllium, and mercury; (c) setting nationwide performance standards for new or modified stationary air contaminant sources. A primary purpose of the 1970 amendments was to prevent the general occurrence of new air contamination problems by requiring the installation of the best controls during initial construction, when the installation of such controls is least expensive. The new standards were not, however, to be applied to existing sources.

In the amendments, "standard of performance" was defined as "a standard for emissions of air pollutants which reflects the degree of emission limitation achievable through the application of the best system of emission reduction which (taking into account the cost of achieving such reduction) . . . has been adequately demonstrated." A "new source" was defined as any stationary source, the construction or modification of which is commenced after the publication of proposed regulations for that source type. "Modification" is defined as any physical change in the method of operation of a stationary source which increases the amount of any air contaminant emitted by the source or which results in the emission of any air contaminant not previously emitted. Examples of the source categories for which EPA has promulgated standards of performance are fossil fuel fired steam generators, municipal incinerators, Portland cement plants, nitric acid plants, and sulfuric acid plants.

Source controls for carbon monoxide, nitrogen oxides, and hydrocarbons have, as previously discussed, been focused on the internal combustion engine. Pre-1968 automobiles, which had no emission controls, gave

Table 8.7 Federal Air-Quality Standards (in $\mu g/m^3$)

	Primary	Secondary
Particulates:		
Annual mean	75	60
24-hr maximum	260	150
Sulfur dioxide:		
Annual mean	80	60
24-hr maximum	365	260
Carbon monoxide:		
8-hr maximum	10,000	10,000
1-hr maximum	40,000	40,000
Photochemical oxidants:		
1-hr maximum	160	160.
Hydrocarbons:		
1-hr maximum	160	160
Nitrogen oxides:		
Annual mean	100	100

Source: U.S. Environmental Protection Agency.

off 8.7 grams per mile (gm/mi) of hydrocarbons, 87 gm/mi of carbon monoxide, and 3.6 gm/mi of NO_x. With the introduction of evaporative controls, positive crankcase ventilation, and exhaust-gas recirculation, the corresponding emissions of 1973-74 automobiles were down to about 3.0, 78.0, and 3.1 gm/mi, respectively. With the addition of oxidation catalysts on the exhaust line, most 1975-model cars had emissions reduced to 1.5, 15.0, and 3.1 gm/mi, respectively. The tighter standards for the 1978-model year specified in 1970 were not met by the auto industry, and were replaced by less-restrictive standards in the 1977 Clean Air Act amendments, as listed in Table 8 – 8.

The 1977 amendments also clarified several issues of concern not specifically addressed in the earlier legislation. One is the prevention of significant deterioration in areas which are presently cleaner than the national ambient air quality standards. Areas having the purest air are designated as Class I; this designation is mandatory for national wilderness sites. Areas where the air is not as pure as the Class I regions, but is still currently cleaner than national standards, will be classified as Class II. Allowable contaminant levels are highest in Class III areas. A state may reclassify any area other than a mandatory Class I area by following a procedure set out in the Act amendments.

The Act also provides for limited allowable increments. An "allowable

Table 8.8 Federal Emission Standards for Automobiles (gm/mi):
Specifications of the Clean Air Act Amendments of 1970 and 1977

Amendment	Beginning in model year	Hydrocarbons	Carbon monoxide	Nitrogen oxides
1970	1977	1.5	15.0	2.0
	1978	0.41	3.4	0.4
1977	1978–1979	1.5	15.0	2.0
	1980	0.41	7.0	2.0
	1981	0.41	3.4*	1.0

*The administrator (of EPA) may waive the 3.4 requirement for carbon monoxide up to 7.0 upon a finding that the technology for control is not available, as determined by cost, driveability, fuel economy, and other factors.

Source: U.S. Environmental Protection Agency.

increment" is the permissible increase in contaminant levels in any Class I, II, or III area. The smallest increments are allowed in Class I, the next largest in Class II, and the largest increments are allowed in Class III areas.

However, a variance above the established Class I increment can be granted by a governor (8% above the allowable increment for low-terrain areas and 15% for high-terrain areas). The President of the United States is made arbitrator regarding approval of a variance in cases where there is a disagreement between the State and the Federal land manager.

Another area of concern addressed by the 1977 amendments is the leeway granted to EPA when air-quality goals are not attained. The new Act endorses EPA's "offset" policy for new or modified major sources of air contaminants in areas that do not meet air-quality standards. The offset policy allows new development if the net effect is an improvement in overall air quality due to decreases from other sources. However, the Act also provides for waivers of offset requirements where the state has an adequate program for incremental reductions in emissions which will assure attainment of the standards by the deadlines (1982 for contaminants other than those which are auto-related; 1987 for those which are auto-related). In order to use the waiver provision, a state must have submitted a revised State Implementation Plan by 1979, showing attainment by the 1982 or 1987 dates.

A third issue addressed by the 1977 Act is coal conversion. It allows for extensions for compliance with emissions limitations for power plants ordered to convert to coal. This extension on meeting the standards is effective prior to the date of the conversion. But sources which are ordered to convert can only begin to actually burn coal when they can do so without causing or contributing to concentrations of any contaminant in excess of primary air-quality standards. This latter feature of the Act is called the "primary standard condition."

The Act also authorizes the state, EPA, or the President to require use of local coal to prevent severe economic disruption or unemployment which might be derived from use of coal other than that locally available.

Water Contamination

Contamination criteria for water have been more narrowly focused than those for air, which is reasonable, considering the basic differences in contaminant dispersion in the two media. The atmosphere is one large and continuous mantle, whose motion distributes contaminants released into it in all directions. Contaminants dispersed in the ambient air will be inhaled by all air-breathing creatures, including man. Surface water, on the other hand, flows only downhill, and generally within confined channels. Thus, the water supplies used for drinking water or other specific purposes can be selected on the basis of their freedom from excessive contamination, or can be purified to specified criteria prior to delivery for the desired use.

Historically, there have been three types of standards established by public agencies for the maintenance of environmental water quality: drinking-water standards, surface-water standards, and effluent-quality regulations. There are also specific industry standards which define the quality factors required of water supplies for cooling and process operations.

In the United States, standards for toxic chemicals in drinking water in the various states have been based on the recommendations of the Federal Government. Some of the limits used in the United States and other countries are summarized in Table 8-9. The 1975 U.S. standards, which also include limits for a variety of chlorinated hydrocarbon pesticides, were set under the Safe Drinking Water Act of 1974. The Act was also designed to protect underground sources of drinking water by prohibiting waste-water disposal in areas which rely upon aquifers as a principal drinking-water source.

Water-quality criteria define levels of contaminants for the maintenance of specified water uses. For example, Table 8-10 lists the maximum levels set by the Federal Government for some elements in water which meets the needs as irrigation waters. Standards for surface-water quality have, however, traditionally been established by state governments, and

Table 8.9 Drinking Water Standards – Chemical Limits in mg/l. (ppm$_w$)

Chemical	United States 1962[a]	1969[b]	1975[c]	Canada[d]	WHO[e]	European region[f]
Arsenic (As)	0.05	0.01	0.05	0.01	0.05	0.05
Barium (Ba)	1.0	1.0	1.0	<1.0	1.0	
Cadmium (Cd)	0.01	0.01	0.01	<0.01	0.01	0.01
Chromium (Cr^{+6})	0.05	0.05	0.05	<0.05	0.05	0.05
Cyanide (CN)	0.2	0.01		0.01	0.2	0.05
Lead (Pb)	0.05	0.05	0.05	<0.05	0.05	0.1
Mercury (Hg)			0.002			
Selenium (Se)	0.01	0.01	0.01	<0.01	0.01	0.01
Silver (Ag)	0.05	0.05	0.05			
Phenolic substances (as (C$_6$H$_5$OH)	0.001	0.001		0.002	0.002	<0.001
Nitrate (as N)	10.	10.	10.	<10.0	10.	
Boron(B)		1.0		<5.0		

Basis:

a. Drinking Water Standards, U.S. Public Health Service, 1962.
b. Manual for Evaluating Public Drinking Water Supplies, U.S. Public Health Service, 1969.
c. Interim Drinking Water Standards – Federal Register, Dec. 1975.
d. Canadian Drinking Water Standards & Objectives – 1968, Dept. of National Health and Welfare, Ottawa, October, 1969.
e. International Standards for Drinking Water, 2nd ed., World Health Organization, Geneva, 1963.
f. European Standards for Drinking Water, 2nd ed., Regional Office for Europe, World Health Organization, Copenhagen, 1970.

Table 8.10 Tolerances for Irrigation Waters (mg/l.)

Element	For continuous use in all soil types	For short-term use, only in finely textured soils
Cd	0.005	0.05
Co	0.2	10.0
Cr	5.0	20.0
Cu	0.2	5.0
Mn	2.0	20.0
Mo	0.005	0.05
Ni	0.5	2.0
Pb	5.0	20.0
V	10.0	10.0
Zn	5.0	10.0

Table 8.11 General Policy, Classification, and Standards of Quality for Interstate Waters: Inland Waters

Water use class and description	Dissolved oxygen	Sludge deposits, solid refuse, floating solids, oil, grease, and scum	Color and turbidity
A — suitable for water supply with treatment by disinfection only, and all other water uses; character uniformly excellent[a, c]	As naturally occurs	None other than of natural origin	None other than of natural origin
B — suitable for bathing and other primary contact recreation. Acceptable for public water supply with appropriate treatment.[c] Suitable for agricultural and certain industrial process and cooling uses. Suitable as an excellent fish and wildlife habitat. Excellent aesthetic value	Minimum 5 mg/l. at any time. Normal seasonal and diurnal variations above 5 mg/l. will be maintained	None allowable	None in such concentrations that would impair any usage specifically assigned to this class
C — suitable for fish and wildlife habitat, boating, fishing, and certain industrial process and cooling uses. Under some conditions acceptable for public water supply with appropriate treatment. Good aesthetic value[c]	Minimum 5 mg/l. any time. Normal seasonal and diurnal variations above 5 mg/l. will be maintained. For sluggish, eutrophic waters, not less than 3 mg/l. at any time. Normal seasonal and diurnal variations above 3 mg/l. will be maintained	None[b]	None in such concentrations that would impair any usages specifically assigned to this class
D — suitable for navigation, power, certain industrial processes and cooling uses. and migration of fish. Aesthetically acceptable	2 mg/l. at any time. Normal seasonal and diurnal variations above 2 mg/l. will be maintained	None[b]	None in such concentrations that would impair any usages specifically assigned to this class

[a.] Class A waters reserved for water supply may be subject to restricted use by state and local regulation.

[b.] Sludge deposits, floating solids, oil, grease, and scum shall not be allowed except in that amount that may result from the discharge of appropriately designed and operated sewage and/or industrial waste treatment plants.

[c.] Waters shall be free from chemical and radiological constituents in concentrations or combinations which would be harmful to human, animal, or aquatic life for the most sensitive and governing water class use. In areas where fisheries are the governing considerations and approved limits have not been established, bioassays shall be performed as required by the appropriate agencies. For public drinking water supplies the raw water sources must be of such a quality that United States Public Health Service or higher appropriate state agency limits for finished water can be met after conventional water treatment.

[d.] Water quality analyses on interstate waters should include tests for both total and fecal coliform supported by sanitary surveys for the development of background data and the establishment of baselines for these two parameters for the specific waters involved.

[e.] The resultant temperature of a receiving water shall not exceed that shown for the most sensitive aspect of the most sensitive species of aquatic life in that water. Normal seasonal and diurnal temperature variations that existed before the addition of heat of artificial origin shall be maintained. In addition:

Coliform bacteria	Taste and odor	pH	Allowable temperature increase
Total coliforms: Not to exceed a monthly arithmetic mean of 100/100 ml[d]	None other than of natural origin	As naturally occurs	None other than of natural origin
Total coliforms: Not to exceed a monthly median of 1,000/100 ml nor more than 2,400/100 ml in more than 20% of samples collected[d]	None in such concentrations that would impair any usages specifically assigned to this class nor cause taste and odor in edible fish	6.5-8.0	Only such increases that will not impair any usages specifically assigned to this class[e]
See footnote d.	None in such concentrations that would impair any usages specifically assigned to this class nor cause taste and odor in edible fish	6.0-8.5	Only such increases that will not impair any usages specifically assigned to this class[e]
None in such concentrations that would impair any usages specifically assigned to this class	None in such concentrations that would impair any usages specifically assigned to this class	6.0-9.0	Only such increases that will not impair any usages specifically assigned to this class[e]

1. For streams, the allowable total temperature rise shall not exceed 5°F above ambient unless it can be demonstrated to the satisfaction of the state regulatory agencies that greater rises at various times will not be harmful to fish, other aquatic life or other uses.
2. For lakes, there shall be no discharge or withdrawal of cooling waters from the hypolimnion unless it can be demonstrated to the satisfaction of the state regulatory agencies that such discharges or withdrawals will not be harmful to fish, other aquatic life or other uses. A heated discharge to a lake shall not raise the temperature more than 3°F at the surface immediately outside a designated mixing zone.

More restrictive requirements may be adopted by a state if deemed necessary to meet the class use of the receiving waters.

Adapted by: Band, R. G. and Straub, C. P. (eds.). *Handbook of Environmental Control*, vol. 3. *Water Supply and Treatment*, Cleveland, CRC Press, © 1973. From: New England Interstate Water Pollution Control Commission. Standards of Water Quality, 1969. Reprinted with permission CRC Press, Inc.

Table 8.12 General Policy, Classification, and Standards of Quality for Interstate Waters: Coastal and Marine Waters[a, c, d]

Water use class and description	Coliform bacteria	Taste and odor
SA-suitable for all sea water use including shellfish harvesting for direct human consumption (approved shellfish areas), bathing, and other water contact sports; excellent aesthetic value	Not to exceed a median MPN of 70/100 ml and not more than 10% of the samples shall ordinarily exceed an MPN of 230/100 ml for a 5-tube decimal dilution or 330/100 ml for a 3-tube decimal dilution[b]	Not allowable
SB-suitable for bathing, other recreational purposes, industrial cooling and shellfish harvesting for human consumption after depuration; excellent fish and wildlife habitat, good aesthetic value	Not to exceed a median of 700/100 ml and not more than 2,300/100 ml in more than 10% of the samples collected in a 30 day period[b]	None in such concentrations that would impair any usages specifically assigned to this class and none that would cause taste and odor in edible fish or shellfish
SC-suitable fish, shellfish and wildlife habitat; suitable for recreational boating, and industrial cooling; good aesthetic value	See footnote b.	None in such concentrations that would impair any usages specifically assigned to this class and none that would cause taste and odor in edible fish or shellfish
SD-suitable for navigation, power, certain industrial processes and cooling uses, and migration of fish; good aesthetic value	None in such concentrations that would impair any usages specifically assigned to this class	None in such concentrations that would impair any usages specifically assigned to this class and none that would cause taste and odor in edible fish

[a.] Coastal and marine waters are those generally subject to the rise and fall of the tide.

[b.] Surveys to determine coliform concentrations shall include those areas most probably exposed to fecal contamination during the most unfavorable hydrographic and pollution conditions. Water quality analyses on interstate marine and coastal waters should include tests for both total and fecal coliform supported by sanitary surveys for the development of background data and the establishment of a baseline for these two parameters for the specific waters involved.

[c.] Waters shall be free from chemical and radiological concentrations or combinations which would be harmful to human, animal, or aquatic life or which would make the waters unsafe or unsuitable for fish or shellfish or their propagation, impair the palatability of same, or impair the water for any other usage. In areas where fisheries are the governing considerations and approved limits have not been established, bioassays shall be performed as required by the appropriate agencies.

[d.] The temperature of a receiving water shall not exceed that shown below immediately outside a designated mixing zone:

1. Coastal waters
The water temperature at the surface of coastal waters shall not be raised more than 4°F over the monthly means of maximum daily temperatures from October through June nor more than 1.5°F from July through September.

pH	Dissolved oxygen	Sludge deposits, solid refuse, floating solids, oil, grease, and scum	Color and turbidity
6.8-8.5	Not less than 5.0 mg/1. at any time. Normal seasonal and diurnal variations above 5 mg/1. will be maintained	None allowable	None in such concentrations that would impair any usages specifically assigned to this class
6.8-8.5	Not less than 5.0 mg/1. at any time. Normal seasonal and diurnal variations above 5 mg/1. will be maintained	None allowable	None in such concentrations that would impair any usages specifically assigned to this class
6.0-9.0	A minimum of 2 mg/1. at any time. Normal seasonal and diurnal variations above 2 mg/1. will be maintained	None except for such small amounts that may result from the discharge of appropriately treated sewage and/or industrial waste effluents	None in such concentrations that would impair any usages specifically assigned to this class
6.8-8.5	None less than 5 mg/1. during daylight hours nor less than 4 mg/1. at any time. Normal seasonal and diurnal variations above 4 mg/1. will be maintained	None except that amount that may result from the discharge from a waste treatment facility providing appropriate treatment	None in such concentrations that would impair any usages specifically assigned to this class

2. Estuaries or portions of estuaries

The water temperature at the surface of an estuary shall not be raised to more than 90°F at any point provided further, at least 50 percent of the cross sectional area and/or volume of the flow of the estuary including a minimum of 1/3 of the surface as measured from water edge to water edge at any stage of tide, shall not be raised to more than 4°F over the temperature that existed before the addition of heat of artificial origin or a maximum of 83°F, whichever is less. However, during July through September if the water temperature at the surface of an estuary before the addition of heat of artificial origin is more than 83°F, an increase in temperature not to exceed 1.5°F, at any point of the estuarine passageway as delineated above, may be permitted.

More restrictive requirements may be adopted by a state if deemed necessary for the protection of the most sensitive aspect of the most sensitive specie of aquatic life in that water and/or to meet the class use of the receiving water.

Adapted by: Band, R. G. and Straub, C. P. (eds.). Handbook of Environmental Control, vol. 3, Water Supply and Treatment. Cleveland: CRC Press, © 1973. From: New England Interstate Water Pollution Control Commission. *Standards of Water Quality*, 1969 Reprinted with permission CRC Press, Inc.

generally on the principle of assigned best usage. The highest classification (typically known as Class A) is for waters used for drinking-water supplies. New York State has two categories of fresh surface waters suitable for drinking water: Class AA can be used without filtration, while Class A requires filtration. Class B generally is suitable for all uses except drinking water; Class C waters are unsuitable for bathing, while Class D waters are unsuitable for fishing. Table 8–11 illustrates the classifications and standards used for fresh surface waters by the New England states; Table 8–12 summarizes the corresponding classification and standards for salt waters. These standards are designed to protect environmental quality, as well as human health. The states classify all surface waters into one category or another and then attempt to regulate discharges and land usage to the degree necessary to maintain the quality factors within the prescribed limits.

The maintenance of water quality depends on the control of noxious discharges. Prior to 1971, the control of contaminant effluents was entirely up to the states. However, in that year, a federal judge ruled that with the enactment of the National Environmental Policy Act, the permits required for discharges into navigable waterways under the 1899 Refuse Act had to be accompanied by an environmental impact statement. As a result, the Army Corp of Engineers had to immediately assume responsibility for evaluating and regulating the impact of toxic effluents on streams. The confusion that resulted was partially ended with the passage of the Federal Water Pollution Control Act of 1972. A major feature of this Act was its emphasis on effluent limitations. The overall goals of the Act were to eliminate discharge of contaminants into navigable waters by 1985, to achieve high water quality by July 1, 1983 (later changed by the 1977 Amendments to 1984), and to eliminate the discharge of specific toxic contaminants. The regulations provided for (a) a National Pollutant Discharge Elimination System (NPDES), which required that point sources which discharge contaminants into waterways obtain discharge permits; (b) Effluent Guidelines and Standards; (c) Pretreatment Standards; (d) Oil and Hazardous Substances Rules; (e) Ocean Dumping Rules; and (f) Toxic-Pollutant Standards. The Environmental Protection Agency requests that NPDES applicants have a preapplication conference with EPA at least 24 months before the discharge starts. This period provides EPA with additional time to prepare an environmental impact statement, if required.

The Act mandated that effluent guidelines and standards would be established for those plants listed in Table 8–13. These standards are end-

Table 8.13 Plants regulated by Effluent Guidelines and Standards

Pulp and paper mills	Organic chemicals mfg.
Paperboard, builders' paper, board mills	Inorganic chemicals mfg.
	Plastics & synthetics mfg.
Meat-product and rendering processing	Soap and detergent mfg.
	Fertilizer mfg.
Dairy-product processing	Petroleum refining
Grain mills	Iron and steel mfg.
Canned and preserved fruits and vegetables processing	Nonferrous metals mfg.
	Phosphate mfg.
	Electric (steam) power plants
Canned and preserved seafood processing	Ferroalloy mfg.
Sugar processing	Leather tanning & finishing
Textile mills	Glass and asbestos mfg.
Cement mfg.	Rubber processing
Feedlots	Timber-products processing
Electroplating	

of-pipe limitations expressed as either pound of contaminant per 1,000 pounds of product, or as milligram of contaminant per liter of effluent.

Regulations apply to "liquid effluents" discharged from "point sources" into "navigable waters." The broad definition applied to these terms makes the Effluent Guidelines and Standards virtually all-inclusive. Liquid effluents include every type of water, from process wastes to uncontaminated stormwater. A point source is a discharge through any type of conveyance (pipe, channel, etc.). A navigable water is any water other than groundwater. Effluent Guidelines and Standards are issued in three increasing levels of restrictiveness: (a) the Best Practical Control Technology Currently Available (BPCTCA), which existing plants had to meet by July 1, 1977; (b) the Best Available Demonstrated Control Technology (BADCT), which new plants had to meet upon startup; and (c) the Best Available Control Technology Economically Achievable, which all plants were to meet originally by July 1, 1983. Thus, these guidelines are technology-based effluent limits which are developed following appraisal of the treatment technologies available to a particular industrial category, and the economic consideration associated with the installation of such technology for the particular industry. Current effluent guidelines are applied for BOD, COD, total suspended solids (TSS), pH, and some other waste water constituents.

The intent of the Act was to use the progressive restriction on the discharge of contaminants to approach the goal of zero discharge by 1985.

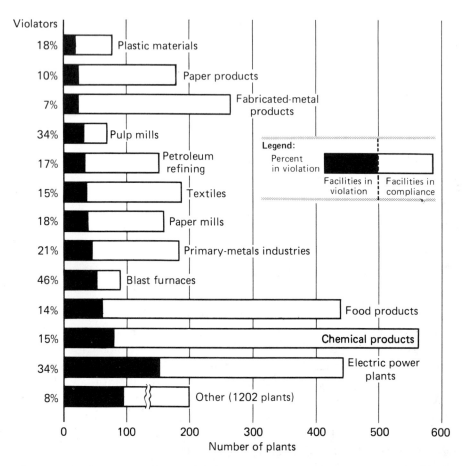

Fig. 8–2   Industry compliance with July 1, 1977, deadline for application of Best Practical Control Technology Currently Available.

Source: U.S. Environmental Protection Agency.

Zero discharge means no discharge of contaminants; therefore, an effluent stream containing substances not harmful to the receiving body of water is compatible with the goal.

The Effluent Guidelines and Standards and the Pretreatment Standards are closely related. Pretreatment Standards apply to discharges into municipal sewers rather than directly into waterways. This sewer discharge could cause the contaminants to interfere with the operation of the waste-treatment facility. These standards eliminate the economic advantages of plants changing their discharge from a navigable water to a municipal system. This economic disincentive is achieved by having the numerical limits virtually identical to the BPCTCA and BADCT levels of the Effluent Guidelines and Standards.

The July 1, 1977, deadline for the application of the Best Practical Control Technology Currently Available was not met by about 15% of the major industrial discharges, nor by about 70% of the municipal sewage treatment jurisdictions (Figure 8–2).

It is likely that the 1984 and 1985 target dates specified in the Act will also prove to be too optimistic. However, it is also clear that the Act has made it possible to achieve substantial progress in water pollution control.

Some chemicals are so highly toxic and impose such severe risks to man or aquatic life that they warrant a special regulatory mechanism to control their presence in waterways. Under the Federal Water Pollution Control Act, the EPA may set up special effluent standards for these toxic contaminants. In 1977, final regulations were prepared by EPA to control direct discharges of DDT, aldrin/dieldrin, endrin, toxaphene, benzidine, and PCB.

Furthermore, pursuant to a 1976 settlement with the National Resources Defense Council, Inc., and Citizens for a Better Environment, the EPA was committed to a program to investigate 65 designated contaminants that represent substantial concern related to their potential health or environmental effects; these include both organic and inorganic agents, many of which are recognized or potential carcinogens, mutagens, or teratogens. The goal established was development of water-quality criteria for each of these by June, 1978, and control of their discharge based upon these criteria.

Food Contamination

The Food and Drug Administration (FDA) is responsible for establishing tolerances, i.e., permissible levels of contamination in food. Table 2–5 lists some tolerance levels for pesticides in food. Tolerance levels have also been established for some carcinogenic contaminants, e.g., 20 ppb_w for aflatoxin in peanuts and corn. On the other hand, the tolerance is zero for carcinogens which are food additives, as specified by the Delaney Clause. An exception was the delay in the implementation of the ban on saccharin, voted by Congress in 1977.

REFERENCES

1. Mitruka, B. M., Rawnsley, H. M., and Vadehra, D. V. (eds.). *Animals for Medical Research*. New York: Wiley, 1976.
2. NCRP. *Basic Radiation Protection Criteria*. NCRP Report No. 39, 1971.
3. NCRP, *Maximum Permissible Body Burdens and Maximum Permissible Concentrations of Radionuclides in Air and in Water for Occupational Exposure*. NBS Handbook No. 69, 1959 USGPO, Washington, D.C.
4. Stokinger, H. E. Modus Operandi of Threshold Limits Committee of ACGIH. *Amer. Ind. Hyg. Assoc. J 25*:589–94 (1964).
5. Toxicology Committee, AIHA. Emergency Exposure Limits. *Amer. Ind. Hyg. Assoc. J. 25*:578–82, 1964.
6. Zielhuis, R. L. Tentative Emergency Limits for Sulphur Dioxide, Sulphuric Acid, Chlorine, and Phosgene. *Ann. Occup. Hyg. 13*:171, 1970.
7. ACGIH. *Threshold Limit Values for Chemical Substances in Workroom Air Adopted by ACGIH for 1977*. American Conference of Governmental Industrial Hygienists, Cincinnati, Ohio, 1977.
8. Brief, R. S. and Scala, R. A. Occupational Exposure Limits for Novel Work Schedules. *Amer. Ind. Hyg. Assoc. J. 36*:467–69, 1975.
9. Mason, J. W. and Dershin, H. Limits to Occupational Exposure in Chemical Environments Under Novel Work Schedules. *J. Occ. Med. 18*:603–06, 1976.
10. Calabrese, E. J. Further Comments on Novel Schedule TLVs. *Amer. Ind. Hyg. Assoc. J. 38*:443–46, 1977.
11. Hickey, J. L. S. and Reist, P. C. Application of Occupational Exposure Limits to Unusual Work Schedules. *Amer. Ind. Hyg. Assoc. J. 38*:613–21, 1977.
12. Elkins, H. B. Excretory and Biological Threshold Limits. *Amer. Ind. Hyg. Assoc. J. 28*:305, 1967.
13. Nelson, K. W. The Place of Biological Measurements in Standard Setting Concepts. *J. Occup. Med. 15*:439, 1973.
14. Hatch, T. F. Permissible Levels of Exposure to Hazardous Agents in Industry. *J. Occup. Med. 14*:134–37, 1972.
15. Zielhuis, R. L. Permissible Limits for Chemical Exposures, pp. 579–88 in *Occupational Medicine*. C. Zenz, (ed.). Chicago: Year Book Medical, 1975.
16. Holmberg, B. and Winell, M. Occupational Health Standards–An International Comparison. *Scand. J. Work, Environ. & Health 3*:1–15, 1977.
17. Stern, A. C. (ed.). *Air Pollution*. New York: Academic Press, 1962.
18. Stern, A. C. (ed.). *Air Pollution*-2nd edition. New York: Academic Press, 1968.
19. Stern, A. C. (ed.). *Air Pollution*-3rd Edition. New York: Academic Press, 1977.

9

Sampling and Measurement of Contaminants

INTRODUCTION

Accurate and reliable qualitative identifications and quantitative determinations of chemical contaminants in the environment are essential for intelligent evaluation of potential hazards to human health and welfare, for identifying and evaluating sources, and for monitoring changes in environmental levels resulting from changes in source characteristics. Unfortunately, the level of available technology and the skills of the analysts have often been inadequate in relation to the needs, and the problems in this field have been compounded by the poor quality of much of the available data. On the other hand, there has been very encouraging progress in recent years in instrument developments and in guidelines and procedures for quality control. There have also been voluntary and mandatory programs for analytical laboratory certifications.

ENVIRONMENTAL SAMPLING AND MEASUREMENTS

A large proportion of the problems in evaluating environmental contaminant levels are not related to the accuracy of laboratory evaluations but, rather, to errors introduced in sample collection and field evaluations. Problems of the latter type are generally the limiting factor in the accuracy of the overall evaluations. Considerations of sampling location, sampling frequency, duration and volume, sampling instruments, and collection substrates all influence the results obtained, and poor choices can invalidate the samples. The investigator generally needs both extensive technical training and professional experience in order to make the appropriate choices. Some of the important considerations in environmental sampling and measurements are discussed in this chapter.

Factors Affecting Sampling Schedules and Locations

Concentrations of contaminants in the environment may be expected to vary continuously and considerably for many reasons. The rates of anthropogenic contaminant release depend upon the level of human activity and the degree of effort expended in limiting discharges. The source strengths of natural contaminants will vary with time, since they depend upon variable biological, geochemical, and meteorological factors. Ambient concentrations will also depend on the degree of dilution which takes place between the source and the sampling site, which in turn depends upon the volume of dilution fluid and its mixing characteristics.

TEMPORAL VARIATIONS

Temporal variations can have several different time constants. There may be rapid fluctuations due to fluid turbulence, which are superimposed on diurnal cycles associated with variations in sources, strengths, and dilution rates. For example, atmospheric dilution depends upon wind pattern and lapse rate which, in the boundary layer, varies with solar flux and radiation from the ground surface. Dilution in aquatic environments depends upon surface-water flow, which varies greatly with the rate of precipitation.

Seasonal variations in contaminant concentrations depend on changes in source strength and mixing characteristics. Because of annual growth cycles, source strength is the dominant factor for natural contaminants such as terpenes from coniferous forests or pollens from plants. The contaminant releases associated with space heating also have an annual surge associated with cold weather. On the other hand, the release of the chemical raw materials involved in photochemical smog formation has little seasonal variability, and the large observed seasonal patterns in oxidant levels are attributable to the differences in solar radiation and atmospheric ventilation at various times of the year.

SPATIAL VARIATIONS

Concentrations of contaminants vary spatially as well as temporally. Thus, it is often necessary to analyze samples drawn from many different locations in order to adequately characterize the average level in, for example, a given airshed, work environment, stream, or food source.

In practice, the choice of sampling sites may be limited by such factors as isolation from local sources and flow disturbances, accessibility for sample changing and sampler maintenance, assurance of sample integrity in relation to tampering, theft, or vandalism, and the need for utilities and/or environmental controls for proper sampler performance. These considerations, and the cost of multiple sampling sites, may lead to dependence on a sampling location not fully representative of the whole. Further evaluations may, therefore, be needed to establish the correspondence, if any, between contaminant levels measured at a fixed site and those in the larger environment it is supposed to represent.

Selection of Materials and Methods

MATERIALS SAMPLED

Environmental samples are generally categorized as air samples, water samples, food samples, soil samples, sediment samples, biota samples (plants and animals), and biological samples (tissues, blood, excreta, etc.).

Most air and some water sampling is done by extractive techniques, whereby the sampled fluid is passed through a collector which extracts the contaminant and passes the fluid. The resulting contaminant samples are much less bulky, and, therefore, are generally easier to package and transport than samples which include the medium as well as the contaminant. Filters are widely used for extractive sampling of suspended particles in both air and water.

Another technique used in both air and water sampling is adsorptive extraction of organics on granular beds of activated carbon or another solid sorbent. Extractions from water samples can also be performed with ion-exchange resins, which exchange and bind ions from the aquatic medium. Other extractive techniques which are primarily limited to air sampling include thermal and electrostatic precipitation and inertial impaction for airborne particles, and scrubbing with liquids for both particles and gases.

In occupational health hazard evaluations, most of the samples collected will be air samples. Biological samples, used to evaluate the degree of chemical exposures or their effects, are also frequently collected, and may include blood, urine, exhaled air, and occasionally other materials, such as hair, fingernails, and feces.

In ambient air evaluations, most of the samples would, of course, be air samples. Other samples which may occasionally be useful are sediment, either as dust fall in a collector or as constituents of surface soil, and biota. Plants can be analyzed for their uptake of contaminant, or for pathological features attributable to exposure.

In water contaminant evaluations, samples of water are most widely used. Sediment samples and biota, including rooted plants, shellfish, and finned fish, are also frequently collected. The biota may be analyzed for contaminant burdens, for pathology, or for changes in population density with level of contaminant.

SAMPLE HANDLING AND PRESERVATION

Certain precautions are necessary in the handling of samples in order to ensure that the characteristics to be analyzed are not altered. Some important considerations are; (a) to minimize the time interval between sample collection and analysis, and (b) to tag or label the sample properly, indicating the exact identification or location of the sampling point and the date and time(s) of sampling, the results of any measurements of ambient conditions at the sampling site, and the identity of the individual who collected the sample.

If the contaminant of interest is produced or affected by biological action, or if the material sampled is chemically or thermally unstable or can be adsorbed on the container surface, the samples will require special handling to prevent changes in content between sample collection and analysis.

Since these preservation techniques may not be sufficient in all cases, some analyses are usually performed only at the sampling site, e.g., pH and DO measurements in aquatic environments.

TYPES OF SAMPLES

There are many kinds of samples which can provide information useful to contaminant evaluations, and it is important not to confuse or misinterpret the contaminant levels found in specific samples. The basic types of samples are: (a) source, (b) ambient, (c) personal, and (d) process.

Source samples include stack samples for airborne effluents and discharge-pipe or channel samples for liquid effluents. They can provide essential information on source strengths, but are of very limited value in evaluating exposures to populations.

Ambient samples provide information on environmental contaminant levels in representative locations within airsheds and waterways. They can be used to indicate whether established standards are being exceeded, and to monitor trends in contaminant levels. The samples are of limited value in identifying or locating specific sources, or for estimating exposure levels for individuals within the population.

Personal samples are used to determine air contaminant exposure levels within the immediate environment of specific individuals. The samplers are generally light in weight and self-contained, so as not to limit the mobility or normal activity of the wearer. They are most widely used to evaluate inhalation exposures of workers, and are also used in some epidemiological studies where indoor exposures are of special interest.

Process samples are samples, collected within process equipment and transfer lines, which are taken to characterize mass flowrates or stream composition. Samples up- and downstream of a pollution-control device belong in this category. The concentrations measured may be very high, but may bear little relation to quantities discharged to the environment, and even less to ambient environmental levels.

SAMPLING EFFICIENCY

It is generally desirable to select the sample collector and its operating conditions so that the fraction of the contaminant which penetrates the collector will be negligible. However, the analyses can be just as reliable when the collection efficiency is much lower, provided that it is known and constant. In this case, a correction factor can be applied. The collection of gases and vapors in air, and dissolved chemicals in water, can generally meet this criterion (provided that the ambient temperature is constant), since each molecule of a particular contaminant is essentially equivalent to every other in terms of capture probability. However, the same consideration does not apply to contaminants present as particles, since an additional variable affecting collection efficiency, i.e., particle size, is involved. Particle samplers should have essentially complete collection to avoid any sampling bias.

SAMPLE SIZE AND ANALYTICAL SENSITIVITY

Every analytical procedure has a lower detection limit at which the accuracy of measurement is low. If the results of the analysis are to be useful, the amount of the specific contaminant in the sample must be as large as, and preferably many times greater than, this limit. Therefore, for a contaminant of interest, the quantity to be sampled depends on two factors: (a) its estimated environmental concentration and (b) the amount of sampled food, air, water, etc.

In extractive sampling, the amount sampled depends on (a) the rate of sample flow and (b) the duration of the sampling interval. Sample size can be increased either by increasing the time interval or the sampling rate. However, freedom to change the rate may be limited in some cases, since sampling efficiency may be rate-dependent. Sampling rates may also be limited by the size, power requirements, or suction capacity of the sampling equipment.

The selection of an appropriate sample size in a given situation must also involve a consideration of the basic reason for collecting the sample. If the purpose is to compare an ambient level to an established standard or specified action level, the sample should be large enough to permit the determination of a specified fraction of that standard or level. The fraction used is generally $1/10$, although equipment limitations may dictate acceptance of a fraction as high as $1/2$.

TIME OF SAMPLING

In the measurement of contaminant levels, it is important to characterize the time and time interval of sampling. In deciding when and how long to sample, the kind and degree of temporal variability likely to occur must be considered, as well as whether it is more important to determine peak levels or average levels. In other words, the individual specifying the sampling frequency and duration should appreciate and anticipate the uses to be made of the resulting data.

Samples which are collected within a brief period are called "grab" samples. Samples collected at a constant rate over a longer period of time are known as integrated samples. While there is no precise demarcation between grab and integrated samples, the former are generally collected in less than a few minutes.

In most cases, grab sampling involves filling a container of a known volume with the fluid being sampled. Examples include filling a one-liter bottle with water from an effluent discharge or stream, or filling a plastic bag with exhaled air for breath analysis. By contrast, integrated samples are usually of the extractive type.

SAMPLING SUBSTRATE

In extractive sampling, the contaminant of interest is deposited onto or into materials which will retain it effectively throughout the balance of the sample collection process, and through any subsequent transport and storage prior to analysis. The material which retains the extracted substance is known as the sampling substrate. A good substrate must not only efficiently retain the sampled material prior to analysis, it must also permit it to be analyzed efficiently in the laboratory. It must either be possible to separate the sampled material from the substrate quantitatively, or it must be possible to perform the analysis in the presence of the substrate. A given substrate may be ideal for some analyses and entirely unsuitable for others where, for example, its presence may preclude the analysis of choice, or where it may contain constituents which would interfere with the analysis.

SPECIFICITY AND INTERFERENCES

One of the most difficult aspects of the analytical chemistry of environmental contaminants is that one is generally applying trace microanalytical techniques to samples of mixed composition. The presence of co-contaminants of unknown composition and concentration may either enhance or depress the response characteristic of the material being analyzed. When the potential for interferences is known, it is generally desirable to achieve greater analytical specificity by performing chemical separations prior to the analyses.

LEGAL REQUIREMENTS

When sampling is performed to demonstrate conformance with legal codes or standards, it may be necessary to use certain specified sampling and analytical procedures rather than, or in addition to, those best suited to scientific investigation.

SAMPLE COLLECTION VS. DIRECT MEASUREMENTS

In this discussion, sample collection refers to the collection of a defined weight or volume of sampled material for a subsequent laboratory analysis or series of analyses. The analyses are performed at a later time, and the results are not generally available until long after the sampling has been completed. In other words, there is no possibility of feedback to guide further sampling.

In direct measurements, on the other hand, the sampling and analysis are performed in the same instrument. This technique allows feedback during the sampling period at the sampling site, so that the site and/or frequency of sampling may be modified.

Some direct instruments are known as continuous analyzers. In these, an intrinsic property of the contaminant is measured as the sampling stream passes through a sensing zone. There is no sample collection, and an uninterrupted output response is a direct function of the concentration of the contaminant of interest. The output can be recorded, and can also be integrated to yield average concentrations.

Other types of direct measurements are based upon extractive sample collection and immediate analysis of the sample. These instruments perform collection and analysis within the same housing, and are known as semicontinuous analyzers. A representative fraction of the contaminant is obtained and analyzed, and the process then repeats itself. The measurement phase is generally performed on a liquid, after controlled dosage of a reagent chemical, using electrochemical or colorimetric techniques. In air sampling, the contaminant is first extracted by a collecting liquid in a gas washer or scrubber. Ideally, the analyzing period is sufficiently short so that no significant chemical changes occur before another sample is measured. However, because of the finite time lag (on the order of a few minutes) between the sampling and measurement steps, and as a result of the mixing and diffusion within the liquid stream, a monitor of this type has a slower response than does a continuous analyzer.

In using direct-reading instrumentation, accurate calibration is essential to the correct interpretation of instrument response. Calibration, which may be defined as the determination of the true values of the scale readings of the instrument, generally involves a comparison of instrument response to that of a reference instrument or to a standardized atmosphere or solution.

There are a number of advantages to direct measurements:

a. Both peak and average concentrations can be determined.
b. Results are immediately available, permitting one to track down sources and initiate corrective actions as appropriate.
c. Continuous and unattended operation is possible under some circumstances.
d. Continuous recordings of concentrations may be useful in retrospective data evaluations and, in industry, as legal documentation of employee exposures.

On the other hand, there are many distinct advantages in sample collection with subsequent laboratory evaluations:

a. Sample collection is a much simpler operation than quantitative chemical analysis, and the hardware needed for collection is less expensive, smaller, and less sensitive to mechanical and thermal stresses.
b. Being smaller and more portable in most cases, the sampling equip-

ment is easier to transport and operate in the field, and can be set up in locations not suitable for continuous monitors.

c. The analyses performed in the laboratory on field-collected samples can be more sensitive, accurate, and reliable. The samples can be analyzed by highly specialized equipment which cannot be incorporated into field monitors. A series of different analyses can be performed on each sample, and the effects of co-contaminants can be eliminated by preliminary chemical separations.

Measurements of Fluid Flow

Characterization of chemical contamination in air or water generally requires data on fluid flow as well as quantitative chemical analyses. For example, when extractive sampling techniques are used, the determination of concentration requires measurement of both the sampled volume as well as the amount of chemical contaminant in the sample. Furthermore, the concentration at the sampling location may be different from that at other locations, and the determination of contaminant transport in a surface stream or airborne plume may require measurements of flow and concentration at many points.

Flowmeters can be divided into three basic groups on the basis of the type of measurement made; there are integral-volume meters, flowrate meters, and velocity meters. In volume meters and flowrate meters, the whole fluid stream passes through the instrument. In this respect, they differ from velocity meters, which measure the velocity at a particular point of the flow cross section. Since the flow profile is rarely uniform, the measured velocity will invariably differ from the average velocity, and it may be necessary to make a large number of measurements in order to determine the average value. However, when the flow field is very large, velocity sensors may be the only indicators that can be used.

INTEGRAL-VOLUME METERS

Some integral-volume meters are used exclusively for measurements in air; these include the wet-test meter and dry-gas meter. Others, which operate by positive displacement, are available for both air and liquid flow metering.

A wet-test meter (Figure 9–1) consists of a partitioned drum half submerged in a liquid (usually water), with openings at the center and periphery of each radial chamber. Air or gas enters at the center and flows into an

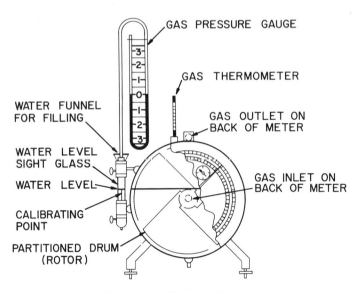

Fig. 9–1 Schematic diagram of a wet-test meter.

Fig. 9–2 Schematic diagram of a dry-gas meter.

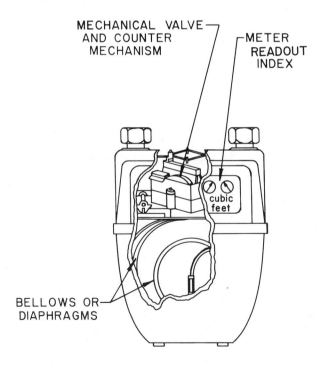

individual compartment causing it to rise, thereby producing rotation. This rotation is indicated by a dial on the face of the instrument. The volume measured will be dependent on the fluid level in the meter, since the liquid is displaced by air.

The dry-gas meter shown in Figure 9–2 is very similar to the domestic gas meter. It consists of two bags interconnected by mechanical valves and a cycle-counting device. The air or gas fills one bag while the other bag empties itself; when the cycle is completed, the valves are switched and the second bag fills while the first one empties. The alternate filling of two chambers as the basis for volume measurements is also used in twin-cylinder piston meters. Such piston meters can also be classified as positive-displacement devices.

Positive-displacement meters consist of a tight-fitting moving element with individual volume compartments which fill at an inlet and discharge at an outlet. Another type of multicompartment continuous rotary meter uses interlocking gears.

VOLUMETRIC FLOWRATE METERS

The integral-volume meters discussed above were all based upon the principle of conservation of mass; specifically, the transfer of a fluid volume from one location to another. On the other hand, the flowrate meters described in this section operate upon the principle of the conservation of energy; they are based upon Bernoulli's theorem for the exchange of potential energy for kinetic energy and/or frictional heat. Each consists of a flow restriction within a closed conduit. The restriction causes an increase in the fluid velocity and, therefore, an increase in kinetic energy, which requires a corresponding decrease in potential energy, i.e., static pressure. The flowrate can be calculated from a knowledge of the pressure drop, the flow cross section at the constriction, the density of the fluid, and the coefficient of discharge, which is the ratio of actual flow to theoretical flow, and makes allowance for stream contraction and frictional effects.

Flowmeters which operate upon the conservation of energy principle can be divided into two groups. One group, which includes orifice meters, venturi meters, and flow nozzles, consists of devices which have a fixed restriction; these instruments are known as variable-head meters, because the differential pressure head varies with flow. A special subclass of this group includes weirs and flumes specifically used to measure flowrates of water. In these, the flow channel is only partially filled with water, and the height of the water column in the restriction varies, providing an indication of the flowrate. The other, smaller, group, which includes rotameters,

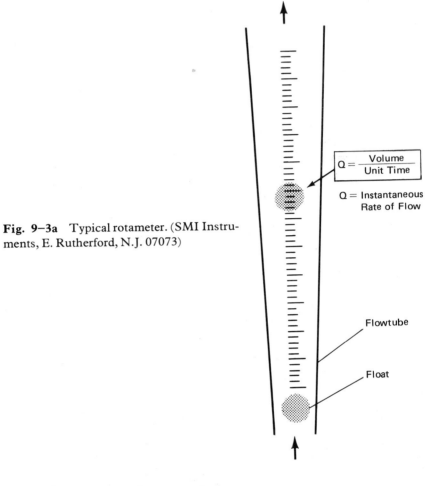

$$Q = \frac{Volume}{Unit\ Time}$$

Q = Instantaneous
Rate of Flow

Flowtube

Float

Fig. 9–3a Typical rotameter. (SMI Instruments, E. Rutherford, N.J. 07073)

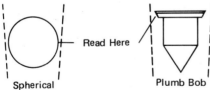

Read Here

Spherical

Plumb Bob

Fig. 9–3b Typical rotameter floats.

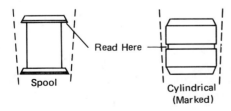

Read Here

Spool

Cylindrical
(Marked)

are known as variable-area meters, because a constant pressure differential is maintained by varying the flow cross section.

A rotameter (Figure 9–3a) contains a "float" which is free to move up and down within a vertical tapered tube which is larger at the top than the bottom. The fluid flows upward, causing the float to rise until the pressure drop across the annular area between the float and the tube wall is just sufficient to support the float. The tapered tube is usually made of glass or clear plastic, and has a flowrate scale etched directly on it. The height of the float indicates the flowrate. Floats of various configurations are used, as indicated in Figure 9–3b. The term "rotameter" was first used to describe variable-area meters with spinning floats, but now is generally used for all types of tapered metering tubes.

The simplest form of variable-head meter is the square-edged, or sharp-edged, orifice illustrated in Figure 9–4. It is also the most widely used, because of its ease of installation and low cost. While the square-edged orifice can provide accurate flow measurements at low cost, it is inefficient with respect to energy loss. The permanent pressure loss for an orifice meter will often exceed 80%.

Venturi meters (Figure 9–5) have converging and diverging angles of about 25° and 7°, respectively; they have high pressure recoveries, i.e., the potential energy which is converted to kinetic energy at the throat is reconverted to potential energy at the discharge, with an overall permanent loss of only about 10%. The characteristics of various other types of variable-head flowmeters, e.g., flow nozzles (Figure 9–6), centrifugal flow elements, etc., are similar in most respects to those of either the orifice meter, venturi meter, or both.

For water flow in open channels, the most common flow measurement techniques utilize weirs or flumes.

A weir is essentially a dam or obstruction placed in the stream. It generally consists of a vertical plate with a sharp crest; the top of the plate is either straight, notched, or rectangular shaped (Figure 9–7). The water level at a given distance upstream from the weir is proportional to the flow rate.

There are a number of different types of flumes. The Parshall flume (Figure 9–8) is a device which is similar to a venturi meter in that it consists of a converging section, a throat, and a diverging section. The level of the floor in the converging section is higher than the floor in the throat and diverging section. The head of the water surface in the converging section is a measure of the velocity through the flume and, therefore, of flowrate.

PRESSURE DISTRIBUTION ACROSS ORIFICE PLATE

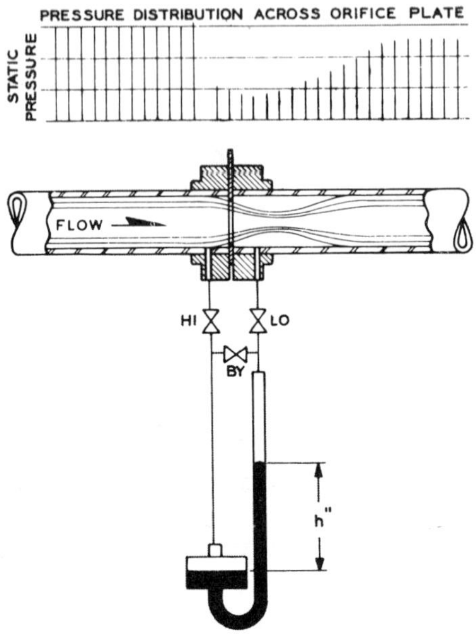

Fig. 9–4 Pipe-line orifice. (Meriam Instrument, Cleveland, Ohio 44102)

The flowrate in a pipe or tube may be strongly dependent on the flow resistance, and flowmeters with a sufficiently low resistance may be too bulky or expensive. A metering element used in such cases is the bypass rotameter, which actually meters only a small fraction of the total flow; however, this fraction is proportional to the total flow. As shown schematically in Figure 9–9, a bypass flowmeter contains both a variable-head element and a variable area element. The pressure drop across the fixed orifice or flow restrictor creates a proportionate flow through the parallel path containing the small rotameter. The scale on the rotameter indicates the total flow.

Flowrates of aqueous discharges from pipes can also be estimated by measuring the distance the flow is projected from the end of an open pipe, as illustrated in Figure 9–10.

For small volumetric flows of water where the stream can be diverted into a vessel of known volume, the rate of flow can be determined by measuring the time required to fill the vessel. This is known as the "bucket-and-stopwatch" technique.

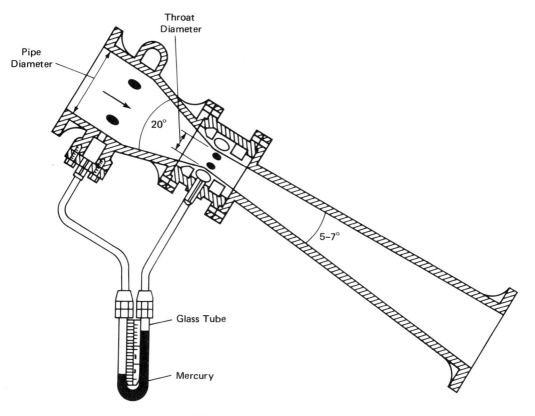

Fig. 9–5 Venturi meter.

Fig. 9–6 Flow nozzle (in a pipe).

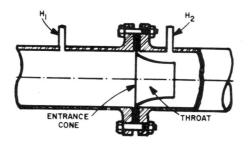

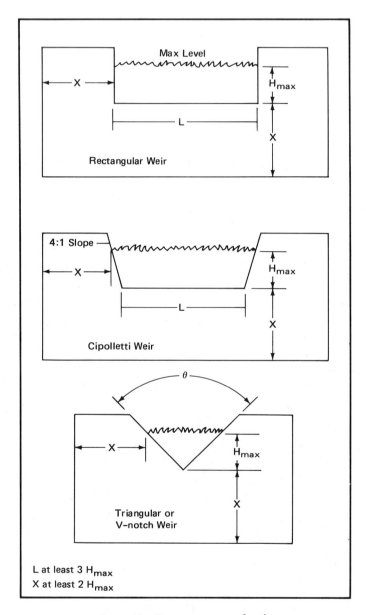

Fig. 9–7 Common types of weirs.

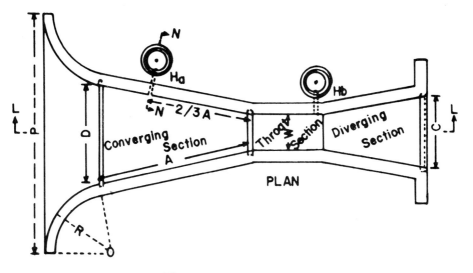

Fig. 9–8   Parshall flume.

Fig. 9–9   Schematic of a by-pass flowmeter. (Brooks Instrument Division, Emerson Electric Co., Hatfield, Pa. 19440)

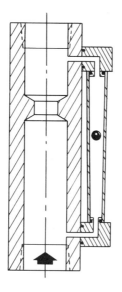

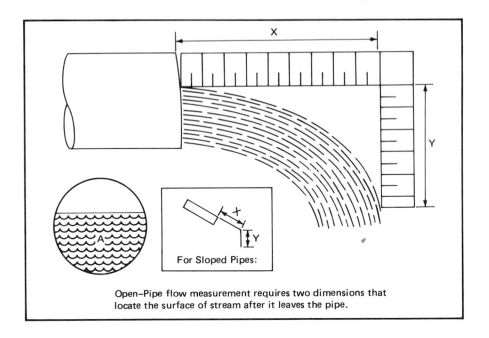

For Sloped Pipes:

Open-Pipe flow measurement requires two dimensions that locate the surface of stream after it leaves the pipe.

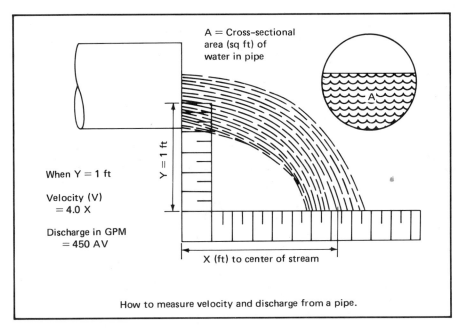

A = Cross-sectional area (sq ft) of water in pipe

When Y = 1 ft

Velocity (V) = 4.0 X

Discharge in GPM = 450 AV

Y = 1 ft

X (ft) to center of stream

How to measure velocity and discharge from a pipe.

Fig. 9–10 Flowrate measurements based upon the distance a stream is projected from the end of an open pipe.

From: Cross, F. L., Jr. *Handbook on Environmental Monitoring.* Westport, Conn.: Technomic Publishing Co., 1974.

FLOW VELOCITY METERS

The Pitot tube (Figure 9–11) is often used as a reference instrument for measuring velocity, and, if carefully made, will need no calibration. It consists of an impact tube whose opening faces axially into the flow, and a concentric static pressure tube with eight holes spaced equally around it in a plane which is eight diameters from the impact opening. The difference between the static and impact pressures is the velocity pressure. Bernoulli's theorem applied to a Pitot tube in a stream simplifies to the dimensionless formula:

$$V = \sqrt{2g_c P_v} \qquad (9.1)$$

where

V = linear velocity
g_c = gravitational constant
P_v = pressure head of flowing fluid (velocity pressure)

There are several other ways beside the Pitot tube to utilize the kinetic energy of a flowing fluid to measure velocity. One technique is to align a jeweled-bearing turbine wheel axially in the stream and count the number of rotations per unit time. Such devices are known as rotating-vane flowmeters. Some are very small and are used as velocity probes. Others are sized to fit the whole duct and become indicators of total flowrate; these latter devices are sometimes called turbine flowmeters.

MASS FLOW AND TRACER TECHNIQUES

A thermal meter measures mass flowrate with negligible pressure loss. It consists of a heating element in a pipe or duct section between two points at which the temperature of the stream is measured. The temperature difference between the two points is dependent on the mass rate of flow and the heat input.

The principle of mixture metering is similar to that of thermal metering. However, instead of adding heat and measuring temperature difference, a tracer is added and its increase in concentration is measured; or clean fluid is added and the reduction in concentration is measured. The measuring device may react to some physical property such as thermal conductivity or vapor pressure.

Mass flow may also be obtained using an ion-flow meter. In this device, ions are generated from a central disc and flow radially toward a collector surface. Airflow through the cylinder causes an axial displacement of the ion stream in direct proportion to the mass flow.

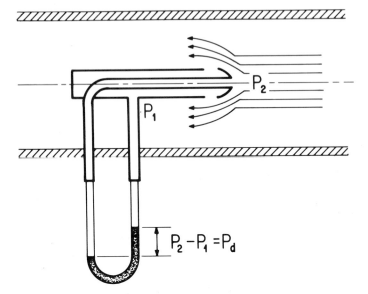

Fig. 9–11 Pitot tube.

Fig. 9–12 Magnetic flowmeter.

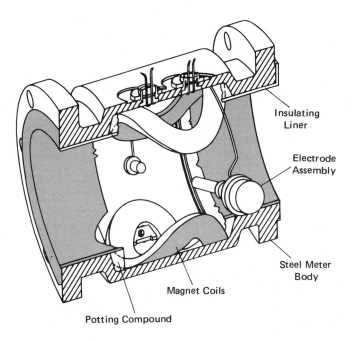

Insulating
Liner

Electrode
Assembly

Steel Meter
Body

Magnet Coils

Potting Compound

An instrument specifically used for measuring water flow is the magnetic flowmeter (Figure 9–12). This apparatus operates according to Faraday's Law of Induction, i.e., the voltage induced by a conductor moving at right angles through a magnetic field will be proportional to the velocity of movement of the conductor through the field. In this device, the water is the conductor and a set of electromagnetic coils in the meter produces the field. The induced voltage is measured to obtain the flowrate.

Factors affecting the selection of flow-metering devices for aqueous streams are rated in Table 9.1.

The surface velocity of water may be measured by placing floating objects (e.g., wood, cork, etc.) in the stream and measuring the time required for the object to traverse a measured distance between two points. The velocity within a given length of channel may also be obtained by the use of dyes.

Air Sampling

INSTRUMENTATION

A description of available instrumentation for air sampling can easily occupy a large reference volume. Therefore, for a more comprehensive review, the reader is referred to other sources, such as *Air Sampling Instruments* (1) or to the appropriate chapters in the multivolume reference works of Stern (2) and Patty (3). The discussion to follow will be limited to the more fundamental considerations in the selection of instruments and sampling techniques for air contaminant evaluations.

ELEMENTS OF AN AIR-SAMPLING TRAIN

There are a number of essential elements which comprise a sampling system, known as a sampling train. These may all be incorporated into a single compact housing or portable instrument, or they may be separate elements assembled into a train.

A sampling probe is needed when sampling in a moving stream, such as in a duct or stack, and, when used, is the first element in the train. It is not needed when sampling from relatively still air in the ambient atmosphere or at the breathing zone of a worker. In such cases, it should not be used, since it would present some opportunity for sample losses without any corresponding benefit.

In sampling quiescent air, the inlet configuration of the sampler is a significant element of the train, and one whose influence is not always recognized, especially in particulate sampling. The inlet size and configu-

Table 9.1 Selection Chart for Water Flow Measuring Devices

	Orifice		Venturi	Nozzle	Pitot	Elbow	Lo-lcss Tube	Magnetic[a] flow meter
	Concentric	Segmental or eccentric						
Accuracy, and amount of empirical data	E	F	G	G	*	P	G	E
Differential for given flow and size	E	E	G	G	F	P	E	None
Pressure recovery	P	P	G	P	E	E	E	E
Use on dirty service	P	F	E	G	VP	P	G	E
For liquids containing vapors	E[b]	E	E	G	F	F	G	E
For vapors containing condensate	E[c]	E	E	G	P	F	G	None
For viscous flows	F	U	G	G	†	U	F	E
First cost small size	E	G	P	F	G	E	P	P
First cost large size	E	G	P	F	G	E	P	P
Ease of changing capacity	E	G	P	F	VP	VP	P	E
Convenience of installation	G	G	F[d]	F	E	E	F[d]	F[e]

All ratings are relative: E excellent, G good, F fair, P poor, VP very poor, U unknown.
* For measuring velocity at one point in conduit, the well-designed pitot tube is reliable. For measuring total flow, accuracy depends on velocity traverse.
† Requires a velocity traverse.
a. Restricted to conducting liquids.
b. Excellent in vertical line if flow is upward.
c. Excellent in vertical line if flow is downward.
d. Both flange type and insert type available.
e. Requires pipe reducers if meter size is different from pipe size.
Source: U.S. Environmental Protection Agency.

ration, in conjunction with the sampling flowrate, establish an actual upper size cutoff for particle acceptance; particles larger than this size will not be aspirated into the inlet.

The next element in the train is the sample collector. It should precede other elements such as flow-measuring devices and pumps, which could remove or add contaminants to the sample stream. The selection of the type of sample collector depends on the nature of the sample to be collected, and will be discussed in the next section.

A pressure sensor should be located downstream of the sample collector, preceding the inlet to the flow-metering element. There will usually be a significant pressure drop between the inlet to the probe or sample collector and the inlet to the flowmeter, due to the flow resistance of the sample collector and the flow path itself. Most flowmeters are calibrated for atmospheric pressure inlets and, therefore, require a correction when used with a reduced inlet pressure. The necessity for, and magnitude of, the correction can be determined by a static pressure measurement at this point.

The accuracy of the concentration measurement is equally dependent on the sampled volume as on the mass of the collected sample. The sampled volume can either be determined directly by an integrating flowmeter, or from the product of the flowrate and the sampling interval. Flowrate measurements are made more frequently than integrated volume measurements, because the meters are generally smaller, lighter, and less expensive. The precision of many flowmeters used in air sampling is, unfortunately, frequently poor, and may act to limit the overall reliability of the concentration determination. Care should be taken to avoid leakage between the inlet of the sampler and the inlet of the flowmeter, so that the volume measured represents only sampled air. Leakage downstream of the flowmeter can reduce the sampling rate, but would not affect the accuracy of the measurement. The most commonly used flowrate meters in air samplers are rotameters and orifice meters.

The final element of the train is the air mover or suction source. It can be an air pump, a blower or fan, or an ejector. An ejector is a device which uses a stream of high-pressure fluid flowing through a jet to create a secondary stream of sampled air. The fluid can be compressed air, steam, or water. The selection of air-mover type and size is dependent on the choice of sampling rate and the pressure drop to be overcome in maintaining the desired flow.

The sequence of the train elements is important, but not sacred. Circumstances sometimes lead to the selection of a different configuration,

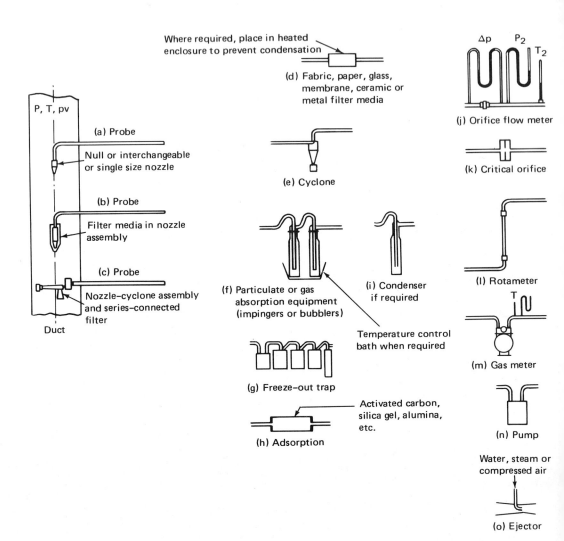

Where required, place in heated
enclosure to prevent condensation

(d) Fabric, paper, glass,
membrane, ceramic or
metal filter media

P, T, pv

(a) Probe

Null or interchangeable
or single size nozzle

(b) Probe

Filter media in nozzle
assembly

(c) Probe

Nozzle–cyclone assembly
and series–connected
filter

Duct

(e) Cyclone

(f) Particulate or gas
absorption equipment
(impingers or bubblers)

(i) Condenser
if required

Temperature control
bath when required

(g) Freeze–out trap

Activated carbon,
silica gel, alumina,
etc.

(h) Adsorption

Δp   P₂
        T₂

(j) Orifice flow meter

(k) Critical orifice

(l) Rotameter

T

(m) Gas meter

(n) Pump

Water, steam or
compressed air

(o) Ejector

Fig. 9–13   Sampling system components.

From: Bloomfield, B. D. *Source Testing*, pp. 487–536 in *Air Pollution*, vol. 2, 2nd Ed. A. C. Stern (ed.).
New York: Academic Press, 1968.

and this is acceptable when it can be demonstrated that excessive losses or errors are not introduced. Some typical sampling-train elements and their normal sequence are illustrated in Figure 9-13.

SAMPLE COLLECTORS FOR GASES AND VAPORS

Gases and vapors can be extracted from an airstream by absorption, adsorption, or condensation. The most common absorption technique involves intimate contact between gas bubbles and an absorbing liquid. Gas washers in various configurations are illustrated in Figure 9-14. Table 9-2 presents examples of some gases commonly collected in liquid sorbents.

The airstream may be broken down into bubbles by a submerged jet, as in impingers (Figure 9-14A) and simple bubblers (Figure 9-14B). Finer bubbles, and, hence, greater gas-liquid contact and removal efficiency, can be obtained by passing the air through a porous plug or frit rather than a single orifice jet. Such a device, known as a fritted bubbler, is illustrated in Figure 9-14D.

Fig. 9-14 Various gas washers: (A) Midget impinger, (B) simple gas washer, (C) spiral absorber, (D) fritted bubbler, (E) glass-bead column.

From: Gaseous Fuels; Coal and Coke; Atmospheric Analysis. *Annual Book of ASTM Standards,* pt 26. Philadelphia: ASTM, © 1976. Reprinted by permission of the American Society for Testing and Materials.

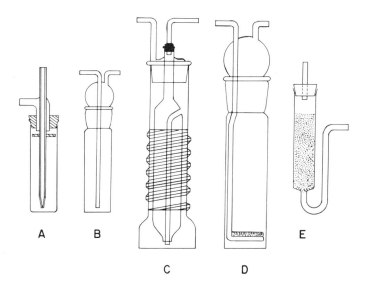

A   B   C   D   E

Table 9.2 Some Gases and Vapors Collected in Liquid Sorbents

Chemical	Sample collector	Sorption medium
Ammonia	Impinger	Sulfuric acid
Carbon dioxide	Fritted bubbler	Barium hydroxide
Formaldehyde	Fritted bubbler	Sodium bisulfite
Hydrogen sulfide	Impinger	Cadmium sulfate
Methyl mercury	Impinger	Iodine monochloride in hydrochloric acid
Ozone	Impinger	Potassium iodide in potassium hydroxide
Sulfur dioxide	Impinger	Sodium tetrachloromercurate

Fig. 9–15 Schematic diagram of personal NO$_2$ sampler.
From: Palmes, E. D., Gunnison, A. F., Dimattio, J., and Tomczyk, C. Personal sampler for nitrogen dioxide. *Amer. Ind. Hyg. Assoc. J. 37*:570–77, 1976.

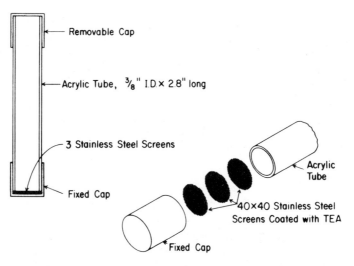

Removable Cap

Acrylic Tube, $\frac{3}{8}$" I.D. × 2.8" long

3 Stainless Steel Screens

Fixed Cap

Acrylic Tube

40×40 Stainless Steel Screens Coated with TEA

Fixed Cap

Exploded View of Sampler Bottom

Absorption can also take place into a liquid film on a solid support. The support can be a screen or filter which has been coated with a reagent chemical. The airstream can be drawn through the screen or filter, or these devices can be used as passive samplers. In the latter, the gas or vapor molecules to be trapped reach the collection surface by molecular diffusion. The NO_2 sampler (4) illustrated in Figure 9–15 utilizes an orifice as a diffusion barrier of known characteristics in order to eliminate the effects of bulk fluid motion on the transport rate of NO_2 from the ambient source to the collection surface at the screen.

Gases and vapors can also be captured by adsorption onto surfaces. The most commonly used adsorption collectors are cylinders filled with granules of an efficient adsorbent, such as activated charcoal or silica gel. As a gas stream is drawn through such a bed, the molecules reach the surfaces of the granules by molecular diffusion. The O_2, N_2, and Ar molecules are not retained, but molecules of the contaminant of interest can be adsorbed by granules of suitable composition. Solid sorbents are generally used for the collection of organic gases and vapors.

Contaminants which condense into liquids or solids below ambient temperature can be removed by drawing them past cooled collection surfaces, which are sometimes called "cold traps." One complication is that the condensate will generally include a large volume of water unless the water vapor in the atmosphere can be removed before the cold trap.

SAMPLE COLLECTORS FOR AEROSOLS

While useful for gas sampling, the impinger, shown in Figure 9–14, was actually designed to sample mineral dusts. The airstream is accelerated as it passes through the orifice nozzle, and particles larger than about 0.75 μm are efficiently collected when they strike the submerged plate and are wetted. The standard formerly used for silica-bearing dusts was based on the particle concentration measured in the suspension within the flask.

In more recent years, almost all particle sampling has been performed using dry-collection techniques. These include other inertial samplers, such as impactors, cyclones, and centrifuges, as well as filtration, electrostatic precipitation, and thermal precipitation.

Inertial Collectors   An impactor is very similar to an impinger, but does not use a trapping liquid. The airstream is accelerated through a nozzle and directed at a collection surface. Particles larger than a certain size (known as the cut-size) strike the surface and may be retained; smaller particles, having less momentum, are carried away with the carrier flow. The

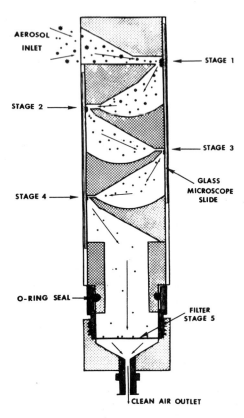

AEROSOL
INLET

STAGE 1

STAGE 2

STAGE 3

GLASS
MICROSCOPE
SLIDE

STAGE 4

O-RING SEAL

FILTER
STAGE 5

CLEAN AIR OUTLET

Fig. 9–16 Diagrammatic cross-sectional view of a cascade impactor while sampling. The particle size is exaggerated.

From: ACGIH. *Air Sampling Instruments,* 3rd Ed. American Conference of Governmental Industrial Hygienists, Cincinnati, Ohio, 1966.

cut-size depends on the velocity in the jet. At ambient pressures, the smallest cut-size readily attainable is about 0.5 μm diameter; lower cut-sizes can be attained by operating the sampler within a vacuum chamber. If the particles and the collection surface are both dry, the particles are more likely to bounce and escape rather than be retained. Impactors can only be used successfully for such applications when the collection plate is coated with an adhesive layer. However, even with the use of an adhesive, collection must be limited to avoid particle bounce. The adhesive can rapidly become saturated, and particles striking those already collected are retained poorly.

In a series or cascade impactor (Figure 9–16), the same volumetric flow is passed through a series of impaction jets, with each successive stage having a smaller jet cross section and particle cut-size. It, therefore, sorts the aerosol into a series of fractions on the basis of aerodynamic particle size. Knowing the proportions of the sample mass on each stage, and the stage constants (i.e., the cut-sizes of each stage), the overall size distribution of the sampled aerosol may be estimated.

Another widely used type of inertial sampler is the cyclone. In the cyclone configuration illustrated in Figure 9-17, the sampled air is drawn into a tangential inlet. The flow follows a spiral path downward along the outer wall and then forms a tighter spiral as it flows up the axial exit pipe. Particles larger than the cut-size strike the outer wall and either remain there or migrate slowly down the wall into the conical section at the bottom. The major application of cyclones as samplers is as the first stage of a two-stage collector. They can also be used for aerodynamic size distribution evaluations in a multicyclone array, with each cyclone having a different cut-size. Their advantage over cascade impactors for such applications lies in the much larger sample masses which can be collected without artifacts due to particle bounce.

More precise separations and collection capabilities for smaller particles can be achieved using aerosol centrifuges. The particles enter at one side of a laminar flow channel and are deposited gently along a collection foil according to their aerodynamic diameter. However, the aerosol centrifuges are basically laboratory instruments, their sampling flowrates are very low, and they are relatively expensive.

Filtration   Filters are currently the most widely used type of aerosol sampler, and by a wide margin. While no single type of filter is suitable for all applications, filters are available in a wide variety of types, ratings, and sizes, so that there is generally one or more well suited for each particular application. Thus, one can generally take full advantage of their inherent advantages: minimal equipment, low cost, and convenience with respect to sample handling.

There are four basic types of filters: (a) fibrous filters, consisting of a mat of randomly oriented fibers of cellulose, glass, asbestos, polystyrene, etc.; (b) membrane filters, consisting of a plastic material with a gel structure having interconnecting pores; (c) Nuclepore filters, which are a solid sheet with uniform parallel holes; and (d) granular beds. The basic structures of Nuclepore and fiber filters are illustrated schematically in Figure 9-18.

In selecting a filter for a given task, the following factors should be con-

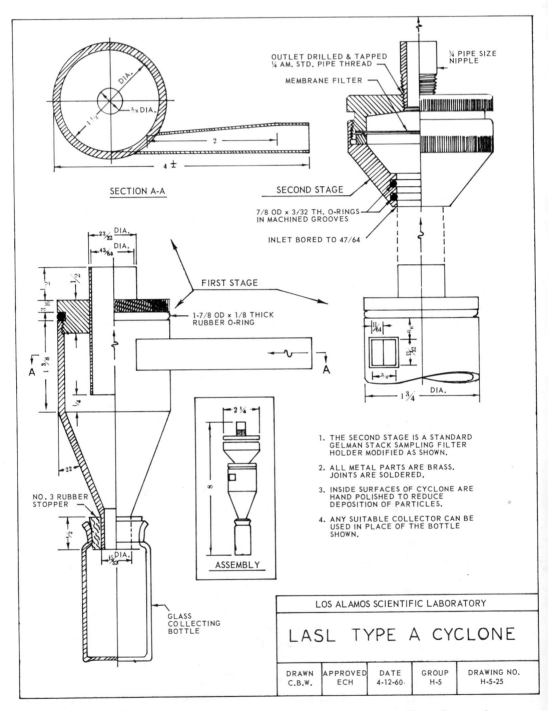

Fig. 9–17 A miniature cyclone-filter unit for two-stage sampling of aerosols. (Los Alamos Scientific Laboratory).

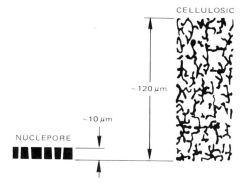

Fig. 9–18 Cross-sectional comparisons of Nuclepore and cellulose-fiber filter thicknesses.

sidered: (a) physical size, composition, and structure of the filter, and whether these would limit the analyses to be performed; (b) collection efficiency and flow resistance, and how they may be expected to change during the sampling interval (e.g., with loading); (c) the characteristics of the available sampling pumps, and whether they can provide the desired suction capacity and sampling rate throughout the sampling interval.

While filters may appear to be very simple devices, the mechanisms by which they capture airborne particles are frequently misunderstood. The most common misconception is that they function like the sieves or screen collectors used to separate large particles from liquid streams, and that particles smaller than the pore or void size will penetrate. This is generally not the case in air filtration, where particles much smaller than the void size can be collected with very high efficiency.

The major mechanisms of particle collection which are operative in air filters are inertial impaction, Brownian diffusion, and interception. In some cases, where the particles or the filter material have high levels of electrical charges, electrostatic precipitation may contribute significantly to the overall collection efficiency.

While the flow path through Nuclepore filters is relatively simple, the flow through all other types is quite complex. The air undergoes accelerations, decelerations, and numerous branchings and directional changes in negotiating its way through the pores or voids. As the aerodynamic particle size and flow velocity increase, the probability of particle retention by impaction increases; with decreasing particle size and flow velocity, the particles' Brownian displacement and retention time within the filter

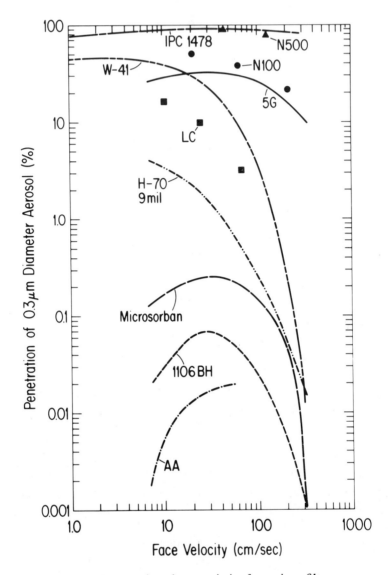

Fig. 9–19 Penetration characteristics for various filters:

AA:	cellulose membrane, 0.8 μm pores
1106BH:	glass fiber
Microsorban:	polystyrene fiber
H–70:	cellulose-asbestos mixed fiber
W–41:	Cellulose fiber
LC:	teflon membrane, 10 μm pores
5G:	thin glass fiber
IPC 1478:	thin cellulose-fiber mat
N100:	Nuclepore, 1 μm pores
N500:	Nuclepore, 5 μm pores

structure increase, increasing the percentage of retention by this mechanism.

The particles most likely to penetrate the sampling filter are those which have both minimal Brownian displacement and minimal momentum. Under most conditions, maximum penetration occurs for particle diameters of about $0.3-0.5$ μm. For some widely used sampling filters, such as most membrane filters with pore sizes below 3 μm and most glass-fiber filters, the maximum penetration will be less than 1%. On the other hand, some commonly used cellulose fiber filters may have particle penetrations as high as 70%, and Nuclepore filters may have even greater penetrations. These considerations are illustrated in Figure 9–19. It can be seen that the collection efficiency of filters having characteristic penetration minima can be greatly improved when they are operated at face velocities (the velocity normal to the filter face) greater or lower than those at which the minima occur.

In fibrous and granular filters, there is defense in depth, i.e., the particles are collected by impaction on and diffusion to the granular or fiber surfaces throughout the depth of the filter. In order to analyze the material collected, one must either extract it from the filter or analyze it on the filter itself. On the other hand, in membrane and Nuclepore filters, which are much thinner, the particles are collected on or close to the upstream surface. This makes it possible to perform some kinds of analyses without removing the sample from the surface. These include electron or optical microscopic examinations of the particles, counting of α-particle emissions without excessive α-particle absorption within the filters, and x-ray fluorescence analysis of elements with low-energy x-ray emissions.

Thermal Precipitation   Particles can be extracted from a sample stream by thermophoresis in a device known as a thermal precipitator. Prior to the mid-1950's, these were widely used in Europe to collect particles for microscopic determinations of particle number concentration and size distribution. A heated wire causes an unequal bombardment of particles by gas molecules, creating a net migration of the particles away from the wire and, in effect, a dust-free zone around the wire. When the temperature is high enough and the sampling flow-rate is low enough, the diameter of the dust-free zone can be larger than the width of the flow channel. In this situation, essentially all of the particles are collected on a linear trace, as illustrated in Figure 9–20.

Electrostatic Precipitation   When an airstream containing particles carrying electrical charges passes through a flow channel having a potential

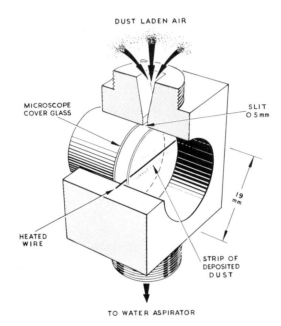

DUST LADEN AIR

MICROSCOPE
COVER GLASS

SLIT
0 5 mm

HEATED
WIRE

STRIP OF
DEPOSITED
DUST

1 9
mm

TO WATER ASPIRATOR

Fig. 9–20   Sampling head of a thermal precipitator.

From: ACGIH, *Air Sampling Instruments*, 2nd Ed. American Conference of Governmental Industrial Hygienists, Cincinnati, Ohio, 1962.

gradient normal to the flow, the charged particles will be deflected toward the electrode closer to ground potential. The efficiency of collection increases with the number of charges on the particle and the potential gradient. Very high collection efficiencies can be achieved in electrostatic precipitator samplers.

Most precipitator samplers utilize a high-voltage corona discharge to generate unipolar air ions which stream toward the grounded electrode from the small area of corona glow along the corona wire or the point of emission. These air ions attach to small particles passing through, elevating their charge level by the process known as diffusion charging. The particles larger than about 2 μm are also charged, by a process known as field or bombardment charging, in proportion to the potential gradient between the corona source and the grounded electrode. Once charged, the particles are accelerated toward the grounded electrode, which also serves as the collection surface.

Electrostatic-precipitator samplers of the type illustrated in Figure

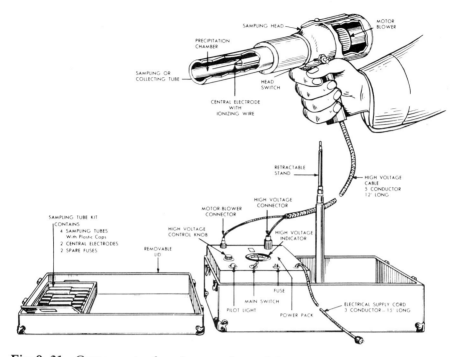

Fig. 9–21   Components of an electrostatic precipitator.

From: ACGIH. *Air Sampling Instruments*, 3rd Ed. American Conference of Governmental Industrial Hygienists, Cincinnati, Ohio, 1966.

9–21 were widely used for particle sampling in the 1930's and 1940's because of their high collection efficiency and their low and constant flow resistance. Their use was limited by the precautions needed to avoid unintended high-voltage discharges. Also, they could not be used in atmospheres containing combustible vapors. With the development of improved filters for air sampling which were simpler to use and much less expensive, electrostatic precipitation has since become limited to special sampling applications.

SPECIAL CONSIDERATIONS IN AMBIENT AIR SAMPLING

Ambient air concentrations of chemical contaminants are greatly influenced by meteorological factors, and their interpretation is, therefore, facilitated by knowledge of wind speed and direction, mixing depth, temperature profile, humidity, etc. It is generally desirable to have appropriate weather instrumentation at each sampling site.

As indicated previously, the selection of sampling sites requires a compromise between an ideal of a relatively large number of sites at strategically located positions and practical limitations imposed by fiscal constraints, availability of sites secure from malicious mischief, accessibility for sample changing and maintenance, and availability of power.

There are some special considerations in the selection of instrumentation for ambient air sampling. For airborne gases and vapors, the trend in recent years has been to use direct-reading continuous instruments having rapid-response gas-phase sensing, with each instrument being specific for a particular contaminant. Direct-reading instruments have also been favored for measurement of particle size distributions. For aerosol composition determinations, the trend has been toward sample collection in discrete aerodynamic size fractions. A size cut is generally made at about 2 μm diameter, on the basis that the smaller "accumulation mode" particles derive primarily from gaseous precursors, while the larger particles derive from mechanical processes and were emitted as discrete particles.

SPECIAL CONSIDERATIONS IN OCCUPATIONAL HYGIENE SAMPLING

Air sampling in occupational environments is generally much more highly focused than ambient air sampling. It is usually designed to determine exposure levels for specific individuals or groups of individuals engaged in a common activity. It is also concerned with only one or, at most, a limited number of specific air contaminants.

There are two basic approaches available for determining an individual's average exposure. The more traditional one involves breaking up the workday into a number of specific subfractions during which the air concentration is reasonably constant or reproducible. Samples are then collected in sufficient numbers in each activity interval to permit a reliable characterization of the exposure in that activity. The product of the exposure concentration at a given activity and the time devoted to that activity during each workday represents the contribution of that activity toward the overall exposure. The time-weighted average exposure is the sum of all the products of concentration and time, divided by the total daily work time. A representative determination of a time-weighted average occupational exposure is illustrated in Table 9–3.

An alternate approach to the determination of average daily exposure became available in the 1960's, with the development of lightweight, self-contained samplers which could be worn by individual workers without restricting their mobility. The sampling rates are low, but the sampled

Table 9.3 Sample Determination of a Time-Weighted Average Occupational Exposure

Given: Workers perform three different operations which involve potential exposure to airborne lead (Pb) during each workday. Breathing Zone (BZ) air samples were collected at each of the operations, and general air (GA) samples were collected which represent the workers' exposure during the balance of the workday.

Sampling rate for all samplers:	15 l./min
Sampling interval for BZ samples:	10 min
Sampling interval for GA samples:	60 min
Sample collections: Operation #1:	20.6, 24.3, and 18.2 μg
Operation #2:	35.2, 33.5, and 39.1 μg
Operation #3:	6.5, 9.7, and 7.8 μg
General air:	3.7, 1.9, and 5.1 μg
Working time/day: Operation #1:	60 min
Operation #2:	90 min
Operation #3:	90 min
General air:	4 hr (240 min)

Calculation:

Operation	Av. sample (μg Pb)	Sampled volume (liters)	Pb concentration (C) (μg/m³)	Exposure time (t) (min)	$C \times t$ (μg-min/m³)
1	21.0	150	140.2	60	8,413
2	35.9	150	239.6	90	21,560
3	8.0	150	53.3	90	4,800
GA	3.6	900	4.0	240	951
			Totals:	480	35,724

Time-weighted average exposure = $\dfrac{35,724}{480}$ = 74.4 μg/m³

Note: Threshold limit value (for 1977) = 150 μg/m³, so the workers in this case are not overexposed. However, if the workers spent more than 300 min per day at Operation #2, their exposure would exceed the TLV.

volume accumulated over a workday is usually large enough to permit an accurate determination of the average exposure for the day. Although there are fewer samples to analyze using this approach, it provides no information on the peak levels of exposure, or which of the activities during the day may have accounted for the bulk of the overall exposure.

The trend in recent years has been toward a greater reliance on personal samplers for routine monitoring of exposures. In sampling airborne particles, there has been a growing utilization of size-selective samplers, where the particles are aerodynamically separated into sample fractions on the basis of their presumed fate in the respiratory tract. The largest application of these is for dusts which produce pneumoconioses following deposition in the alveolar region of the lungs. The dust which penetrates to this region is simulated by the second stage of a two-stage sampler having a first-stage collector removing that fraction which would be deposited in

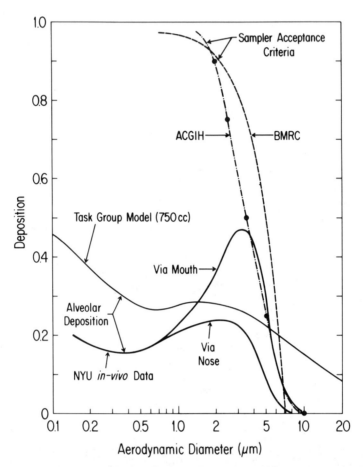

Fig. 9–22 A comparison of "respirable" dust-sampler acceptance criteria of the British Medical Research Council (BMRC) and the American Conference of Governmental Industrial Hygienists (ACGIH) with alveolar deposition in humans according to the International Commission on Radiological Protection (ICRP) Model and median human in vivo data in studies conducted at New York University.

From: Lippmann, M. Regional Deposition of Particles in the Human Respiratory Tract, pp. 213–232 in *Handbook of Physiology* – Section 9 – Reactions to Environmental Agents. D. H. K. Lee, H. L. Falk, S. D. Murphy (eds.). American Physiological Society, Bethesda, Md., 1977.

the head and tracheobronchial tree, and a second stage which collects all of the particles penetrating the first stage. The overall size-penetration characteristics specified for such two-stage samplers are illustrated in Figure 9–22.

SPECIAL CONSIDERATIONS IN SOURCE SAMPLING

The air flowing in discharge ducts and stacks generally has much higher concentrations of contaminants than the ambient air or workroom air. As a result, it is important to use sampling substrates that (a) can retain relatively large sample masses without losses due to resuspension in the sample stream and (b) do not exhibit changes in sampling rate due to clogging.

When the flow channel is large and/or where there are branch entries or elbows near the sampling port, the flow pattern and contaminant concentration across the channel are likely to be far from uniform. Thus, in order to determine the average concentration or total mass discharge rate, it is necessary to measure both the flowrate and concentration at numerous points across the channel. Such a series of measurements is known as a traverse; standardized traverse locations for circular and rectangular cross sections are available, and are shown in Figure 9–23.

It is important that the stream drawn into the sampler probe at each traverse point be representative of the stream at that point. No special precautions are needed when sampling most gases, or particles with diameters smaller than about 2μm. However, for larger particles moving in high-velocity airstreams, the particle momentum may lead to a biased sample. The bias may be either for or against the larger particles, depending on the velocity and orientation of the sampling probe with respect to the stream flow. Representative samples of aerosols containing larger particles can only be taken using the technique of isokinetic sampling, where the probe inlet faces into the airstream, and the velocity in the probe is equivalent to that of the stream flowing past it. Isokinetic sampling is illustrated in Figure 9–24.

In some stacks, such as vents for hot industrial processes or furnace discharges, the discharge stream may be much warmer than ambient temperature, and may also be high in moisture content. When hot, moist air passes through a tube whose wall temperature is lower than the dew point, the moisture will condense on the inside walls of the tube. If such condensation takes place before the stream reaches the sample collector, some or all of the sample can be lost. Insulated or heated sampling lines may be needed to avoid such complications.

When the stream is hot, it may also be difficult to obtain an accurate

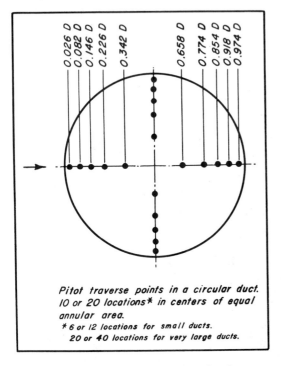

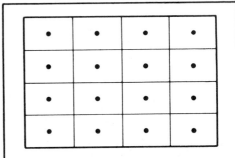

Fig. 9–23 Traverse points for flow measurements in ducts.

From: ACGIH. *Industrial Ventilation: A Manual of Recommended Practice,* 15th Ed. American Conference of Governmental Industrial Hygienists, Lansing, Mich., 1978.

measurement of the sampling rate or sampled volume because of the change in air density with temperature and humidity. Elevated temperatures may also enhance chemical reactions between gas-phase constituents and the walls of the sampling line, resulting in line losses.

Water Sampling

The evaluation of water-quality parameters requires characterization of both the quantity of flow in the waterway as well as its contaminant load. Since water generally is confined to pipes, channels, and defined streams, it is possible to establish mass flowrates at critical sampling points. Total waste load is then obtained by multiplying mass flowrate and the concentration of contaminant. This knowledge should help to identify both source locations and strengths for contaminants, and also the effectiveness of installed controls.

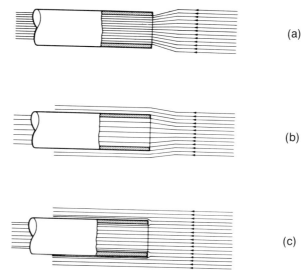

(a)

(b)

(c)

Fig. 9–24 Various cases of aerosol sampling. (a) Isokinetic sampling. (b) Flow-rate less than isokinetic rate. Flow streamlines diverge outward near the edge of the probe, causing larger particles to diverge into probe, and smaller particles to follow streamlines; thus, sample has too many large particles. (c) Flowrate greater than isokinetic rate. Streamlines diverge inward; sample has too few large particles.

From: Crawford, M. *Air Pollution Control Theory*. New York: McGraw Hill, 1976. Used with permission of McGraw-Hill Book Co.

SAMPLE COLLECTORS

GRAB SAMPLING

A grab sample of water is usually obtained with a very simple instrument, such as an empty jar or a bottle with a tight-fitting removable cap. The sample of water containing the contaminants of interest has a volume equal to the internal volume of the bottle, and the sample is collected within a short time interval. The sample may be obtained at the surface, or at various depths by use of weighted bottles which are opened and closed at the selected collection depth.

INTERMITTENT SAMPLING

The temporal pattern of contaminant levels can be determined by analyzing a series of manually collected grab samples. However, since it is usually less expensive to automate such a simple task, it is more common to use devices which collect samples periodically according to a preselected

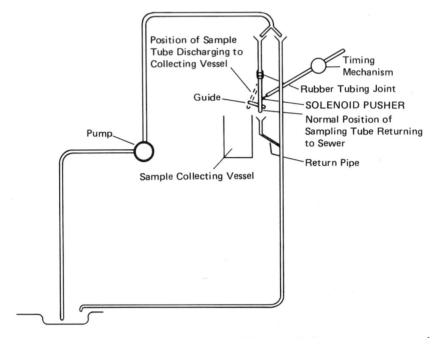

Fig. 9–25 An intermittent water sampler. This type is known as a pump-solenoid-diverter-timer sampler.

schedule. A series of individual samples can be collected, or a composite sample may be preferred, representing a mixture of equal-volume increments collected throughout a day or duty-cycle. One type of intermittent water sampler is illustrated in Figure 9–25.

CONTINUOUS SAMPLING

Intermittent sampling at fixed time intervals can miss surges in contaminant discharge, and, therefore, it may be desirable to draw the sample continuously. Continuous samplers are generally operated at a very low flowrate to avoid processing larger volumes than those needed for the laboratory analyses. A very simple continuous sampling system is shown in Figure 9–26.

PROPORTIONAL SAMPLING

Samples of constant volume collected on a fixed time schedule and samples collected continuously at a constant sampling rate can be used to determine the flux of contaminant directly only when the stream being sampled moves at a constant rate. When the stream flow varies, both the

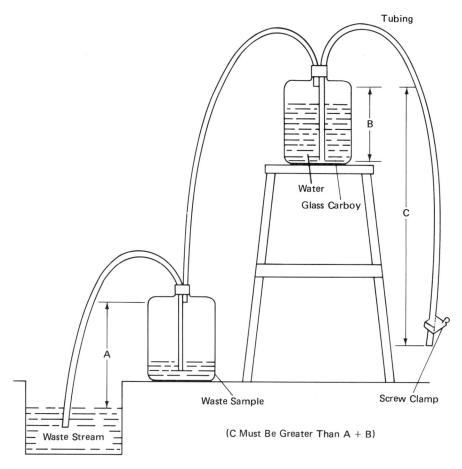

Fig. 9–26   A simple continuous sampler. As water drains from the upper bottle, a vacuum is created which siphons stream water into the lower bottle.

flowrate and the concentration must be determined in order to obtain the average or total mass flux of contaminant. Alternatively, the samples can be collected at variable rates which are proportionate to the flow.

There are two basic types of proportional samplers. One type collects a definite volume at irregular time intervals; the other collects a variable volume at equally spaced time intervals. Both are flow dependent: in one, flow dictates the time interval; in the other, flow regulates the sample volume. In most of these, the pressure differential needed to extract the sample is provided by the moving stream itself, ensuring the required propor-

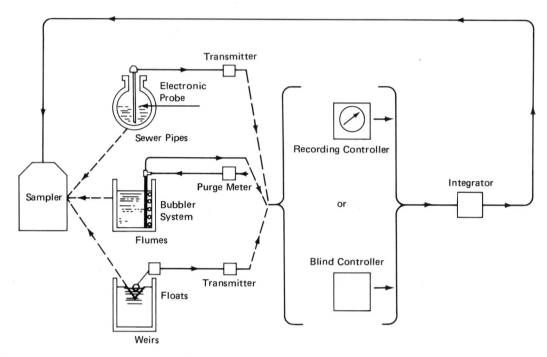

Fig. 9–27   "Flow-proportional" sample-control systems.

tionality. A schematic diagram of a proportional-sampler control system is presented in Figure 9–27.

A variety of approaches have been used in flow proportional sampling. In some, a constant sampling flow is pumped through a pipe. After a predetermined volume has passed through the flowmeter, a diverter is activated and a sample is taken. An air-lift automatic sampler can be used to obtain samples when a pump cannot be used. When the compressed air supply is shut off (the air-control valve is connected to a timing relay from the flow-measuring device), a spring in the sampler raises a piston (normally kept closed by the air) which opens an inlet so that water enters the sampler and goes to the sample container. The air valve is then opened and the piston is forced down, closing the inlet. The air passes through the air-escape port into the main chamber of the sampler and forces the liquid up the sample line into the collecting container. The cycle is then repeated.

In a vacuum-type sampler, which is very often used in water with high suspended-solids content or corrosive properties, a signal from a flow-meter activates the vacuum system, which lifts liquid through a suction line into the sample chamber. When the chamber is filled, the vacuum line is automatically closed. The pump then turns off, and the sample is drawn into the sample container. A float check prevents any liquid from reaching the pump.

Another type of automatic sampler consists of a cup or dipper on a chain. When activated by a signal, the cup is carried down through the stream and returns filled with a sample, which is emptied into a container.

In theory, most automatic-type samplers may be connected to flow-measuring devices in order to obtain proportionate samples.

EXTRACTIVE SAMPLING

Extractive techniques have many advantages for water sampling, corresponding to those previously discussed for air sampling. The samplers are much lighter, more compact, and are easier to preserve and ship, and they allow the concentration of contaminants from large volumes of water in a practical manner.

Unfortunately, effective and reliable extractive techniques have not been developed for a large number of water contaminants of interest, including many of the soluble gases and salts, and most colloidal solids.

Effective extractive techniques are available for suspended solids, many organics, and some of the toxic metallic ions. Membrane and glass-fiber filters are widely used for collecting suspended solids. In liquid filtration, sieving is an important mechanism of collection, and a major fraction of the particles smaller than the void or pore size can penetrate the filter. Membrane filters are available with pore sizes as small as 0.01 μm, but the smaller-pore-size filters have large pressure drops and, therefore, have very low flow capacity, limiting the sample size that can be drawn through them.

Adsorption is the most widely used extractive sampling technique for the collection of water-soluble organic material. Activated carbon and various organic resins (e.g., XAD, Tenax) are commonly used in pipes or columns of varying capacities and configurations through which the water percolates.

Ion exchange is a widely used technique for extractive sampling of contaminants in ionic form; it is applicable to both organic and inorganic contaminants. This technique involves the reversible transfer of ions between the water solution and a solid material capable of binding the ions.

The resins used in ion exchange are graded powders or beads of porous compounds, packed into columns.

A number of materials have ion-exchange properties. Among the solid inorganic materials commonly used are natural aluminum silicate minerals (zeolites); hydrous TiO_2 or ZrO_2; metal phosphates, e.g., zirconium phosphate, microcrystalline ammonium molybdophosphate; oxides of Al(III), Si(IV), Fe(III), and Mn(IV). The solid organic resins include cellulose, lignin, cation-exchange resins such as natural clay, chelating resins (with amine carboxylates as functional groups), and anion resins (usually in OH^- or Cl^- forms). The selectivity for specific ions depends largely upon the chemical structure of the resin and its physical configuration (e.g., degree of cross-linking).

SPECIAL CONSIDERATIONS IN STREAM SAMPLING

It is especially difficult to collect representative samples of the water in a free-flowing stream or river. The velocity profile in a stream is very variable, depending as it does on the cross section for flow, the volumetric rate of flow, the disturbances introduced upstream by bends in the stream, tributary streams, surface winds, etc. The stream flow varies in the short term with the recent history of precipitation, and in the longer term with seasonal variations in runoff. Variations in flow affect both the levels of contaminants attributable to surface runoff and the resuspension of contaminants from the stream sediments.

Since no one point in the stream cross section can be considered representative of the overall stream, it follows that contaminant flux can only be determined accurately from an analysis of samples collected at a series of representative locations across the flow cross section. Unfortunately, the appropriate traverse locations are not as easily defined as those for the circular and rectangular cross sections of Figure 9–23; it may be necessary to sample at only one or a few points and to accept the uncertainties introduced by the circumstances. Although, ideally, the sample should be taken from mid-stream where the velocity is greatest and the possibility that any solids have settled is minimal, in practice most samples are collected at the edge of the stream and near the upper surface. One potential error introduced by such a practice is illustrated in Figure 9–28.

SPECIAL CONSIDERATIONS IN EFFLUENT SAMPLING

The collection of representative samples of effluent wastewaters is relatively simple in comparison with sampling of process lines or free-flowing streams. In most cases, the number of outfalls of discharge points will be

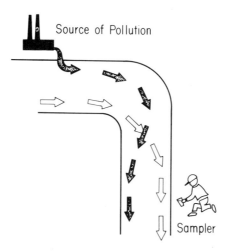

Fig. 9–28 An illustration of a potential problem encountered by sampling near the edge of a stream.

Fig. 9–29 Flow variations of water and wastewater and variation in the strength of wastewater.

From: Fair, G. M., Geyer, J. C. and Okun, D. A. *Water and Wastewater Treatment,* vol. 2. Water Purification and Wastewater Treatment and Disposal. New York: John Wiley and Sons, 1968.

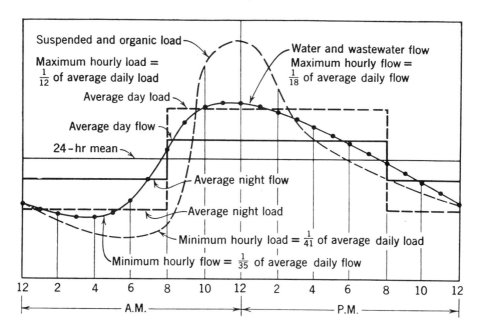

limited, their volumetric rate of flow will generally be known, or at least definable, and the flow will be relatively well mixed. The major variables will usually be in the temporal aspects of the waste discharges. Industrial wastes may be expected to vary greatly in both composition and volume with time, reflecting the batch nature of many production operations, the periodicity of routine cleanup and maintenance operations, and the consequences of accidental, unanticipated releases.

The first step in designing an effluent sampling program is to define all of the significant discharge points. The next step is to determine their likely variability in discharge, and to select the best means of characterizing the discharges from each. If the total amounts discharged are of primary concern, it may be satisfactory to continuously collect an integral sample whose volume is proportional to the overall flow. On the other hand, if the objectives include identifying the contributions of the component sources to the overall discharge, it may be necessary to collect a series of effluent samples at appropriate times throughout the production or other cycle. Some of these considerations are illustrated in Figure 9–29.

SPECIAL CONSIDERATIONS IN SAMPLE HANDLING
AND PRESERVATION

Some special precautions are necessary in the handling of water samples. Water which is contaminated with chemicals is generally contaminated with microorganisms as well. Since the metabolism of these microorganisms may add or subtract materials of interest between the time of collection and the subsequent laboratory analyses, water samples are often refrigerated to minimize such complicating effects.

Many other time-dependent changes may occur in water samples after their collection. These may include changes in pH, loss of some dissolved gases and absorption of others from the atmosphere, change in valence state of metal ions via oxidation-reduction reactions, and precipitation of metal cations due to chemical reactions. Procedures for sample preservation for the analysis of various contaminants are discussed in detail in *Standard Methods* (5). It is important that the procedures used to stabilize one contaminant of interest do not affect the determination of another.

Some precautions which may be needed in a water-sampling train to preserve contaminants of interest are illustrated in Figure 9–30.

Sampling of Biota, Sediments, and Soil

A complete evaluation of the extent and significance of environmental contamination by a given material will frequently require the analysis of

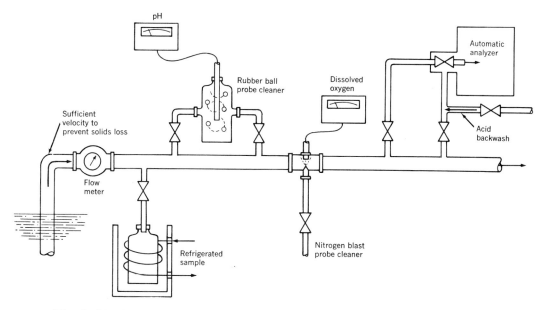

Fig. 9–30 Examples of possible precautions needed in a sampling train to prevent changes in composition in the water sample.

From: Ripley, K. D. Monitoring Industrial Effluents. *Chemical Engineering 79* (No. 10):119–122, 1972. ©McGraw-Hill. Used with permission of the publisher.

soil and aquatic sediments and biota, since they can act as temporary or permanent sinks for the material. As long as the contaminant is available for ingestion by mobile life forms, absorption by roots or leaves of plants, or chemical leaching into the surrounding medium, it retains a potential for adverse effects.

The collection of sediment samples is generally much more difficult than collection of representative water samples; the former must be transported through the overlying water without loss or exchange of material with the water. Dissolved gases and reduced chemicals may be altered or released during transport through more oxygenated waters to the surface.

One popular sampling technique is collection of a flocculent sample with a dredge. However, there may be considerable loss of material to the water. A better method utilizes a core-sampling device. This technique minimizes both material loss to the water and oxidation of reduced materials during sample ascent. Furthermore, an intact core sample may be used to study the contaminant deposition pattern over a period of time, such as several seasons.

Contaminant evaluations are easier when the pathways within the environment are well established. In such cases, a sample from an established indicator organism may provide a sufficient index of the extent of the contamination. When less is known about contaminant behavior, a larger range of materials and a greater number of samples will be required.

An indicator organism is a plant, animal, or microbe which provides an index of environmental conditions by its presence, absence, or specific characteristics. The narrower the tolerance of the indicator organism for a specific condition, the greater is the accuracy in describing ecologic conditions. For example, presence of an aquatic organism which only survives in a pH from 7.0 to 8.0 is an accurate field monitor of mild alkalinity in the waterway.

Indicator plants may be used in evaluating the extent of chemical contamination of the ambient air by characteristic pathological changes. For example, intercostal markings on the leaves of violets indicate high SO_2 levels; banding on snowstorm petunia leaves indicate aldehydes; white flecks on tobacco leaves indicate high ozone levels (6).

Other plants may accumulate chemicals, and are thus an index of ambient levels. For example, samples of Spanish Moss growing on trees downwind of highways may be collected in order to determine their Pb content, on the basis that this content is proportional to traffic density.

Some of the best biological indicators of chemical contamination can be found in the aquatic environment. Some filter-feeding crustacea extract organic contaminants and metallic ions very efficiently from the water. Aquatic plants may also greatly concentrate some of the trace metals which accumulate in the bottom sediment of streams and lakes. Certain algae, such as the green and blue-green types, are indicators of high-nutrient content in waterways.

Samples of aquatic biota may be obtained with nets, dredges, or an artificial substrate, such as glass slides for microorganisms. Guidance on the sampling of rooted aquatic plants, algae, phytoplankton, crustacea, finned fish, and bottom sediments for the evaluation of the extent of chemical contamination of the aquatic environment can be found in Cross (7).

Certain plants on land are associated with the presence of particular chemicals in the soil. For example, the growth of Eastern Colombine in the eastern United States is indicative of high calcium carbonate content, and plants of the genus *Astragalus* in the western United States grow in association with selenium in the soil.

Food Sampling

Concern about the extent of chemical contamination of food may lead to several kinds of sampling programs. One involves identification of specific batches of contaminated, adulterated, or spoiled food products which may reach the market. The random- and spot-sampling programs of the FDA and various state agencies fall into this category. When the samples indicate that the quality of the food is substandard, the batch of food product represented by the sample will be destroyed or diverted to another usage where its quality would be acceptable.

A second type of program focuses on the definition of pathways and accumulation rates for materials of interest which do not present immediate hazards but may be worthy of concern and continued observation. One example was the worldwide food-sampling program and analysis for ^{90}Sr content during and after the period of atmospheric testing of thermonuclear weapons in the 1950's and 1960's. Another is the continuing monitoring of pesticides in foods.

In much of this large-scale food sampling, the extent of the problem is determined from a relatively limited number of composite samples rather than from large numbers of analyses of individual foods. The approach is generally known as market-basket sampling. In it, a representative diet is selected and the food products needed to prepare this diet are purchased from regular retail outlets. The foods are prepared for cooking as they would be in the home, i.e., wrappings, bones, trimmings, etc., are discarded. The parts that would be consumed are then analyzed for their content of the contaminant of interest.

Market-basket sampling has several important advantages for routine monitoring of overall contaminant levels and their trends. These are: (a) fewer samples need to be analyzed; (b) the samples are relatively large, avoiding problems of too limited analytic sensitivity; and (c) the results are directly relatable to an average population exposure. On the other hand, there are important limitations: (a) there is no information on the major sources within the overall diet and (b) there is no information on maximum levels of exposure within the population attributable to unusual dietary patterns.

Biological Sampling .

The sampling and analysis of ambient air, drinking water, and foods can indicate the contaminants to which people are exposed. However, environmental data cannot define the fractions retained after inhalation or ingestion either initially or at later times. For most toxicants, the amount retained is very variable among individuals, and with time in a given individual. Furthermore, the basic metabolic pathways are poorly defined for many contaminants. For these reasons, it is usually a good practice to use biological samples as supplements to environmental samples in evaluating the health significance of environmental contaminants.

Biological samples have, in the past, been most widely used in occupational health evaluations, as discussed in Chapter 8, and Biological Limit Values have been suggested for various chemicals, their metabolites, or the alterations they produce in biological constituents or physiological functions. The kinds and sizes of the samples needed for such analyses depend on the contaminant and its metabolism, and the sensitivity of the assay.

The easiest kinds of samples to collect are urine and exhaled air. Hair and fingernails may also be useful and can also be collected readily in many cases. It is usually more difficult to obtain the cooperation needed for the collection of blood or fecal samples, but they frequently are needed and can be successfully collected. Table 9–4 lists some biological indices of contaminant exposure.

An ever-present problem in the collection and handling of all of these types of samples is contamination, and extreme caution must be exercised in the selection and handling of the sample container. The material to be analyzed must not be extractable from the container itself, and anyone handling the sample must be aware of all the possible means of introducing artificial material, and must avoid them. For example, industrial workers usually have much more of the chemical of interest on their hands than in their urine and, if they haven't first washed thoroughly, can easily contaminate a urine bottle while filling it.

Another problem is the collection of a representative sample. Most biological samples are short-interval (grab) samples containing a material of interest whose body concentration varies considerably with time. Proper interpretation may depend on the availability of supplementary information, such as the times of exposure relative to the times of sampling, and the characteristic metabolic time constants.

For some contaminants, especially radioactive materials, the overall retention in the body and within various organs can sometimes be determined by noninvasive, external measurements of emitted radiation.

Estimates of individual and population burdens of contaminants can also be made by analysis of tissues taken at autopsy from accident victims.

Table 9.4   Some Biological Samples Used as Exposure Indices

Contaminant	Specimen for evaluation	Chemical indicator
Aldrin; dieldrin	blood	dieldrin
Arsenic	urine, blood, hair	arsenic
Benzene	urine	phenol
	blood, exhaled breath	benzene
Cadmium	hair, nails	cadmium
Carbon monoxide	blood	carboxyhemoglobin
Cyanide	blood	cyanmethemoglobin
DDT	urine	DDA (dichlorodi-phenylacetic acid)
	blood	DDE (dichlorodiphenyl dichloroethylene)
Fluoride	urine, hair	fluoride
Hydrogen sulfide	blood	sulfhemoglobin
Lead	hair, blood	lead
	urine	lead, delta amino-levulinic acid
Nitrite	blood	methemoglobin
Organic mercury	urine, hair	mercury
Parathion	urine	p-nitrophenol
Polycyclic aromatic hydrocarbons	urine	analysis for parent compound
Vinyl chloride	exhaled breath	vinyl chloride

However, the great difficulties associated with obtaining adequate-sized samples of this type generally precludes the acquisition of a significant body of data through this approach.

MEASUREMENT OF CONTAMINANT CONCENTRATIONS

Concentrations of contaminants may be measured by sample collection and subsequent laboratory analysis, or by use of direct-reading instrumentation in the field. Many of the instruments used in the field are merely more compact and/or rugged versions of those used in the laboratory.

Laboratory Analysis of Environmental Samples

All of the samples of air contaminants, water, soil, biota, food, and biological materials brought to the analytical facility can be analyzed by similar techniques. While they may differ considerably in their nature, such samples generally have a number of common characteristics. These include: (a) the likely presence of a variety of co-contaminants at concentrations equal to or greater than the contaminant of interest and (b) the presence of a collection substrate or matrix from which the contaminant must be extracted prior to analysis. Thus, the first task in a laboratory evaluation is generally to separate the contaminant of interest from its co-contaminants and matrix, so that it can be further analyzed unambiguously.

SEPARATION AND PRECONCENTRATION

Separation is often combined with preconcentration procedures. Preconcentration improves the sensitivity of the analytical method and increases the precision and accuracy of analysis. The necessity to concentrate prior to analysis is dependent upon the specific analytical method which is to be used and the lower limit of detection of the procedure. For example, organic constituents of water are sufficiently numerous and usually present at such low concentrations that they require preconcentration and separation. Many inorganics are present at concentrations above the sensitivity limits of analytical methods; however, numerous interferences can arise from other agents which should be removed prior to analysis. Many of the current analytical methods are subject to well-characterized interferences, which are noted in standardized procedures for contaminant analysis (5,9,10).

Contaminants collected by extractive sampling techniques must be separated from the collection matrix. Following adsorption on activated carbon, the organic chemicals are extracted from the carbon using a chloroform wash, sometimes followed by methanol. Following evaporation of the solvent, the weights of the organic extract residues are expressed in units of $\mu g/l$. as carbon-chloroform-extract (CCE) and carbon-alcohol-extract (CAE). These extracts may be further separated into groups, e.g., bases, weak acids, strong acids, etc., via differential solubility techniques.

Recovery of contaminants from ion-exchange resins is performed by elution with a solvent appropriate for the particular material being analyzed. Ion exchange is a particularly useful separation technique for selective removal of ions which may interfere with the analysis of other ions.

A number of specific separation techniques are used in contaminant analyses. These are presented in Table 9–5. Often, more than one procedure is required prior to final analysis. Many of these separation techniques can also be used for the preconcentration of contaminants.

Table 9.5 Some Common Separation Techniques

Technique	Principle	Typical separation applications
Coprecipitation	Removal of an ion from solution via adsorption on a carrier precipitate	trace metals
Liquid-liquid extraction (solvent extraction)	Physical separation based upon the selective distribution of a substance in two immiscible solvents (e.g., water and a water-immiscible organic solvent)	many organic and inorganic compounds, especially water-immiscible organics and metal ions
Freeze concentration	Water in sample is frozen, yielding pure ice crystals and leaving water-soluble impurities in a liquid phase with a reduced volume	wide range of inorganic and organic solutes
Adsorptive bubble separation	Passage of gas bubbles through a solution or suspension in a vertical column; material of interest may be adsorbed at bubble-solution interface or onto bubble surface	surface active agents; colloids and some other particles
Distillation, evaporation, sublimation	Separation from water via removal of a component as a gas or vapor	organic compounds; dissolved inorganic gases
Centrifugation	Decrease time required for particles to sediment by increasing the gravitational forces affecting them	macromolecules; suspended and colloidal particles
Chromatography	Based upon the differential migration of substrates due to their selective retention in a fixed phase while subjected to movement by a flowing bulk phase (gas or liquid)	wide applicability; generally used for organics
Gel filtration	Lodging of colloidal particles and high-molecular-weight organic compounds in pores of a cross-linked gel packed in columns	colloids; organics
Chelation separation	Complexing of metals by certain reagents	metals
Liquid-anion exchange (ion-pair formation)	Separation of metal complexes by formation of an ion pair with an onium cation (e.g., trialkyl ammonium) formed by use of a strongly basic solvent	metals
Dry ashing	Combustion of material (e.g., filter, biological tissues) and recovery of contaminants of interest in unburned residue or in gaseous emissions	metals
Wet ashing	Oxidation of material in acid to separate contaminant of interest from an organic matrix	metals

Table 9.6 Some Common Methods of Contaminant Analysis

Method	Principle of operation	Instrumentation
1. Gravimetric	Isolation of component of interest in the form of one of its compounds which shows insolubility under test conditions and has a known chemical composition; weigh the product	Conventional lab equipment
2. Titrimetric (e.g., acid-base; oxidation-reduction; precipitation)	Reaction of material in predictable manner with a standard solution of known concentration	Conventional lab equipment
3. Absorption Spectrophotometry		
a. Visible	Measure of selective absorption or transmission of visible light, usually following reaction with a reagent	Colorimeter; Spectrophotometer
b. Ultraviolet	Absorption of UV light	UV spectrophotometer
c. Infrared	Absorption of IR (from heated filament)	IR spectrophotometer
4. Emission Spectroscopy		
a. Flame Photometry	Spectral-emission analysis following excitation by flame (arcs or sparks are also used)	Flame photometer; Spectrograph (arc)
b. X-ray Fluorescence spectrometry	Spectral-emission analysis following x-ray excitation	XRF spectrometer
c. Spectrofluorimetry	Reemission of radiation absorbed by dissolved molecules	Recording spectrofluorimeter
d. Chemiluminescence	Emission of spectral radiation resulting from a chemical reaction	Chemiluminescence analyzer
5. Atomic Absorption Spectrophotometry	Absorption of radiation by free atoms in a vapor state	AA spectrometer
6. Chromatography (gas; liquid; thin-layer)	Differential migration of agents due to selective retention in a fixed phase while subjected to movement by a flowing bulk phase	Gas chromatograph, liquid chromatograph
7. Electrochemical		
a. Conductivity	Measurement of electrical conductivity following absorption of an agent through a solution	Conductivity meter
b. Anodic stripping voltammetry	Electrolysis of solution and electrodeposition followed by stripping at various potentials	DC-pulse polarograph
c. Coulometry	Quantitative electrochemical conversion from one oxidation state to another	Coulometer
d. Polarography	Electrolysis of solution and measurement of current-voltage relation	Polarograph
8. X-ray Diffraction	Recording of scattered X-rays from sample subject to X-ray beam	X-ray diffractometer

Sample[a]	Specificity[b]	Sensitivity[c]	Typical applications
SLG	Good	$1-10\ \mu g$	Numerous analyses for inorganic and organic materials
L	Good	$10^{-6}-10^{-7}$ M in solution	Numerous analyses, e.g., acidity, alkalinity, water hardness (Ca^{+2}, Mg^{+2}), Cl^-, S^{-2}, transition metals, DO
SLG	Fair	0.005 ppm	Metals (e.g., Fe, Cu); nutrients (NO_2^-, NO_3^-, PO_4^{-3}), NH_3, phenol, SO_2, NO_x, oxidant gases (O_3), COD, TOC (total organic carbon)
SLG	Fair	0.005 ppm	Organic compounds, some inorganic gases, e.g., SO_2, O_3, NO_2
SLG	Fair	1 ppm	Organic compounds, some inorganic gases, e.g., CO, CO_2, SO_2, NO_x, NH_3, O_3; TOC
SL	Good Excellent	0.001 – 0.1 ppm	Metals, halogens, phosphorus, sulfur (depending on specific method)
SL	Good	10 ppm	All elements having atomic numbers above 11
SL	Good	0.001 ppm	Organic compounds; some inorganic constituents
G	Excellent	ppb	O_3, NO_x
SL	Excellent	0.001 ppm	Most metals
LG (gas chr) SL (liquid chr) SL (thin-layer)	Excellent (gas) Good (liquid)	ppb– ppm (TLC)	Organic compounds, trace metals, anions
LG	Poor	0.01 ppm	Any gas which forms electrolytes in aqueous solution, e.g., SO_2, NO_2, Cl_2, H_2S, NH_3
L	Good	0.001 ppm	Some metals, e.g., Cd, Cu, Fe, Pb, Zn,
L	Good	1 ppm	SO_2, NO_2, O_3, H_2S, olefins, F^-, mercaptans
L	Good	0.1 ppm	Some organic compounds and some metals, e.g., Sb, As, Cd, Cr, Ni, Zn
S	Good	–	Solids of crystalline structure, e.g., asbestos, quartz

cont.

Table 9.6   cont.

Method	Principle of operation	Instrumentation
9. Microscopy (Polarizing; Fluorescence)	Optical analysis of enlarged images	Various types of microscopes
10. Mass Spectrometry	Determination of material by creation of ionic current via electrostatic or electromagnetic field and separation of the ions according to their mass. Generally used in conjunction with gas chromatographs	Mass spectrometer
11. Neutron Activation Analysis	Measurement of emitted ionizing radiation following production of radionuclides by neutron bombardment	–

[a.] Sample: Most usual form of sample suited to the specific method; S = solid; L = liquid; G = gas.
[b.] Specificity: This is an indication of the general ability of the method to measure the material in the presence of matrix interferences. The classifications are very broad and use of preconcentration or separation methods may increase specificity.
[c.] Sensitivity: Typical lower limits of detection are presented; these may differ depending upon the specific chemical analyzed and the background concentrations present.

ANALYTICAL INSTRUMENTATION

The full range of analytical instrumentation that can be used for the qualitative and quantitative analyses of environmental samples could not be adequately described in a whole volume, let alone a small part of this chapter. Thus, the reader is referred to several of the more concise descriptions in the standard references (5, 8, 9, 10) readily available in most technical libraries. An overview of some of the more commonly used techniques and their typical applications are presented in Table 9–6.

The selection of the best analytical techniques in a specific situation is often quite difficult. Factors to consider in any choice are: (a) any sample preprocessing required, (b) specificity with regard to the chemical being analyzed, (c) influence of the matrix, (d) amounts needed for determination of the chemical, (e) accuracy for the concentration ranges expected, and (f) time and cost.

Direct Measurement of Environmental Concentrations

As mentioned earlier, many instruments may be used for direct measurements of contaminants in the field. This discussion of equipment for the direct determination of air and water concentrations will be brief. It will

Sample[a]	Specificity[b]	Sensitivity[c]	Typical applications
S	Excellent	–	Nature and size of particles, e.g., fibers, dusts
SLG	Excellent	ppb	Organic compounds
SL	Excellent	ppb to ppm	Hg, As, Pb, Fe, and about 66 other elements which can become radioactive when bombarded with neutrons

be limited to an illustration of some of the more common or innovative approaches to the general task. For a more comprehensive description, the reader is once again referred to the references listed at the end of the chapter.

Bear in mind that similar instrumentation may be used for both water and air sampling. The determining factor is the specific state in which the sample must be introduced to the instrument.

AIRBORNE CONTAMINANTS

DIRECT-READING INSTRUMENTATION

Gases and Vapors   Any characteristic physical property of a chemical can be utilized in its detection and measurement. A variety of instrumental principles can be employed to maximize sensitivity, specificity, and precision. A widely used instrument for monitoring airborne carbon monoxide (CO) is based on the capacity of CO to absorb infrared radiation (IR). Thus, CO molecules in the path between an IR source and an IR sensor will reduce the incident flux at the sensor. A higher concentration or longer path length will increase the attenuation. Other gases and vapors present in relatively high concentrations in the air, including methane and water vapor, will also absorb IR. However, they do not absorb at the same wavelength as does the CO, and by using an IR beam of an appropriate wavelength, the instrument response can be limited to CO. It also follows

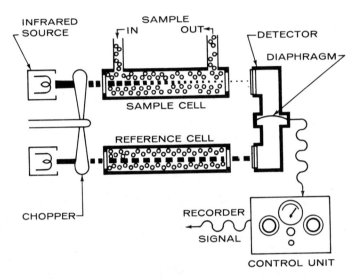

Fig. 9–31   Schematic of an infrared analyzer (Beckman Instruments).

From: ACGIH. *Air Sampling Instruments*, 3rd Ed. American Conference of Governmental Industrial Hygienists, Cincinnati, Ohio, 1966.

that the same basic instrument can be used to make specific measurements of the concentration of other airborne molecules by using appropriate wavelengths.

A schematic diagram of an IR instrument is shown in Figure 9–31. It uses both a sensing cell and a sealed reference cell, and the signal is proportional to the difference in the heat absorbed by the two detectors which, in turn, varies with the CO content of the sample cell. In this configuration, small variations in IR output by the source affect both detectors equally, and do not significantly affect the output signal.

Instruments have also been developed for airborne contaminants, such as mercury vapor and benzene, which strongly absorb ultraviolet light. More recently, instruments have become available which detect molecules which fluoresce upon excitation by a radiant source. In this case, the detector is tuned to the fluorescence frequency, rather than the source fre-

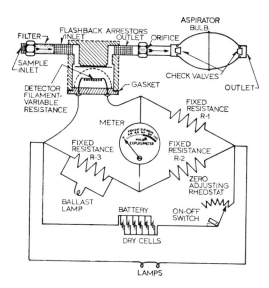

Fig. 9–32   Schematic of a combustible-gas indicator flow system with circuit diagram (Mine Safety Appliances, Co.).

From: ACGIH. *Air Sampling Instruments*, 3rd Ed. American Conference of Governmental Industrial Hygienists, Cincinnati, Ohio, 1966.

quency. Instruments of this type are currently being used for carbon monoxide and sulfur dioxide analysis.

Of still greater complexity are some relatively new instrument designs which analyze patterns in the radiant spectra of excited molecules rather than just total absorption over a given wavelength band. These second derivative spectrometers can be used to monitor a wide variety of gases and vapors, including sulfur dioxide, ammonia, nitrogen dioxide, and benzene.

Another approach to continuous measurement of gas and vapor contaminants is to react them with other chemicals in the stream, or at surfaces in contact with the stream, and to measure the rate of reaction or the presence of reaction products. One of the earliest applications of this approach, and one still widely used, is the combustible gas indicator. A gas stream is passed over a catalytic filament at which combustible vapors burn. The heat of combustion raises the temperature of the filament, reducing its electrical resistance. The change in resistance provides an indication of the concentration of combustible material. A schematic diagram

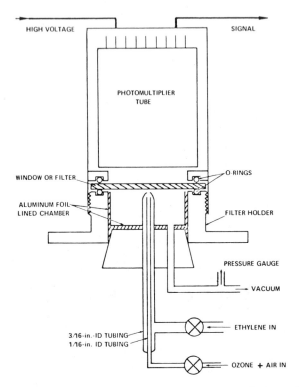

HIGH VOLTAGE

SIGNAL

PHOTOMULTIPLIER
TUBE

WINDOW OR FILTER

O-RINGS

ALUMINUM FOIL
LINED CHAMBER

FILTER HOLDER

PRESSURE GAUGE

VACUUM

3/16-in.-ID TUBING
1/16-in. ID TUBING

ETHYLENE IN

OZONE + AIR IN

Fig. 9–33 Schematic of ozone monitor utilizing ozone-ethylene chemiluminescent reaction and photometric detection.

From: ACGIH. *Air Sampling Instruments*, 5th Ed. American Conference of Governmental Industrial
Hygienists, Cincinnati, Ohio, 1978.

of this type of instrument, in which the catalytic filament is one leg of a
Wheatstone bridge, is shown in Figure 9–32.

A slightly more sophisticated approach, which has also been used for
many years, is utilized in the halide meter. It employs the "Beilstein" reaction, in which halogen atoms in contact with a copper element at high
temperature emit an intense green light. In the halide meter, halogen-
containing hydrocarbons are burned in a copper arc, and the intensity of
the spectral emission indicates the concentration of the vapor.

Neither the combustible gas meter nor the halide meter can provide
accurate or specific determinations when there are mixtures of air contaminants. The former responds to all combustible vapors, and its response
depends on the heat of combustion. The halide meter responds only to

halogens, but has a different degree of response for different halogen compounds.

Some of the newer instruments utilizing gas-phase conversions achieve specificity by employing highly individual chemical reactions. For example, the ozone monitor illustrated in Figure 9–33 utilizes the chemiluminescent reaction between ozone and ethylene. An excess of ethylene is used, and the luminescense is limited by the ozone concentration. None of the other airborne gases react with ethylene, so there are no significant interferences.

Aerosols   There are very few methods for measuring particle concentrations in airstreams without sample collection, and almost all of these utilize the light-scattering properties of the particles. Some measure the integral scatter of an aerosol. Since the amount of light scattered by each particle varies greatly with particle size, these instruments have little value unless the size distribution is known and constant. Other instruments measure the scattered light from individual particles as they pass through the sensing zone, and sort the pulses by size. Thus, they can accumulate number concentrations within defined size intervals for particles larger than about 0.3 μm diameter. Instruments of this type, such as the one illustrated in Figure 9–34, are widely used in industrial and ambient air studies. They are calibrated for reflective spherical particles, and, therefore, may not be accurate for size analyses of particles with other shapes and optical properties.

SAMPLE COLLECTION WITH ON-LINE MEASUREMENTS

In on-line systems, the analysis instrumentation is combined with the sampling instrumentation. A sample is collected as a grab, composite, or continuous sample, and then analyzed by a direct-reading instrument. Most sample-collecting instruments with direct-measurement capabilities combine one of the extractive-sampling techniques previously described with a simplified version of a laboratory-type analytical instrument.

Many of the older on-line instruments used wet chemical techniques following extraction of the air contaminant by a bubbler or gas washer. A widely used SO_2 instrument passed the scrubbing solution through an electrical conductivity cell. However, it was not specific for SO_2, since other ionizable contaminants in the air would also contribute to the conductivity. Other two-stage instruments used colorimetric reagents to collect the contaminants, and could give more specific analyses, provided that co-contaminants in the air did not produce or quench color changes at the wavelengths used in the built-in colorimeter.

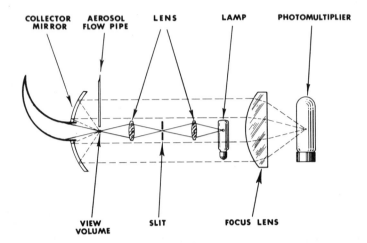

Fig. 9–34 Sketch of optical system of a light-scatter particle analyzer (Bausch and Lomb).

From: ACGIH. *Air Sampling Instruments*, 3rd Ed. American Conference of Governmental Industrial Hygienists, Cincinnati, Ohio, 1966.

Most of the newer instruments use dry collection techniques. For example, the gas chromatographic flame photometric analyzer illustrated in Figure 9–35 collects various sulfur gases simultaneously on a chromatographic column. When the sampled gases are eluted from the column into the flame photometric detector, they each give essentially the same signal per mole of sulfur. However, since each compound has a different retention time on the column, they give distinct peaks, and the quantity of each can thus be determined.

Perhaps the simplest example of sample collection with a rapid readout of concentration is that provided by detector tubes. Glass tubes of the type illustrated in Figure 9–36 are filled with adsorbing granules coated with a colorimetric chemical reagent which changes its color after reacting with a specific gas or compound class.

A specific volume of air, typically 100 cm³, is drawn through the tube from one end toward the other, usually with a hand-powered pump; the contaminant of interest reacts with the adsorbed chemical, changing its color. The capacity of the granules to take up the chemical is limited, so that the penetration of contaminant into the tube before its molecules are captured is dependent on concentration. The length of color change is, therefore, an index of the concentration. Detector tubes were developed for occupational health and safety evaluations, and are currently available

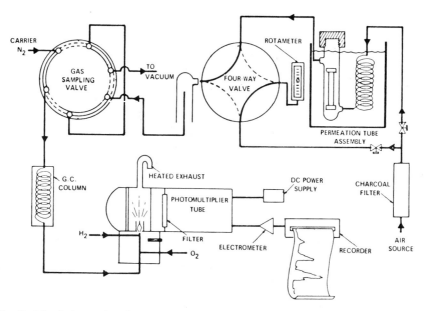

Fig. 9–35 Schematic of an automated gas chromatograph/flame photometric detector system for analysis of SO$_2$ and other sulfur compounds.

From: ACGIH. *Air Sampling Instruments,* 5th Ed. American Conference of Governmental Industrial Hygienists, Cincinnati, Ohio, 1978.

Fig. 9–36 Enlarged view of a detector tube for CO.

From: ACGIH. *Air Sampling Instruments,* 3rd Ed. American Conference of Governmental Industrial Hygienists, Cincinnati, Ohio, 1966.

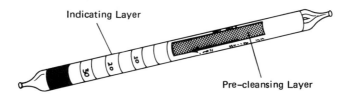

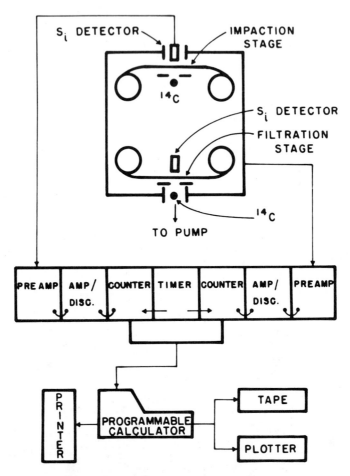

Fig. 9–37 Schematic diagram of the Two-Mass aerosol sampler, including the data acquisition and analysis system.

From: Macias, E. S. and Husar, R. B. A Review of Atmospheric Particulate Mass Measurement via the Beta Attenuation Technique, pp. 536–564 in *Fine Particles. Aerosol Generation, Measurement, Sampling and Analysis*, B. Y. H. Liu, (ed.). New York: Academic Press, 1976.

for about 200 gases and vapors. However, their accuracy is generally not very good; only 39 tube types for 14 gases have been demonstrated to have accuracies equal to or better than $\pm 25\%$, according to tests conducted by NIOSH.

Some detector tubes have been calibrated for long-period sampling using mechanical sampling pumps, and can, therefore, be used to determine lower airborne concentrations in industrial and community environments.

While there are many direct-reading instruments which are specific for gases and vapors, there are virtually none for airborne particles. There are, however, several nonspecific instruments which are used to measure particle mass concentrations. One example, the Two-Mass sampler, is illustrated in Figure 9–37.

The Two-Mass sampler collects particles on paper tapes in two size fractions. The particles larger than about 3.5 μm are collected in the first stage by inertial impaction, while the smaller particles are collected in the second stage by filtration. The attenuation of β-particles between the source and the radiation detector on the opposite side of the tape is measured both before and after the sample-collection interval. The increased attenuation, attributable to the collected sample, indicates the particulate mass collected, since β-attenuation per unit mass is almost the same for all elements except for hydrogen, and hydrogen is usually a minor mass constituent of particulate matter. The collection tapes can be advanced between samples to expose fresh collection surfaces, and to store the collected samples so that other analyses can be performed if desired.

WATERBORNE CONTAMINANTS

DIRECT-READING INSTRUMENTATION

Direct-reading instruments are commonly used for the measurement of water-quality parameters other than contaminant concentrations, such as temperature, pH, and dissolved oxygen. Examples of these instruments are presented in Table 9–7.

One of the more commonly used methods for the direct measurement of dissolved contaminants is the ion-selective electrode. A variety of electrodes are available which respond rather selectively to specific ions; their response depends upon the potential of a measuring electrode vs. a reference electrode. Among the contaminants for which selective electrodes exist are NH_4^+, CN^-, SO_4^{-2}, F^-, Na^+, and Ag^+. Spectrophotometric methods utilizing the IR- or UV-absorption characteristics of specific contaminants are also widely used. Some of their applications are for mea-

Table 9.7   Direct Measurements for Some Nonspecific Water-Quality Parameters

Parameter measured	Typical instrumentation	Principle of operation
pH	pH meter	Voltage difference between a measuring electrode (a glass tube having a special membrane which responds to H^+ at one end) and a reference electrode
DO	DO electrode or DO probe (gas membrane electrode)	Gas dissolved in the water diffuses through a semipermeable membrane into an electrochemical cell compartment, which is comprised of a sensing electrode, reference electrode, and supporting electrolyte. DO reacts at sensing electrode to produce a current flow
Turbidity	1. Turbidimeter	1. Measures amount of light passing directly through the water
	2. Nephelometer (low turbidity ranges)	2. Measures light-scatter in water
Temperature	1. Thermometer	1. Expansion and contraction of a liquid
	2. Thermocouple	2. Measures voltage generated at junction of two wires made of different metals
	3. Thermistor	3. Electrical resistance varies with temperature in special resistors

surement of total organic carbon and certain dissolved inorganic gases. Electrochemical methods, such as coulometry, potentiometry, and polarography, are used for the measurement of a wide range of metals in water.

SAMPLE COLLECTION WITH ON-LINE MEASUREMENTS
The water-quality instrumentation which measures a rate or product of reaction between a contaminant and a reagent is essentially the same as that described earlier for air contaminants, except that a reactor vessel replaces the scrubber, and the sampled stream contributes water as well as contaminant to the mixture in the reactor.

Quality Assurance

The accurate determination of the concentrations of trace contaminants in environmental media is no small or simple task, as should be evident from the preceding discussion. It should, therefore, not be surprising that many of the data in the literature are at least partially suspect. One unfortunate result is that there is a great deal of confusion and controversy about the status and significance of many chemicals in the environment.

Table 9.8 Some Recommended and Standard Methods Relating to Environmental Sampling and Instrument Calibration

Organization		Types of methods
American Conference of Governmental Industrial Hygienists	(ACGIH)	Analytic methods for air contaminants
American Industrial Hygiene Association	(AIHA)	Analytical guides
American National Standards Institute, Inc.	(ANSI)	Sampling airborne radioactive materials
Air Pollution Control Association	(APCA)	Recommended standard methods for continuing air monitoring for fine particulate matter
American Society for Testing and Materials	(ASTM)	Sampling and analysis of atmospheres and waters; recommended practices for sampling procedures, nomenclature, etc.
Environmental Protection Agency	(EPA)	Reference and equivalent methods for air and water contaminants
Intersociety Committee on Methods for Air Sampling and Analysis	(ISC)	Recommended methods of air sampling and analysis
National Bureau of Standards	(NBS)	Standard reference gases and liquids

Fortunately, there has been a much greater awareness of the problems of quality assurance of environmental data in recent years, and significant progress is being made in the availability of standard reference materials and calibration instrumentation and techniques for analysis of specific contaminants. There are an increasing variety of laboratory certification programs, and programs for interlaboratory evaluation of analytical methods. Some of these important activities are summarized in Table 9–8. More complete discussions of standardization and calibrations are presented in a variety of standard reference works (1,2,3,5,8,9,10).

REFERENCES

1. ACGIH. *Air Sampling Instruments* — 5th Edition. American Conference of Governmental Industrial Hygienists. Cincinnati, Ohio, 1978.
2. Stern, A. C., (ed.) *Air Pollution* — 3rd Ed., vol. 3. New York: Academic Press, 1977.
3. Patty, F., Clayton, G., and Clayton, F. *Industrial Hygiene and Toxicology* — 3rd Ed., vol. 1. New York: Wiley, 1977.
4. Palmes, E. D., Gunnison, A. F., DiMattio, J., and Tomczyk, C. Personal Samplers for Nitrogen Dioxide. *Amer. Industr. Hyg. Assoc. J.* 37:570–577, 1976.

5. APHA, AWWA, and WPCF. *Standard Methods for the Examination of Water and Wastewater* — 14th Ed. American Public Health Assoc., Washington, D.C., 1975.

6. APCA. *Recognition of Air Pollution Injury to Vegetation: A Pictorial Atlas.* Air Pollution Control Association, Pittsburgh, 1972.

7. Cross, F. L., Jr. *Handbook on Environmental Monitoring.* Technomic Publishing Co., Westport, Conn: 1974.

8. ISC. *Recommended Methods of Air Sampling and Analysis.* Intersociety Committee on Methods for Air Sampling and Analysis, New York, 1976.

9. AOAC. *Official Methods of Analysis* — 12th Ed. Association of Official Analytical Chemists, Washington, D.C., 1975.

10. ASTM. *Annual Book of ASTM Standards*-Part 31, Water, and Part 26, Atmospheric Analysis. American Society for Testing and Materials, Philadelphia, 1977.

10
Control of Contamination

INTRODUCTION

Implicit in the discussions of the previous chapters was that ways and means existed to control contamination when it was deemed to be excessive. One may expect that the administrative and technical skills and ingenuity displayed in creating the technologies which generate contaminants could also be employed in controlling them. By and large, this expectation is realistic. When the motivation is provided, whether by enlightened self-interest or legislative mandate, it has been possible to achieve substantial control over ambient levels of chemical contaminants.

Unfortunately, there has been, in practice, too little application of existing control technology. The installation and maintenance of controls on automobile exhaust, process equipment in industry, and liquid and gaseous waste streams from industrial operations may be very costly. In the absence of clear-cut standards and regulations for effluent discharges, or the lack of evidence for damage to human health or environmental quality associated with such discharges, there is too little incentive for voluntary application of contaminant controls. One of the major benefits of the recently developing consciousness of the impact of environmental contamination on human health and welfare, exemplified by the substantial body of recent environmental legislation (Table 10-1), has been an impetus toward the application of existing technology, as well as the development of new and improved technologies, for the control of contaminant effluents.

Responses to damaging conditions in the environment will always continue to generate needs for controls. In future years, alleviation of pre-existing conditions should become a diminishing part of control activities.

Table 10.1 Major Federal Legislation Pertaining to Control of Chemicals

Air Contaminants

1955 Air Pollution Control Research and Technical Assistance Act
Provided temporary authority and funding for five years for a federal program of research in air contamination, and technical assistance to local governments. (Extended in 1959 for an additional four years.) Regulatory agency: HEW

1963 Clean Air Act
Granted permanent authority for federal air pollution control activities. Provided for (a) federal grants to state and local air pollution control agencies to establish and improve their control programs, (b) federal action to abate interstate air contamination through a system of hearings, conferences, and court actions, (c) an expanded federal research and development program with particular emphasis on motor-vehicle exhaust and sulfur oxide emissions from coal and fuel-oil combustion. Regulatory agency: HEW

1965 Amendments to Clean Air Act
Provided for (a) the promulgation of national standards relating to motor-vehicle exhaust, (b) cooperation with Canada and Mexico to abate international air contamination.

1967 Air Quality Act (amending the Clean Air Act)
Declared a national policy of air-quality enhancement and provided a procedure for designation of air-quality control regions and setting of standards by cooperation between federal and state governments; provided for registration of fuel additives.

1969 Amendments to Clean Air Act
Extended authorization for research on low-emission motor vehicles and cleaner fuels.

1970 Amendments to Clean Air Act
Provided for (a) the establishment of national ambient air quality standards and their achievement through the implementation plans of air-quality control regions and states, (b) a 90% reduction from 1970 levels of hydrocarbon and carbon monoxide emissions from automobiles by the 1975-model year, and 90% reduction from 1971 levels in nitrogen oxide emissions by the 1976-model year (with one-year extensions if necessary); (c) studies of aircraft emissions. Regulatory agency: EPA

1977 Amendments to Clean Air Act
(see text)

Water Contaminants

1899 "Refuse Act of 1899" (part of River and Harbor Act of 1899)
Prohibited discharges into navigable rivers.

1912 Public Health Service Act
Authorized investigation of water contamination in relation to human diseases.

1924 Oil Poilution Act
Prohibited discharge of oil by any means, except in emergency or by accident, into navigable waters of the United States.

1948 Water Pollution Control Act
Provided five-year authorization to fund research studies, low-interest loans for construction of sewage- and water-treatment works, and a Federal Water Pollution Control Advisory Board. Authorized the Department of Justice to bring suits against indi-

viduals or firms, but only after notice, hearing, and consent of the state involved. (Extended in 1952 for an additional three years.)

1956 **Amendments to Water Pollution Control Act**

Provided for (a) permanent authority, (b) grants for construction of sewage-treatment plants, (c) abatement of interstate water contamination by federal enforcement through a conference public-hearing court-action procedure.

1961 **Federal Water Pollution Control Act**

Permitted the Secretary of HEW, through the Department of Justice, to bring court suits to stop contamination of interstate waters without seeking permission of the state. Extended pollution abatement procedures to navigable intrastate and coastal waters, with permission of state. Authorized seven regional laboratories for research and development in improved methods of sewage treatment and control. Authorized funds for grants to local communities for sewage-treatment plants.

1961 **Oil Pollution Act**

Enacted to implement provisions of the International Convention for the Prevention of the Pollution of the Sea by Oil, 1954.

1965 **Water Quality Act**

Declared a national policy of water-quality enhancement. Established the Federal Water Pollution Control Administration (FWPCA). Provided for the states to adopt water-quality standards for interstate waters and plans for implementation and enforcement, to be submitted to the Secretary of HEW (later to Secretary of Interior after FWPCA was transferred to the Department of the Interior) for approval as federal standards; authorized the Secretary to initiate federal actions to establish standards if the state criteria were inadequate. Authorized grants for research and development to control storm water and combined sewer overflows and authorized funds for sewage treatment plant grants.

1966 **Clean Water Restoration Act**

Provided for grants for research and development of advanced waste-treatment methods for municipal and industrial wastes. Authorized grants for construction of treatment plants. Amended the Oil Pollution Act of 1924 by transferring responsibility to Secretary of the Interior and provided for suits against "grossly negligent, or willful spilling, leaking, pumping, pouring, emitting or emptying of oil."

1970 **Water Quality Improvement Act**

Strengthened federal authority to deal with sewage discharges from vessels, hazardous contaminants, and contamination from federal and federally related activities. Provided for liability for oil spills from onshore- and offshore-drilling facilities and from vessels.

1972 **Marine Protection, Research, and Sanctuaries Act (Ocean Dumping Act)**

Regulation of the ocean dumping of materials which may be hazardous to human health and welfare. Regulatory agency: EPA

1972 **Ports and Waterways Safety Act**

Regulations relating to prevention or mitigation of damage to the marine environment due to transport of oil and other hazardous materials. Regulatory agency: Department of Transportation

1972 **Water Pollution Control Act (Clean Water Act)**

Extended national water pollution program to intrastate waters. Established (a) a system of national effluent limitations with goals of best practicable water pollution con-

cont.

Table 10.1 cont.

Water Contaminants (*continued*)

trol technology by mid-1977 and best available technology by mid-1983, (b) national performance standards for new industrial and publicly owned waste treatment plants, (c) a national system of permits for discharge of contaminants. Authorized funds for treatment facilities. Regulatory agency: EPA

1974 **Safe Drinking Water Act (amendment to Public Health Service Act)**
Authorized EPA to establish federal drinking-water standards for protection from all harmful contaminants, applicable to all public water supplies in the United States. Established a joint federal-state system for compliance with standards and protection of underground sources of drinking water.

1977 **Amendments to Water Pollution Control Act of 1972**
Delayed installation of best available control technology until 1984. Continued federal grants for sewage-treatment plants. Added 65 chemicals and classes of chemicals to list of toxic agents.

Solid Wastes

1965 **Solid Waste Disposal Act**
Began a national research, development, and demonstration program for solid wastes. Provided financial assistance to interstate, state, and local agencies for planning and establishing solid-waste disposal programs. Regulatory agency: EPA

1970 **Resource Recovery Act**
Provided for (a) research into new and improved methods to recover, recycle, and reuse wastes, and (b) financial assistance to states in the construction of solid-waste disposal facilities.

1976 **Resource Conservation and Recovery Act**
Promulgates regulations for treatment, storage, and ultimate disposal of hazardous solid wastes. Regulatory agency: EPA

Pesticide Contamination

1938 **Federal Food, Drug, and Cosmetic Act**
Provided for the establishment of tolerances for pesticide residues in food.

1947 **Federal Insecticide, Fungicide, and Rodenticide Act**
Provided for (a) registration of "economic poisons" by the U. S. Department of Agriculture for products marketed in interstate commerce and (b) seizures of adulterated, misbranded, unregistered, or insufficiently labeled pesticides.

1954 **Miller Amendment to the Federal Food, Drug, and Cosmetic Act**
Provided for condemnation of raw agricultural commodities containing pesticide residues in excess of tolerances as fixed by the Secretary of HEW.

1972 **Federal Insecticide, Fungicide, and Rodenticide Act**
Gave EPA authority to regulate uses of pesticides so as to prevent environmental damage.

Occupational Environment

1936 **Walsh-Healey Act**
Enabled Federal Government to set standards for safety and health in work places engaged in activities relating to Federal contracts.

Occupational Environment *(continued)*

1970 **Occupational Safety and Health Act**
Established National Institute for Occupational Safety and Health to conduct research and training, develop exposure standards, and make inspections relevant to the protection of the health and safety of workers. Regulatory agency: OSHA

Radioactive Wastes

1954 **Atomic Energy Act (plus amendments)**
Regulates release of radioactive wastes into the environment. Regulatory agency: Nuclear Regulatory Commission (NRC)

General

1969 **National Environmental Policy Act**
Declared a national policy for the environment to encourage harmony between man and his environment, to promote efforts that will prevent or eliminate damage to the environment, and to enrich the understanding of ecological systems and natural resources important to the nation. Established a three-member Council on Environmental Quality, in the Executive Office of the President, charged with making studies and recommendations to the President and with preparing an annual Environmental Quality Report. Directed that all agencies of the Federal Government include in every recommendation or report on proposals for legislation and other major Federal actions significantly affecting the quality of the human environment, a detailed statement on: (a) the environmental impact of the proposed action, (b) any adverse environmental effects which cannot be avoided should the proposal be implemented, (c) alternatives to the proposed action, (d) the relationship between local short-term uses of the environment and the maintenance and enhancement of long-term productivity, and (e) any irreversible and irretrievable commitments of resources which would be involved in the proposed action should it be implemented.

1976 **Toxic Substances Control Act**
Provided for regulation of new chemicals or new uses for existing chemicals to prevent adverse effects upon human health or the environment. Provides for chemical testing, as well as authority to delay use of or ban specific chemicals or uses of chemicals. Regulatory agency: EPA

The major application will be those which are basically preventative in nature.

Such an "anticipatory" control attitude has been developed in recent decades by some more enlightened industries. They sought to recognize potential problems affecting industrial workers, neighboring communities, and consumers or users of their products, at the design stage of the process or product. Evaluations in terms of toxicity tests and/or environmental modelling are performed before the plant is built or the process is installed. This type of approach can be economically advantageous, since the costs of controls are invariably much less when they are designed into the process than when they have to be retrofitted onto the process at some later date.

While the pattern of problem anticipation is relatively highly developed, it has not been, unfortunately, too widely practiced. Occupational diseases among uranium miners, asbestos-insulation workers, coke-oven workers, and personnel manufacturing the pesticide Kepone, are classic examples of conditions which were readily preventable through the application of existing knowledge and technology. Similarly, the damage to aquatic life and the restrictions of commercial seafood harvests resulting from releases of mercury, PCB, Kepone, and Mirex to surface waters, could also easily have been prevented. Much of the recent environmental legislation has, in fact, been spurred by public reaction to the failure of industry to recognize their own interests in the application of preventive controls.

APPROACHES TO CONTROL

When a decision has been made that an imminent or serious potential hazard exists due to the presence of a chemical contaminant in the environment, and there is a need to control its further release or usage, choices must be made concerning the means by which control is to be achieved.

One approach involves outright bans, specification of permissible usages, permissible times for operation and discharge of effluents, and permissible concentrations and amounts of effluents. These are all examples of control by the regulatory process.

Another approach, which is largely restricted to the occupational environment, is to control access to potentially hazardous conditions and environments. Management may be able to restrict access of personnel to contaminated areas, or to obtain the cooperation of its employees in using personal protective devices, so as to limit the duration and/or the intensity of exposures.

A third approach is known as engineering control. In this approach, controls are built into the process itself in order to reduce the formation and release of chemicals into the environment. This may be achieved through appropriate design and selection of materials, processes, and equipment; the intent is to depend as little as possible on the individual actions of plant personnel or consumers. Engineering controls can be subdivided into process controls and source (or effluent) controls. The former emphasize the reduction or diminution of contaminant formation, while the latter act to reduce or prevent the release to the environment of any contaminants which are formed.

A fourth approach involves control via dilution in the ambient air and water.

Regulatory Controls

A total ban on usage of a material may be appropriate when (a) such usage is not essential, but rather one of marginal economic consequence or personal preference, such as fluorocarbon propellants in spray cans of household products, or (b) less hazardous materials are available which are equal to or better than the hazardous material for the purpose. A ban on lead-based white pigments in indoor paint was feasible because of the availability of titanium-based pigments to replace them.

When a material has unique properties and cannot readily be replaced without introducing or increasing other hazards, its usage must be rigorously controlled. An example is the use of asbestos for fireproofing steel beams. Asbestos had been sprayed onto the beams at the construction site, with resulting excessive exposures of construction workers and passersby to airborne fibers. Now it is applied in a matrix which assures coverage without significant airborne dispersion.

There are several levels of regulatory controls. In some areas, states and local governments retain significant regulatory authority. However, to an increasing extent, the basic authority for governmental regulation comes from a series of federal acts which have created new programs or expanded existing ones, and directed that certain objectives be achieved and maintained. The key federal legislation affecting the control of chemical contamination in the environment is briefly summarized in Table 10-1.

Under the enabling legislation passed by Congress, each of the federal agencies charged with environmental-control responsibilities promulgates regulations which are published in the Federal Register. These regulations have the force of law and, unless overruled by court actions, are enforced by the appropriate agencies.

As discussed in Chapter 8, regulations may be of several different types as appropriate to the specific task. They can specify standards or permissible levels of contamination in community and workroom air, in drinking water supplies and streams, or in foods and packaging materials. They can also specify effluent standards and work practices. The regulatory standards involving numerical concentration limits may also require specific types of analytical measurement techniques, as discussed in Chapter 9.

Federal regulations which establish numerical standards provide a uniform frame of reference and goals to be achieved and maintained, putting all parties involved on an equal basis and giving them due notice of what is required. These regulations free industry and local authorities from the need to make technical judgements as to the levels which are toxic or may produce environmental damage. Such judgements, as previously discussed, are very difficult ones, and when made for a particular situation on a local basis, are likely to differ substantially from place-to-place and time-to-time.

Governmental regulations may also specify requirements for work practices and the maintenance of records on effluent discharges and the exposures and medical histories of employees. Such regulations may, therefore, affect the manner in which industry achieves control of exposures through both administrative and technical controls.

Administrative Controls

Administrative controls of contaminant emissions and exposures depend upon the ability of management to motivate and control the personnel whose actions affect contaminant release, dispersion, and accessibility of themselves and others to contaminated areas.

The most effective means of administrative controls are those achieved through programs of education and motivation. Personnel whose actions affect their own exposures, or the extent of contaminant releases to the community environment, should be aware of the potential consequences of the resulting exposures and of the effects that their own work practices have on the extent of the releases. They should be thoroughly familiar with the intended functions, proper operating parameters, and maintenance procedures for the process equipment in their unit and for any control devices associated with its operation, so that substandard equipment performance may be recognized and controlled. These people should also be trained to cope with malfunctions and emergencies, including safe procedures for shutting down process operations, and to be thoroughly familiar with evacuation routes and procedures.

Personnel training programs are also needed for the proper use of personal protective devices such as gloves, aprons, coveralls, eye-protectors, hearing protectors, and respiratory protective devices. These programs should include the criteria for the selection of the particular devices provided, when and how they should be worn or used, the degree of protection that they provide, their useful life and replacement schedules, and the conditions under which they should be maintained and stored between periods of use.

Management's role goes beyond education and encouragement of employee participation. Other aspects include human engineering of equipment, incentives for positive efforts, and disincentives or penalties for poor performance. In terms of human engineering (known as ergonomics), the provision of process controls and protective equipment within easy reach, which can be readily operated with natural motions, will usually increase effective performance. On the other hand, the lure of special incentives or the threats of penalties are not usually very effective in achieving a uniformly high level of employee performance.

Management also has a responsibility to limit the access of personnel to contaminated areas when unlimited access would result in overexposures. This type of control can take several forms. One involves the designation of personnel into those who shall have access to certain areas and those who shall not. The former category may include employees who have been fitted with personal protective devices. It may also include personnel without protective equipment, but whose work-time within the contaminated area will be sufficiently brief that their cumulative exposures will still be well within the permissible daily limits.

Among those for whom management is obligated to restrict access to contaminated areas are people who may be hypersensitive, have preexisting conditions, or accumulated prior exposures which would place them at special risk. The information needed for the identification of such personnel can be obtained at preemployment and periodic medical examinations and, for some conditions and materials, by special screening tests.

Engineering Controls

The most effective controls of contaminant release are those which are built into the process itself, and do not normally require the attention or assistance of operating personnel. There are four basic approaches to engineering control: material substitution, process modification, and isolation, which are all techniques of process control, and source control.

MATERIAL SUBSTITUTION

Substitution of materials can ensure against exposure and contamination due to a particular chemical by eliminating it from the process entirely and

Table 10.2   Applications of Control by Substitution

I. SUBSTITUTION OF LESS-TOXIC MATERIALS		
Original material	Less-toxic substitute(s)	Applications
Yellow phosphorus	Red phosphorus	Safety-match heads
Carbon tetrachloride	Perchloroethylene	Cleaning solvent
	Methylene chloride	Degreasing solvent
Benzene	Toluene, cyclohexane	Chemical raw material, solvent
Lead	Titanium, zinc	Interior paint pigments
Mercury nitrate	Various materials	Caroting fur for hats
Beryllium	Calcium phosphate	Phosphors for fluorescent lamps
Asbestos fibers	Glass fibers	Insulating materials
Sand	Silicon carbide	Abrasive cutting and shaping
Radium	Tritium	Luminous phosphors in dials

II. SUBSTITUTION OF EQUIPMENT OR PROCESSES WHICH DISPERSE LESS CONTAMINANT		
Original Equipment or Process	Less-Contaminating Equipment or Process	Applications
Spraying	Dipping	Painting of parts and products
Sandblasting	Hydroblast; chemical etching	Cleaning of parts
Soldering or welding	Riveting, crimping, or adhesive bonding	Joining of parts
Batch charging	Continuous feeding	Addition of materials
Diesel engines	Battery-powered electric engines	Fork-lift and delivery vehicles
Flanged piping	Welded pipe	Transfer lines
Mercury gauges	Mechanical gauges	Instrumentation

replacing it with another chemical which is less toxic or offensive. This technique is usually practical where the material in question is used as a solvent, pigment, thermal or electrical insulator, or for other applications where there are a number of alternate materials having similar properties which may be used. Some notable examples of materials substitution which have substantially reduced occupational disease incidence and/or environmental contamination are presented in Table 10-2.

PROCESS MODIFICATION

The control of contaminant discharges by modification of process procedures and equipment involves using process elements which provide a greater degree of enclosure, and/or the combination of a series of process operations in a single step. Both of these techniques are generally used in switching from batch-processing operations to continuous processing. The loading, unloading, and transfer of materials from one batch-processing vessel to another provides ample opportunity for uncontrolled releases, and these are minimized in continuous operations. Some notable examples of process substitutions which have substantially reduced contaminant release are also presented in Table 10-2.

ISOLATION

When the generation of contaminants cannot be completely eliminated, the next best thing is to prevent contact of the contaminants with people and the environment. This can be done by isolating the contaminant source or, occasionally, by isolating the people to be protected from the source. It is generally easier and less expensive to isolate the source; usually this is done by enclosing it within a box or shell which will eliminate or reduce the escape of the material. On the other hand, when the sources are numerous and the people to be protected are few, it may be less expensive to isolate the people and create a clean artificial environment for them. This is frequently done, for example, in foundries, where the crane operator works within a cab provided with clean breathing air; the operator moves within the cab through dusty air above the foundry operations.

SOURCE CONTROLS

When processes generate contaminants that must be controlled, the most economical and effective controls are those that exert their action closest to the source. It becomes more difficult and expensive to collect a chemical from a waste stream as process volume increases and the concentration decreases, such as following dilution by ambient air or water, and as the number of other constituents increases. Co-contaminants may produce interferences in the separation procedures and may also increase the cost of, or limit the freedom for, ultimate disposal of the extracted material. The use of process controls may act to concentrate contaminants in a small value of air or water, thus decreasing the cost of any source cleaning devices, as well as increasing their efficiency.

Effluent controls include the removal of contaminants from waste streams, and recycling of the extracted materials back into the process, or disposal in an environmentally acceptable manner. When recycle or sale of the extracted material is feasible, it will usually be necessary to segregate waste streams according to their content and to treat some separately in order to avoid mixing and dilution of materials. On the other hand, where the material is simply to be disposed of, it may be possible to combine various waste streams in ways favoring the neutralization of one waste with another. For example, an acidic liquid waste can be mixed with an alkaline waste to produce a neutral salt. It may also be possible to bubble boiler-flue (smokestack) gases containing carbon dioxide to neutralize alkaline liquid wastes.

Control by Dilution

It is frequently stated that "dilution is the solution to pollution." As time goes on, this old adage becomes harder to justify. There will, however, always be some suitable application of control by dilution, such as for contaminants with short residence times in the environment and for those whose degradation products are innocuous. For contaminants which persist, or which are degraded in the environment into persistent chemicals, control by dilution is inappropriate. Also, what was acceptable when there are few sources may become unacceptable when the number of sources increases.

In the air environment, toxic chemicals from power plants and industrial operations may be vented through high stacks. Such stacks can penetrate normal inversion layers and, as discussed in Chapter 4, the maximum ground-level concentrations decrease approximately as the square of the effective stack height. Some power plants and industrial operations depend on dilution via tall stacks most of the time, but switch fuels or process operations to control effluent levels on those days when stack dispersion will not be adequate.

In the water environment, dilution rates depend on stream flow, and industrial operations frequently utilize detention basins on site to hold some of the wastes for periodic discharge during times of high stream flow.

In the occupational environment, dilution ventilation is sometimes used to reduce air concentrations from levels slightly above to slightly below acceptable limits, especially when there are multiple small sources, and where the material is considered more of a nuisance or housekeeping problem than a toxic hazard. Where the sources are few and well defined, or where toxicity is involved, source control by local exhaust ventilation is usually preferred over dilution ventilation.

CONTROL IN THE OCCUPATIONAL ENVIRONMENT

Local Exhaust Ventilation and Dilution Ventilation

Local exhaust ventilation is one of the key techniques for controlling air contaminant levels within the occupational environment. The principle is to create sufficient suction in the zone of active contaminant release to draw essentially all of the contaminant into a hood enclosure, along with a minimum volume of workroom air. The contaminant stream is then generally guided through a duct system to an air-cleaning device, fan, and discharge stack. Various exhaust-hood designs for specialized industrial and laboratory operations have evolved over the last 40 years, and many of them are illustrated in detail in the ACGIH publication *"Industrial Ventilation."* This frequently updated design manual also provides guidance on ventilation fundamentals, duct-system design, fan and air-cleaner selection, and measurements of ventilation systems. Examples of some local exhaust hood designs are illustrated in Figures 10-1 and 10-2.

A special consideration in ventilation control of air contaminant concentrations in industry is the cost of supply make-up air. For each volume of contaminated air exhausted, it is necessary to supply an equivalent volume of clean air, and this latter air will usually have to be cooled or heated to maintain the desired temperature in the workroom. However, with the increasing costs of energy, the relatively large volume requirements of dilution ventilation make it increasingly less attractive as a contaminant control technique. This is, however, still a major technique used in underground mines.

Wet Operations

A widely used technique for source control of industrial dusts is to wet the dust generated in operations such as crushing, grinding, cutting, and drilling. Dust generation may be greatly reduced by spraying a water mist or a stream of water onto the working face of the machine or tool.

Personal Protective Devices

Personal protective devices play an important role in occupational health protection. Chemical contaminants in the workplace can be systemically absorbed through the skin, eyes, lungs, and gastrointestinal tract, or can affect these portals directly.

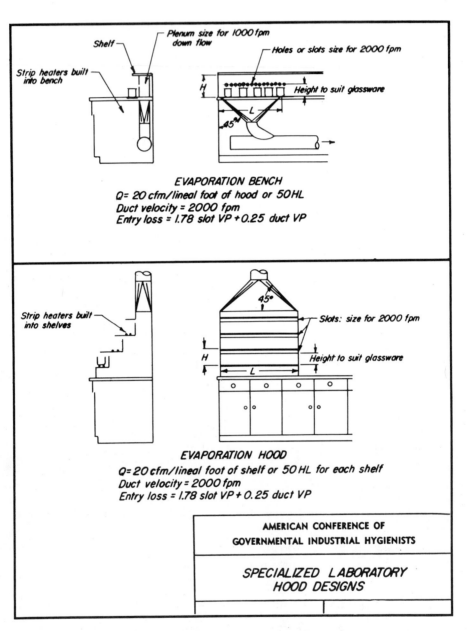

Fig. 10–1 Specialized laboratory exhaust-hood designs.

From: American Conference of Governmental Industrial Hygienists. *Industrial Ventilation*, 15th ed. ACGIH, Lansing, Mich., 1978.

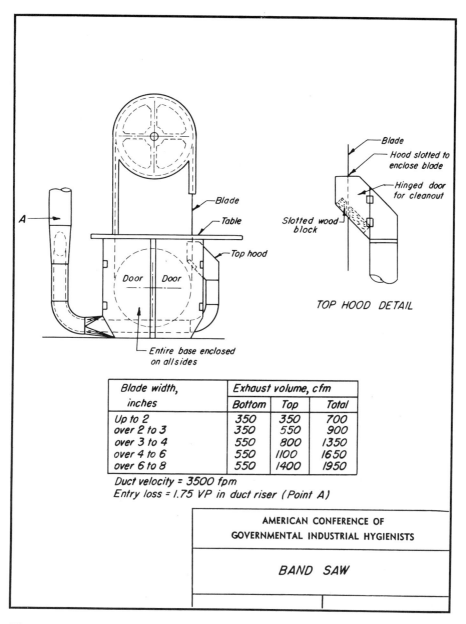

Blade width,	Exhaust volume, cfm		
inches	Bottom	Top	Total
Up to 2	350	350	700
over 2 to 3	350	550	900
over 3 to 4	550	800	1350
over 4 to 6	550	1100	1650
over 6 to 8	550	1400	1950

Duct velocity = 3500 fpm
Entry loss = 1.75 VP in duct riser (Point A)

AMERICAN CONFERENCE OF
GOVERNMENTAL INDUSTRIAL HYGIENISTS

BAND SAW

Fig. 10–2 Hood design for a band saw.

From: American Conference of Governmental Industrial Hygienists. *Industrial Ventilation*, 15th ed. ACGIH, Lansing, Mich. 1978.

Gloves, aprons, face shields, goggles, and various kinds of protective clothing can provide effective barriers against skin and eye contact, provided the proper materials are selected for their construction.

The direct ingestion of chemicals in industry usually occurs from poor personal hygiene practices, such as touching food or cigarettes with contaminated hands. When contaminated cigarettes are burned, the chemicals on them, or their thermal degradation products, can be inhaled.

Respiratory protective devices provide a last resort defense against the inhalation of airborne contaminants. There are two basic types of devices: air-supplying and air-purifying respirators. In the former type, an excess of clean air is introduced into or around the nose or mouth. The air can be supplied by a portable tank under pressure, as in scuba diving, or by an air hose from a fixed tank or pump located out of the contaminated air. Portable tanks can only contain enough air or oxygen for brief periods, while air hoses limit mobility and freedom of action.

In air-purifying respirators, the contaminated air passes through a canister containing a filter and/or bed of adsorbing granules which extract the contaminants and pass the air. One type uses a battery-powered fan to push the air through the canister and thereby deliver it at positive pressure to the interior of a face mask or hood. It, therefore, resembles the air-supply type of device, in that leakage is outward.

By contrast, in most air-purifying respirators, the driving force to move the air through the canister is provided by the wearer's lungs, the pressure inside the mask is lower than that outside, and leakage is inward. Thus, even with complete contaminant removal from the air passing through the canister, the inhaled air will be partially contaminated if there is any leakage through or around the seal between the mask and the face. There is always resistance to flow through the canister, and this resistance generally increases with increasing collection efficiency and increasing flow.

For maximum protection, the correct air canister must be used. This is one which is rated for the particular contaminant in the exposure atmosphere and within its concentration and time limits. If it is used at higher concentrations, or if its adsorptive capacity is exceeded by prolonged usage, it will fail to be protective. Also, no canister can compensate for a deficiency of oxygen. Selection of respirators and canisters should be limited to those which have been tested and approved for the particular duty by NIOSH.

With a well-fitted mask, the appropriate air-purifying canister, and the motivation to use it properly, an individual can be protected from the inhalation of contaminated air, and respirators are, therefore, very useful for protection during infrequent brief operations and during unanticipated emergencies.

CONTROL OF COMMUNITY AIR QUALITY

The effective control of ambient air quality ultimately depends upon the control of source emissions. For the criteria contaminants, i.e., total suspended particulates (TSP), SO_2, CO, NO_x, hydrocarbons, and oxidants, the first step in their control was to construct an inventory of sources, source strengths, and patterns of dispersion so that control priorities could be set up. The inventories established that TSP and SO_2 were primarily attributable to stationary fossil-fuel combustion sources, that CO, hydrocarbons, and oxidants were primarily derived from mobile sources, and that NO_x was attributable in similar measure to both of these sources.

Most noncriteria contaminants which are regulated are primarily of industrial origin, and have been controlled by the specification and enforcement of effluent standards.

While source controls are preferable in most cases, there still remain times when satisfactory levels of air quality can be maintained most economically by adequate dilution of the contaminants in the atmosphere. Since the mixing characteristics of the atmosphere are quite variable, dilution is frequently coupled with intermittent discharge.

Source Controls

Source controls will be discussed first in terms of air-cleaning techniques. This will be followed by a discussion of the application of available technology to the specific control of air contamination from mobile sources and stationary combustion sources.

AIR-CLEANING TECHNIQUES

The process of removing contaminants from an air stream is known as air cleaning. It can involve extraction of the material from the air, or its conversion to less-objectionable forms. An example of the latter method is the burning of toxic organic compounds, producing CO_2 and H_2O.

Gases and vapors can be extracted from an airstream by condensation, adsorption, and absorption in gas washers and scrubbers. Particles can

Brownian Separator

The Brownian Separator removes low-viscosity liquid particulates in the particle size range from 0.01 to 0.05 microns. This patented device consists of filaments, usually glass, arranged so that the space between them is less than the mean free path of the particle to be collected. Thus, the particle cannot escape; it must collide with one of the filaments where it is captured. Eventually, liquid builds up on the filaments and runs down to a collection point.

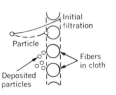

Filtration

The panel-type filter is limited to relatively light dust loadings because cleaning is time-consuming and costly. The baghouse, on the other hand, can be used with heavy grain loadings. These units are made up of a large number of cylindrical bag filters, which may be cleaned automatically to provide continuous operation. The baghouse should not be used for oily, hygroscopic or explosive dusts. Gas temperature is limited by the composition of the filter medium.

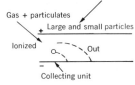

Electrostatic Precipitator

In this common collecting device, a high voltage is imposed on a relatively slow-moving gas stream. Charged particles are attracted to the grounded electrode, where they are periodically removed by rapping, vibration or washing. Conventional precipitators are quite bulky because of low air velocities. They have the advantage of providing dry collection, generally down to the 0.2 to 0.5 micron range. Theoretically there is no minimum limit to the size of particles that can be collected.

Centrifugal Collectors

Available in a number of different designs, cyclone separators rely on centrifugal force to drive particles to the wall of the chamber where they drop out of the gas stream into a collector. They are useful for relatively coarse separations. A large-diameter cyclone will remove particles 15 microns and larger. Small-diameter units, usually connected in parallel, are effective down to the 10-micron range.

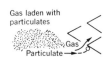

Inertial Separators

A variety of dry mechanical collectors are based on inertial impaction of particles on baffles arranged in the gas stream. Because there is a practical limit on how closely together the baffles can be placed, these devices are best for separations above 20 microns. However, some new designs can achieve 5 to 8 micron separations.

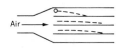

Gravity Settling

The oldest and simplest means of particle separation is the gravity settling chamber, in which particles fall onto collecting plates. These are useful for removing coarse dusts above 40 microns, and may be placed ahead of other separation equipment.

Fig. 10–3   Basic methods for dry-particle collection.

From: Teller, A. J. Air Pollution Control. *Chemical Engineering 79* (No. 10):93 – 98, 1972.
© McGraw-Hill. Used with permission of the publisher.

Fig. 10–4   Basic equipment which may be used for both particle and gas collection.

From: Teller, A. J. Air Pollution Control. *Chemical Engineering 79* (No. 10):93 – 98, 1972.
© McGraw-Hill. Used with permission of the publisher.

Spray Towers

In spray towers the gas stream passes at low velocity through water sprays created by pressure nozzles. These units are simple but only moderately effective. They will remove particles down to the 10 to 20 micron range and absorb very soluble gases.

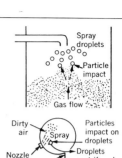

Wet Cyclones

In this version of the cyclone separator, swirling gas flows through water sprays. Droplets containing dusts and absorbed gas are separated by centrifugal force and collected at the bottom of the chamber. Wet cyclones are more effective than spray towers. They will absorb fairly soluble gases and remove dusts down to the 3 to 5 micron range.

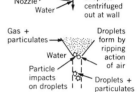

Venturi Collectors

The venturi design relies on high gas velocities on the order of 100 to 500 ft./sec. through a constriction where water is added. The impact breaks the water into droplets, with the fineness of spray determined by gas velocity. Venturis collect particles to the 0.1 micron level and recover soluble gases.

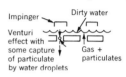

Perforated Impingement Trays

Here, gas flows through small orifices at 40 to 80 ft./sec. and hits a liquid layer to form spray. The liquid captures particles and gases and is collected on impingement baffles. Performance is relatively good with particle collection possible down to 1 micron and absorption of fairly soluble gases.

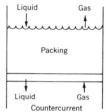

Packed Towers

Packed towers were primarily created for gas absorption. With some new designs, they can also be used for dust removal. Gas-liquid flow may be concurrent, cocurrent or crossflow. Two kinds of mass and particulate transfer are possible, depending on the packing. Extended surface packings provide absorption by spreading the liquid surface, and collect particles by cyclonic action. Recovery is limited to 10-micron sizes and larger. With filament packing, absorption by surface renewal is twice as effective and particles to 3 microns are separated by inertial impact. A cross-flow tower gives the highest solids handling capability with the lowest pressure drop.

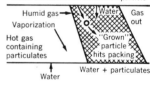

Nucleation Scrubber

This process grows submicron particles, down to 0.01 micron, by condensation. It collects grown particles on filament-type packing by inertial impact at low energy. At the same time, absorption occurs as in a conventional cross-flow packed tower.

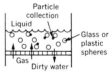

Turbulent Contactor

With this device gas flows through spherical packing made up of 0.5 to 2-in. balls, which oscillate or bounce in liquid. Particles down to the 1 to 5 micron range collect by baffle impact on the spheres. Gases will absorb in the turbulent liquid and absorption efficiency will depend on the number of stages.

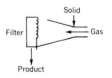

Wet Filters

This is an open filter whose efficiency is improved by spraying with water. Its absorption ability is limited to those gases that are very soluble.

Chromatographic Recovery

The chromatographic unit (proprietary) contains inexpensive extended surface solids with a mono-to-termolecular coating of an active reagent. Thus the solute gas need only diffuse to the surface of the solid where it reacts either by ionic or molecular reaction. Where the reaction is ionic, the absorption is generally irreversible. Where the reaction is molecular, the absorption can be reversible and the solute gas can be stripped off essentially pure.

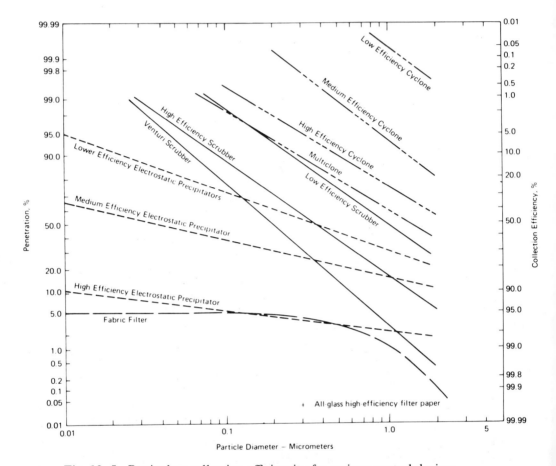

Fig. 10-5 Particulate collection efficiencies for various control devices.

From: Billings, C. E. Technological Sources of Air Pollution. pp. 350–408 in *Industrial Pollution*, N. I. Sax (ed.). New York: Van Nostrand, 1974. © 1974 by Litton Educational Publishing, Inc. Reprinted by permission of Van Nostrand Reinhold Co.

also be efficiently captured in some scrubbers. They can also be captured in dry collection systems on the basis of their diffusional and inertial displacements during passage through filters or specialized collectors, and by electrostatic precipitation. The basic mechanisms of dry-particle collection are illustrated in Figure 10-3, and those used to collect gases and vapors, and in some cases particles by scrubbing, are illustrated in Figure 10-4. Figure 10-5 presents particulate collection efficiencies for various control devices.

The selection of any particular cleaning method depends on the contaminants, the degree of removal required, the volumetric rate of flow to be processed, and the availability of suitable and economical means of disposal of the materials extracted from the air or of the secondary products formed during the control process.

COMBUSTION

The simplest and generally the most economical way to dispose of organic wastes or other combustible contaminants in air is to burn them. If combustion is complete, all wastes are converted into CO_2 and H_2O. The burning of so-called "sour" waste gases, i.e., containing H_2S and organic sulfur compounds, in refineries results in SO_2 in the effluent, but this SO_2 is generally much less of a problem than the original compounds. In most cases, the amounts of SO_2 and NO_x in the effluent of a waste-gas combustion system will be negligible in comparison to the other anthropogenic sources of these contaminants.

Two methods are widely used for the combustion of effluents. These are the flame method and the catalytic method. In the former, an external gas supply is generally needed to maintain a flame temperature sufficiently high to ensure complete combustion. An example is a refinery flame used to dispose of hydrocarbon vapors. The other method is based on catalytic oxidation, and is generally used for low-concentration streams. Such devices are used on most post-1975 cars to reduce the carbon monoxide and hydrocarbon content of exhaust gases.

ADSORPTION AND CONDENSATION

Contaminant gases and vapors can be removed by passing the airstream through a bed of granules whose surfaces will adsorb and retain the contaminants which reach them by diffusion. For example, activated carbon is often used to adsorb organic vapors. Since the surfaces eventually become saturated, the adsorption bed must either be periodically replaced or reconditioned, and provisions must be made for the disposal of the bed or the materials collected on and stripped from the bed during its reconditioning.

Adsorbants may be impregnated with various materials (Table 10–3) to convert the contaminant to less-harmful forms (e.g., CO into CO_2) or to promote a reaction which results in a form which may be more readily collected.

Some gas-phase contaminants may be removed by use of cold surfaces within or lining the flow path. Similar considerations concerning reconditioning apply here also. Organic vapors are often controlled using this technique.

Table 10.3   Some Adsorbent Impregnations for Gaseous Contaminants

Adsorbent	Impregnant	Contaminant	Action
Activated alumina	Potassium permanganate	Easily oxidizable gases, e.g., formaldehyde	Oxidation
	Sodium carbonate or bicarbonate	Acidic gases	Neutralization
Activated carbon	Bromine	Alkenes, e.g., ethylene	Conversion to dibromide
	Iodine	Mercury	Conversion to HgI_2
	Lead acetate	Hydrogen sulfide	Conversion to PbS
	Oxides of Cu, Cr, V, etc.; noble metals (Pd, Pt)	Oxidizable gases, including H_2S and mercaptans	Catalysis of air oxidation
	Phosphoric acid	Ammonia; amines	Neutralization
	Sodium carbonate or bicarbonate	Acidic vapors	Neutralization
	Sodium silicate	Hydrogen fluoride	Conversion to fluorosilicates
	Sodium sulfite	Formaldehyde	Conversion to addition product
	Sulfur	Mercury	Conversion to HgS

Compiled from: Turk, A., Adsorption, pp. 329–63 in *Air Pollution*, 3rd Ed. vol. 4. A. C. Stern (ed.). New York: Academic Press, 1977.

SCRUBBING

Liquid streams and droplets can be used to extract gases and vapors from airstreams in devices known as scrubbers. The gases and vapors are removed by absorption into the liquid phase. The removal efficiency depends on both the solubility and the degree of saturation within the liquid, and also on the contact time between the contaminant in the gas phase and the surfaces of the liquid phase. Table 10–4 summarizes contaminants commonly removed by scrubbing and the scrubbing media used.

Scrubbers can also be used to collect particulates. Particles which contact a liquid surface are collected, with removal due to impaction, diffusion, and in some cases, electrostatic precipitation. Not surprisingly, collection efficiency is strongly dependent on particle and droplet sizes (Figure 10–5).

Scrubbers such as the Venturi type, which break up the liquid into very small droplets, may have relatively high collection efficiencies for small particles. However, these devices consume much more energy in the process than do gas washers such as the spray tower or packed tower.

Table 10.4 Some Widely Used Scrubbing Media for Gaseous Contaminants

Contaminant	Scrubbing media
Chlorine (Cl_2)	H_2O, NaOH, NH_3
Hydrogen chloride (HCl)	H_2O, NaOH, NH_3, $Ca(OH)_2$, $Mg(OH)_2$
Hydrogen sulfide (H_2S)	NaOH, Na_2CO_3, KOH, K_2CO_3
Nitrogen oxides (NO_x)	$Ca(OH)_2$, $Mg(OH)_2$, Na_2SO_3 – $NaHSO_3$, urea
Sulfur dioxide (SO_2)	NaOH, Na_2CO_3, Na_2SO_3, KOH, K_2CO_3, K_2SO_3, CaO, $Ca(OH)_2$, $CaCO_3$, NH_3

The contaminants contained within the scrubbing liquids will generally require special handling in terms of treatment and disposal. Considerations in the handling of such contaminated liquids will be discussed later in this chapter under water-treatment techniques.

BAG FILTERS

Fibrous-filter media formed into cylindrical sleeves or bags are the most widely used type of dry-particle collector for air cleaning. By appropriate selection of fiber type, diameter, and packing density, fiber-mat thickness, and total surface, a high degree of collection efficiency may be achieved under a wide variety of operating and ambient conditions, and for aerosols having a broad range of characteristics.

Filtration in bag filters is similar in some respects and different in others to that occurring in air-sampling filters. The face velocity across a bag filter is generally only a few centimeters per second and, therefore, collection seldom depends on impaction deposition. Furthermore, the filtration occurs in the accumulated dust layer as well as in the filter mat itself. By contrast, in air sampling, the dust loadings are generally much lower, and a major portion of the sampled volume may be drawn through the filter before a significant dust layer accumulates. The clean filter must, therefore, be an efficient particle collector.

As dust accumulates on a bag filter, the flow must pass through the dust layer; the resistance to flow, therefore, increases, and will eventually impose a great burden on the air mover used. An analogous situation occurs in canister-type home vacuum cleaners which use a permanent-type felt bag, which must periodically be emptied. When the bag is emptied, some dust clings to the felt, and, therefore, the pressure drop and collection efficiency never return to the original new-bag levels. The same considerations apply to industrial bag filters, which are emptied and reused many times before they develop tears or other defects which lead to their replacement.

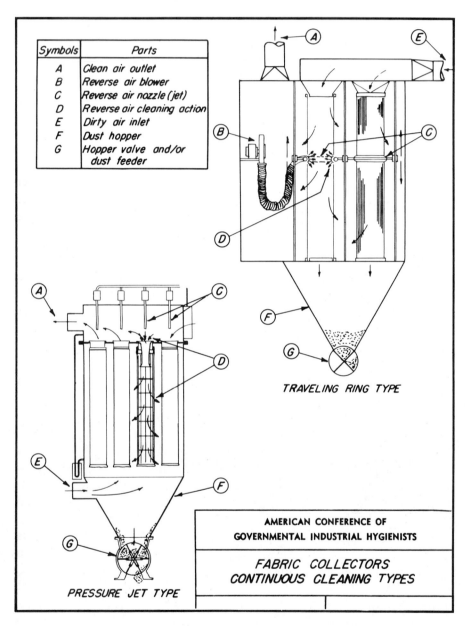

Symbols	Parts
A	Clean air outlet
B	Reverse air blower
C	Reverse air nozzle (jet)
D	Reverse air cleaning action
E	Dirty air inlet
F	Dust hopper
G	Hopper valve and/or dust feeder

TRAVELING RING TYPE

PRESSURE JET TYPE

AMERICAN CONFERENCE OF
GOVERNMENTAL INDUSTRIAL HYGIENISTS

FABRIC COLLECTORS
CONTINUOUS CLEANING TYPES

Fig. 10-6 Two continuous-cleaning fabric collectors.

From: American Conference of Governmental Industrial Hygienists. *Industrial Ventilation*, 15th Ed. ACGIH, Lansing, Mich., 1978.

Two typical configurations for bag-type industrial air cleaners are illustrated in Figure 10-6. In these devices, numerous identical bags hang vertically within an enclosure called a bag house. The bags are closed at one end. Particle-laden air enters the housing at the top or bottom, and flows through the felt surfaces of the bags and out of the center of each bag into the clean-air plenum. The particles accumulate on the outsides (pressure-jet type) or the insides (traveling-ring type) of the bags, until the pressure drop across the bag reaches the preselected action level. In the pressure-jet device, the dust cake is periodically removed by a pressure pulse created by air-jet discharges. The pulse makes the bags expand, and the dust layer falls off and down into the collection hopper. Between cleaning cycles, the dust reaccumulates on the bags. In some bag filters, the cleaning is done by mechanical shaking. In the traveling-ring type, a doughnut-shaped ring with small air jets on its inner surface travels slowly up and down each bag to dislodge the dust.

Bag-filter control devices are available with an enormous variety of design features, numbers, sizes and shapes of bags, efficiency ratings, and operating conditions, and can be used for many, but not all, air-cleaning applications. They cannot be used effectively at very high temperatures without special high-temperature fiber mats, nor for dusts which tend to gum, or accumulate moisture excessively.

ELECTROSTATIC PRECIPITATORS

Electrostatic precipitators used for air-cleaning applications are generally single-stage types, in which both the charging and collection of the particles take place in the same electrode configuration.

A negative high voltage is applied to the corona wires, creating a bipolar glow region around each wire, and a unipolar stream of negative air ions which migrate towards the collecting plates on each side. Particles in the airstream passing through are charged and accelerated toward the collecting plates.

The collecting plates must be periodically cleaned in order to maintain a high overall collection efficiency, since the resistivity of the dust cake affects the voltage gradient and corona current and, thereby, the performance of the precipitator; the resistance to air flow does not change appreciably with dust accumulation. The plates can be cleaned by any of several techniques, as appropriate to the precipitator and the dust. They can be mechanically shaken, or periodically washed. Alternatively, the plates can be continually wetted by a falling film of water, and thereby maintain a more constant operating characteristic.

High-voltage power supplies and electrodes for precipitators are very costly to install, and precipitators are generally not economically competitive with bag filters, except for very large installations. However, for high-temperature, high collection efficiency air-cleaning applications, precipitators may be the only types of air cleaners available. They are widely used to clean coal-fired power-plant effluents.

INERTIAL SEPARATIONS

When particles larger than about 1 to 2 μm in diameter are suspended in a moving stream of air, their motion will deviate less than that of the carrier air when the stream is turned or deflected. If particle displacement within the stream is sufficient to bring them into contact with a surface, they can be collected. The probability of collection increases rapidly with increasing particle size, air velocity, and sharpness of the directional change in the flow. Two types of inertial air cleaners, the baffle type and the cyclone, are illustrated in Figure 10–3.

In the cyclone, dusty air enters tangentially, and the particles are projected onto the outer walls as the stream follows a descending spiral. As dust accumulates in an air-cleaning cyclone, it migrates in flocs along the wall toward the collecting hopper at the bottom. Part of the way down the conical section, the descending air spiral containing the fine particle fraction turns inward and upward, and leaves the axial exit pipe on a tighter spiral.

Their simplicity of construction and operation, and their relatively low power requirements, make inertial collectors relatively inexpensive to purchase and operate. However, they are of limited value for controlling the release of particles having health significance, because of their inability to capture particles smaller than about 5μm. These devices are, however, frequently used to collect nuisance-type dusts having large particle sizes. They are also used as pre-cleaners upstream of more efficient and expensive particle collectors, such as bag filters, electrostatic precipitators, or Venturi scrubbers. When the mass median particle size is large, an inertial precollector can remove a substantial fraction of the total mass at low cost, reducing the loading on the more expensive second-stage air cleaner.

MOBILE-SOURCE CONTROLS

The control of mobile-source (automobile) emissions was mandated by Congress in stages, as indicated in Table 8–8. The sequence of steps was intended to permit adequate lead time for the development and testing of control technology. The entire industry adopted essentially the same tech-

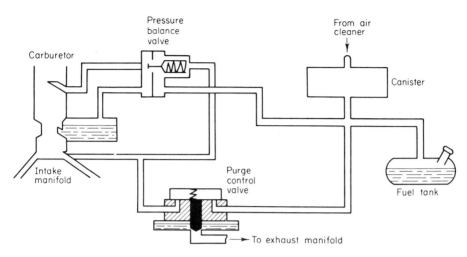

Fig. 10–7 Fuel-evaporation loss-control system. Vapors from the gasoline tank and carburetor are vented through a canister containing activated charcoal. The adsorbed vapors are dissolved and metered into the intake manifold during engine operation when intake air demand is high.

From: Wohlers, H. C. Air – A Priceless Resource. pp. 189–294 in *Environmental Health*. P. W. Purdom (ed.). New York: Academic Press, 1971.

nology for the control of evaporative losses of fuel (Figure 10–7) and gaseous leakage around the pistons (Figure 10–8), but chose several different approaches to the control of tailpipe emissions of CO and hydrocarbons. Most adopted the catalytic converter to more completely oxidize exhaust gases; others elected to achieve more complete combustion within the engine, some with electronic ignition systems, and others with different combustion-chamber designs such as the stratified-charge engine.

STATIONARY COMBUSTION SOURCE CONTROLS

The effluents from industrial and utility boilers fired by coal and heavy oil contain substantial quantities of ash, sulfur dioxide, and nitric oxide. Up until the 1960's, the only exhaust-gas cleaning that was commonly done was removal of most of the ash, using either electrostatic precipitators or bag filters. Neither type of air cleaner removed any significant amount of the gas-phase contaminants.

The initial focus of stationary-source control under the Clean Air Act of 1970 was the reduction of SO_2 emissions by (a) reducing the average sulfur content of the fuel burned and (b) extracting SO_2 from the effluent

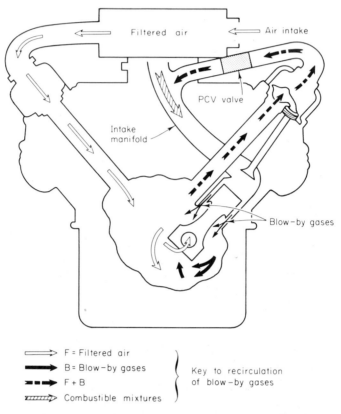

Filtered air   Air intake

PCV valve

Intake
manifold

Blow-by gases

F = Filtered air
B = Blow-by gases
F + B
Combustible mixtures

Key to recirculation
of blow-by gases

Fig. 10–8 Positive crankcase ventilation (PCV). Crankcase (or blow-by) gases (primarily unburned fuel) are routed through a control valve to regulate the flow to the fuel-intake system under varying operating conditions of the engine.

From: Wohlers, H. C. Air – A Priceless Resource. pp. 189 – 294 in *Environmental Health*. P. W. Purdom (ed.), New York: Academic Press, 1971.

gas before it leaves the stack. Most utilities having an adequate supply of low-sulfur fuel chose the former approach. The increased fuel costs could be passed directly to the power consumers. The alternative was to incur the additional expenses required for land acquisition and plant construction and operation for a stack gas-cleaning system. Some utilities sought permission for a third alternative, i.e., dilution via high stacks, but EPA refused to approve control by dilution, and this position has been upheld in the courts. Since there is not enough low-sulfur fuel to accomodate the

Table 10.5   Flue-Gas Desulfurization Processes

Process	No. of boilers	Total controlled generating capacity (MWe)
Lime-slurry scrubbing	9	2,702
Limestone-slurry scrubbing	6	2,037
Alkaline-fly ash scrubbing	2	720
Sodium carbonate scrubbing	3	375
Magnesium oxide scrubbing	1	120
Wellman-Lord	1	115
Total:	22	6,069

Source: "Summary Report – Flue Gas Desulfurization Systems," Jan., Feb., Mar., 1977, PEDCo Environmental, Inc.

demand, an increasing number of utilities have installed or are committed to installing flue-gas desulfurization (FGD) processes to clean their effluents.

In the United States, there are currently six types of FGD systems in use for controlling full-scale utility boilers. Table 10–5 summarizes these systems by type and extent of application, and shows that most of the operating capacity involves lime- or limestone-slurry, or sodium carbonate scrubbing techniques. These are known as "throwaway" processes, since the SO_2 is removed in a form that is discarded. On the other hand, the magnesium oxide and Wellman-Lord processes produce salable sulfur products, such as elemental sulfur, sulfuric acid, or liquid sulfur dioxide.

In concept, lime- or limestone-slurry scrubbing processes are very simple. In practice, however, the chemistry and the system design for a full-scale operation can become quite complex.

The overall absorption reaction taking place in the scrubber and the hold-tanks for a limestone-slurry system produces hydrated calcium sulfite:

$$CaCO_3 + SO_2 + \tfrac{1}{2}H_2O \rightarrow CaSO_3 \cdot \tfrac{1}{2}H_2O + CO_2 \qquad (10-1)$$

With a lime-slurry system, the overall reaction is similar, but yields no CO_2:

$$CaO + SO_2 + \tfrac{1}{2}H_2O \rightarrow CaSO_3 \cdot \tfrac{1}{2}H_2O \qquad (10-2)$$

(The actual reactant in Eq.$(10-2)$ is $Ca(OH)_2$, since CaO is slaked in the slurrying process.)

In practice, some of the absorbed SO_2 is oxidized by oxygen which is also absorbed from the flue gas; the resultant product appears in the slurry as either gypsum ($CaSO_4 \cdot 2H_2O$) or as a calcium sulfite/sulfate mixed crystal. The slurry is recycled around the scrubber to obtain the high liquid-to-gas ratios required.

Lime- and limestone-slurry scrubbing systems can be engineered for almost any desired level of SO_2 removal. Commercial utility systems are generally designed for 80% to 95% removal; however, some systems have, at times, achieved removal efficiencies as high as 99%.

Some low-sulfur coals contain large amounts of alkaline metal oxides (e.g., CaO, MgO) in the ash which results from combustion. These coals appear particularly suitable to SO_2 control by scrubbing the waste gases with a slurry of the alkaline fly ash. There are two ways to add the ash to the system: (a) collecting the fly ash in an electrostatic precipitator upstream of the scrubber and then slurrying the dry fly ash with water so that it can be pumped into the scrubber circuit, and (b) scrubbing the fly ash directly from the flue gas by the circulating slurry of fly ash and water.

"Throwaway" FGD systems depend, of course, on having a suitable place to throw the waste away. The enormous amounts of impure gypsum slurry produced by lime- and limestone-process plants must be piped to large settling and evaporation ponds. These FGD processes are, therefore, economically feasible only where land for such ponds is available close by and at low cost. The sodium carbonate process produces a solution of Na_2SO_4, which has no commercial value and cannot be readily disposed of except in some desert areas.

The magnesium oxide process differs from the above techniques in that it is a "regenerable" or "salable product" process; the SO_2 removed from the flue gas is concentrated and used to make marketable H_2SO_4 or elemental sulfur. This process employs a slurry of MgO or $Mg(OH)_2$ to absorb SO_2 from the flue gas in a scrubber, and yields magnesium sulfite and sulfate. When dried and calcined, the mixed sulfite/sulfate produces a concentrated stream (10–15%) of SO_2 and regenerates MgO for recycle to the scrubber. Carbon added to the calcining step reduces any $MgSO_4$ to MgO and SO_2

In commercial applications, the scrubbing and drying steps would normally take place at the power plant. The regeneration, and the sulfur or H_2SO_4 production steps, might be performed at a conventional sulfuric acid plant. Alternatively, a central processing plant could produce sulfur from mixed magnesium sulfite/sulfate brought in from other desulfurization locations.

The Wellman-Lord (W–L) process is also a regenerable system. When coupled with other processing steps, it can make salable liquid SO_2, H_2SO_4, or elemental sulfur.

The W–L process employs a solution of Na_2SO_3 to absorb SO_2 from waste gases in a scrubber or absorber, converting the sulfite to bisulfite:

$$Na_2SO_3 + SO_2 + H_2O \rightarrow 2\,NaHSO_3 \qquad (10-3)$$

Thermal decomposition of the bisulfite in an evaporative crystallizer can regenerate sodium sulfite for reuse as the absorbent:

$$2\,NaHSO_3 \xrightarrow{\triangle} Na_2SO_3 + SO_2 + H_2O \qquad (10-4)$$

The evaporative crystallizer produces a mixture of steam and SO_2 and a slurry which contains sodium sulfite/sulfate plus some undecomposed $NaHSO_3$ in solution. As water condenses from the steam/SO_2 mixture, it leaves a wet SO_2-enriched gas stream to undergo further processing for recovery of salable sulfur.

Dilution and Intermittent Discharge

Dilution is a control technique only in the sense that it depends on turbulent diffusion in the atmosphere to reduce the concentration of the effluents to acceptable levels. It may also depend on the ability of the normal processes in the atmosphere to degrade, transform, or remove the contaminants, so that they do not accumulate excessively within the atmosphere. Consideration of the acceptability of control by dilution may even extend to the effects of the accumulation of airborne contaminants on the land, in surface waters, and in the biosphere. The technical considerations in determining the extent of dilution of contaminants discharged into the atmosphere were covered in some detail in Chapter 4.

Intermittent discharge is, as previously discussed, a useful adjunct to dilution in the atmosphere. Since effective dilution varies with the atmospheric lapse rate and wind velocity, control by dilution is better justified when there are means of stopping or reducing effluent discharges during periods when dilution factors are unfavorable. The most notable example of voluntary effluent reductions during periods of poor atmospheric dispersion is the use of reserve supplies of low-sulfur fuel for industrial and utility boilers, as a means of limiting the build-up of ambient sulfur dioxide levels.

CONTROL OF SURFACE-WATER QUALITY

Background

By the turn of the century, the concentration of people in metropolitan areas and the increasing use of sewer systems which discharged directly into surface waters resulted in the frequent appearance of floating fecal matter and attendant problems of odor and visual blight. Thus, until quite recently, water-pollution control was primarily concerned with sewage and other wastes having similar properties. The installation of primary-treatment facilities was generally sufficient to alleviate the immediate problem. This stage of treatment involves removal of larger suspended and floating solids, and oil and grease, generally by mechanical means.

Primary treatment, however, only removed part of the oxygen demand of the waste, and the continuing increase in the total amount of sewage led to frequent depletion of the dissolved-oxygen levels in surface waters around major metropolitan areas. Oxygen depletion, in turn, led to fish kills and odors associated with putrefaction. The general solution to this problem was the installation, beginning on a large scale in midcentury, of second-ary-treatment facilities to supplement the primary treatment. Secondary treatment generally is synonomous with biological treatment. In this process, the liquid waste is contacted with bacteria, which consume the organic material in the waste during their metabolism. The actual reduction in oxygen demand achieved depends upon the ability of the bacteria to use

Table 10.6 The Concentration of Contaminants That Make Primary Treatment Necessary

Contaminant	Limiting concentration
Acidity	Free mineral acidity
Alkalinity	0.5 lb alkalinity as $CaCO_3$/lb BOD removed
Suspended solids	> 125 mg /l.
Oil, grease	> 50 mg/l.
Dissolved salts	> 16 gm/l.
Organic-load variation	> 4 : 1
Trace metals	> 1 mg/l.
Sulfides	>100 mg/l.
Phenols	>70 – 160 mg/l.
Ammonia	>1.6 gm/l.

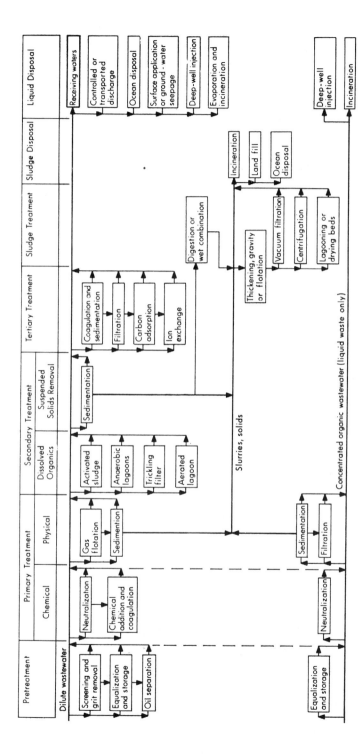

Fig. 10-9 Wastewater-treatment sequence. Very often, one process may be substituted for another, or a variety of different processes may be needed for treatment of specific contaminants. (Pretreatment is usually considered to be part of the primary treatment of wastewater.)

From: Eckenfelder W. W., Jr., and Ford, D. L. Economics of Wastewater Treatment. *Chemical Engineering 76* (Aug. 25):109–18, 1969. © McGraw-Hill. Used with permission of the publisher.

Table 10.7   Typical Applications of Wastewater Treatment Processes

Process	Contaminant					
	Acids	Alkalies	Total dissolved solids	Suspended solids	Nitrate	Phosphate
Filtration				X		
Sedimentation				X		
Clarification				X		X
Flotation				X		
Neutralization	X	X				
Biological treatment					X	X
Lagooning				X	X	X
Chemical oxidation or reduction						
Emulsion breaking						
Adsorption				X		
Ion exchange			X		X	X
Reverse osmosis			X	X	X	X
Electrodialysis			X		X	X

Compiled from: Paulson, E. G. Water Pollution Control Programs and Systems, pp. 611–27 in *Industrial Pollution Control Handbook*, H. F. Lund (ed.). New York: McGraw-Hill, 1971; and Metcalf and Eddy, Inc. *Wastewater Engineering: Collection, Treatment, Disposal*. New York: McGraw-Hill, 1972.

the organics in the waste, and the contact time of the contaminants with the bacteria.

Depending upon their biodegradability, nonsewage chemical contaminants in waste streams may be partially removed by the bacteria in secondary-treatment facilities. Some chemical contaminants, however, present serious problems to sewage treatment plant operations in that they are toxic to the bacteria. If too many bacteria are killed, the operation of the treatment plant can be seriously disrupted, and it may take many days for the reestablishment of an effective bacterial colony in the facility.

Primary-treatment steps are frequently needed to modify the waste to the extent that it will be amenable to biological treatment. Conditions calling for such preliminary treatments are listed in Table 10–6.

It is only recently that concern has shifted from the oxygen demand to

| | | Contaminant | | | |
Trace Metals	Organic material (BOD)	Oils	Persistent pesticides	Color	Taste and odor
X					
		X			
X	X	X			
		X			
	X			X	X
	X	X		X	X
				X	X
		X			
	X		X	X	X
X					
	X				

the toxicity and carcinogenicity of contaminants in surface waters. This concern has led to an acceleration in the development of a variety of advanced physical and chemical treatment techniques (also known as tertiary treatment) for the removal of many chemicals from waste streams. Much of this modern technology has been based on adaptations of the process technology of chemical engineering, having been used for many years to recover materials from industrial-process waste streams. However, applications of these techniques to large-volume municipal-waste streams of extremely heterogenous composition is a relatively new concept.

The kinds of wastewater-treatment processes used, and their sequence, are illustrated in Figure 10–9. The various treatment schemes accomplish different levels of removal, and as the degree of treatment increases, the cost also rises. Note that there is some overlap in techniques used in the different stages of water treatment. The contaminants for which the various processes are effective are presented in Table 10–7.

Control of Chemical Contamination from Point Sources

STREAM MANIPULATION

SEGREGATION OF WASTE STREAMS

When there are numerous waste streams, with many having similar characteristics, it may be desirable to have two or more separate waste systems, e.g., one for oily wastes, one for inorganic chemicals, one for storm drainage. In some cases, it may not be necessary to clean the storm drainage at all, and keeping it out of the chemical-waste system will help to avoid flow surges that can overwhelm the capacity of the waste-treatment facilities.

MIXING OF WASTE STREAMS

Under some circumstances, it is better to mix waste streams than to segregate them for separate treatment. If, for example, one stream is acidic and another is basic, one waste can neutralize the other, producing a salt solution which may require no further treatment. Even when the salt is insoluble and requires collection and disposal, the net effect may still be beneficial, and the purchase of chemicals for intentional neutralization can be minimized. The economic benefits of complementary wastes frequently lead neighboring industries to exchange waste streams or to share in the costs of operating a combined waste-treatment facility.

The mixing of streams can also be beneficial in terms of equalization of the strength of the waste, especially when one or more of the major sources are intermittent. The mixing of streams within the sewer system or preliminary-treatment steps may prevent concentration peaks which could adversely affect subsequent biological treatment or advanced chemical processes.

REDUCTION OF WASTE VOLUMES

Reduction of waste volumes involves both decreases in the amount of contaminant and the amount of dilution water; both can generally be achieved by a good preventive maintenance program which detects and corrects leakage. Improvements in process design, equipment, and operating procedures may also make it possible to maintain or improve process output with reduced amounts of waste materials.

WATER-TREATMENT TECHNIQUES

SCREENING

Screening is a mechanical process that separates suspended particles on the basis of size. There are several types of devices which may have static,

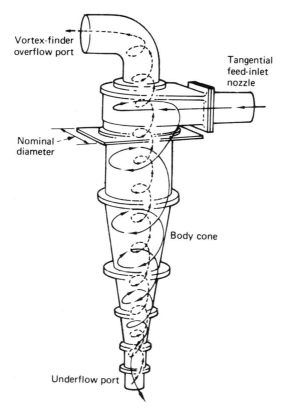

Fig. 10-10 Hydrocyclone.

From: Lash, L. D. and Kominek, E. G. Primary-Waste-Treatment Methods. *Chemical Engineering 82*(No. 21):49–61, 1975.
© McGraw-Hill. Used with permission of the publisher.

vibrating, or rotating screens. Openings in the screening surfaces range from several centimeters down to about 20 μm.

Screens are used for treating raw water, storm water, and wastewaters from certain industries which produce large volumes of solids, such as food-process plants, textile mills, and pulp and paper mills. The amount of solids removal depends upon screen size and characteristics of the water or wastewater being treated. For example, fine-mesh microstrainers, with screen apertures of about 20 μm, remove solids, such as various forms of plankton and general microscopic debris; coarse bar screens which have metal bars spaced approximately 25 to 50 mm apart are used to remove the large suspended solids. The screens are periodically cleaned, generally by automatic equipment.

Comminutors, or large circular garbage grinders, may be used to grind into smaller pieces those solids which penetrate the coarse screens.

CENTRIFUGAL SEPARATION

Hydrocyclones are used to remove relatively dense suspended solids from a liquid stream; a common application is for separating grit from storm water or wastewater. Figure 10–10 shows their operation. Fluid-pressure energy is used to create rotational fluid motion in a cyclone. This motion causes relative movement of materials suspended in the fluid, and permits separation of these materials from the fluid. The bulk of the fluid is removed through a top outlet, and the separated solids are discharged through a second outlet located in an axial position at the bottom.

Smaller particles can be separated using centrifuges, but such separations are much more expensive, both in terms of capital outlay and operating costs.

NEUTRALIZATION

When there is an excess of acid or base in the wastes, it may be necessary to treat the waste stream with sufficient basic or acidic chemicals to produce neutralization. For small volumes of acidic wastes, sodium hydroxide is used, and the resulting salts will usually not present any significant disposal problem. Lime and ground limestone are much less expensive per unit of basicity, but produce insoluble neutralization products which will usually require separate disposal.

Alkaline wastes can be neutralized with sulfuric or hydrochloric acids. Relatively inexpensive neutralizations can be achieved by bubbling flue gas containing CO_2 through the waste stream.

PRECIPITATION

Precipitation reactions are often carried out in water and waste treatment. Lime is frequently used to precipitate calcium carbonate and magnesium hydroxide in water treatment. In waste-treatment operations, the precipitation of metallic hydroxides from plating, metallurgical, and steel mill wastes and also from acid mine drainage is commonplace. Wastes that contain phosphates may also be treated to precipitate calcium or aluminum phosphate. Lime or limestone is used for the precipitation of calcium sulfite or sulfate from stack-gas scrubber water.

Essentially, every reaction is carried out with the stoichiometric amount of precipitant and at the optimum pH. If the solubility of a compound exceeds the required effluent concentration, an excess of the precipitant may be added to reduce the solubility.

Precipitation reactions are usually carried out in the presence of previously formed sludge. This minimizes supersaturation, and also results in more rapidly settling precipitates, because precipitation occurs on existing slurry particles.

FLOCCULATION AND COAGULATION

Flocculation is the process of bringing together fine particles, e.g., colloids, so that they agglomerate, forming larger particles (flocs) which will settle more rapidly. Mechanical stirring or air injection may be used to cause the suspended particles to collide. Alternatively, a chemical coagulant may be used; these cause suspended solids to agglomerate by reducing colloidal charge levels. When a metallic cation such as aluminum, e.g., as aluminum sulfate (alum), or iron, e.g., as ferric chloride, is used, particulate metallic hydroxides are initially formed, which adsorb the colloids.

Synthetic polymers are often used as coagulants or as coagulant aids. Polymers may be cationic, and adsorb a negatively charged particle; anionic, which permit bonding between a colloid and polymer; or nonionic, which adsorb flocculates by hydrogen bonding between the solid surfaces and the polar groups of the polymer.

SEPARATION BY DENSITY DIFFERENCES

Separation by density differences involves the removal of suspended solids or oils when the specific gravity difference from water causes settling or rising of the solids or oils during passage through a tank under suitably quiescent conditions.

Static conditions that affect such separations are (a) water temperature, which affects both viscosity and density of the water, (b) specific gravity of the oil or suspended solids, (c) size and shape of suspended oil droplets and of particles, and (d) solids concentration, which results in free or hindered settling.

Sedimentation Sedimentation is the removal of suspended solids via gravity settlement, and is usually performed in tanks (known as settling tanks) having a continuous flow and a detention time of several hours. The settling velocities of particles will change with time and depth, as particles agglomerate and form larger floc sizes.

Settling tanks are divided into four zones: (a) inlet zone, to provide a smooth transition from the influent flow to a uniform steady flow desired in the settling zone; (b) outlet zone, to provide smooth transition from settling zone to the effluent flow; (c) sludge zone, to receive and remove set-

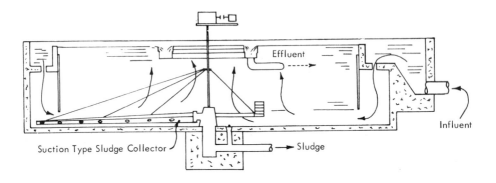

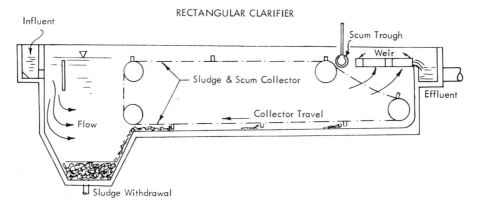

Fig. 10–11 Typical clarifier units.

From: Ross, R. D. (ed.). *Industrial Waste Disposal*. New York: Reinhold, 1968.
© 1968 by Litton Educational Publishing, Inc. Reprinted by permission of Van Nostrand Reinhold Co.

tled material and prevent it from interfering with the sedimentation of particles in the settling zone; and (d) settling zone, to provide tank volume for settling, free of interference from the other three zones. Settling tanks which follow preliminary treatment such as screening are known as primary clarifiers. Typical primary-clarifier units are shown in Figure 10–11.

The surface area of a settling tank is one of the most important factors that influence sedimentation. The tank should be designed so as to produce a clarified effluent at minimum water temperature, and to allow for separating floc particles that may not be of maximum size or density. This latter problem may be due to either partial deflocculation in the inlet zone, or to partially ineffective coagulation. Tanks are generally about 30 m in diameter and about 4–5 m deep.

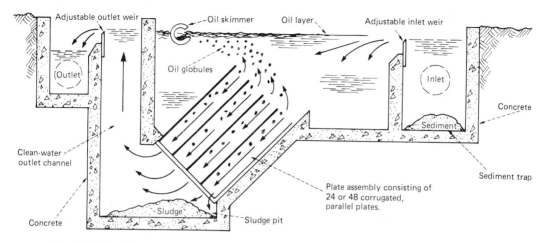

Fig. 10–12 Corrugated-plate interceptor separator.

From: Ford, D. L. and Elton, R. L. Removal of Oil and Grease from Industrial Wastewaters. *Chemical Engineering 84*(No. 22):49–56, 1977.

© McGraw-Hill. Used with permission of the publisher.

If the release of entrained air might cause floating floc, or if floating solids may be present, a scum baffle plus a scum-skimmer mechanism are usually installed. Sludge removal from the tank bottom may be performed manually, or by a continuous belt-drive system.

An oil/water separator is used to separate free oil from refinery wastewater. This type of unit will not separate soluble substances, or break emulsions.

A corrugated-plate interceptor for oil is shown in Figure 10–12. Wastewater enters a separator bay and flows downward through corrugated plates arranged at an angle of 45° to the horizontal. Oil collects on the underside of the plates and rises to the surface where it is skimmed. Solids settle into a sludge compartment, with the now-clarified waste discharging into an outlet channel.

Air Flotation In dissolved air flotation, illustrated in Figure 10–13, air is intimately contacted with an aqueous stream at high pressure, dissolving the air. The pressure on the liquid is reduced through a back-pressure valve, thereby releasing micron-sized bubbles that sweep suspended solids and oil from the contaminanted stream to the surface of the air-flotation unit. Applications of this technique include treating effluents from oil refineries, metal-finishing processes, pulp and paper mills, cold-rolling

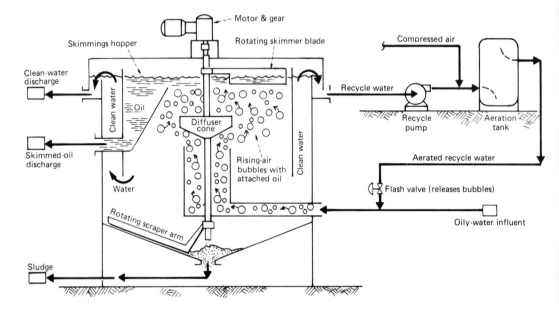

Fig. 10–13 Dissolved-air flotation unit.

From: Ford, D. L. and Elton, R. L. Removal of Oil and Grease from Industrial Wastewaters. *Chemical Engineering 84*(No. 22):49–56, 1977. © McGraw-Hill. Used with permission of the publisher.

mills, poultry processing, grease recovery in meat-packing plants, and cooking-oil separation from french-fry potato processing. An increasingly important application is the thickening of sludge.

The attachment of gas bubbles to suspended solids or oily materials occurs by several mechanisms. The suspended-solids/gas mixture is carried to the vessel surface after (a) precipitation of air on the particle; (b) collision of a rising bubble with a suspended particle; (c) trapping of gas bubbles, as they rise, under a floc particle; and (d) adsorption of the gas by a floc formed or precipitated around the air bubble.

Flocculants, such as synthetic polymers, may be used to improve the effectiveness of dissolved-air flotation. Also, coagulants, such as alum, may be used to break emulsified oils and to agglomerate materials for improved flotation recovery.

Flotation can also be achieved by induced air, as illustrated in Figure 10–14. Air drawn into the flotation cell by action of the rotor is mixed with the water and transformed into minute bubbles. Oil particles and suspended solids attach themselves to the gas bubbles and are borne to the surface

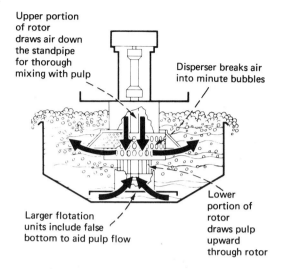

Upper portion
of rotor
draws air down
the standpipe
for thorough
mixing with pulp

Disperser breaks air
into minute bubbles

Larger flotation
units include false
bottom to aid pulp flow

Lower
portion of
rotor
draws pulp
upward
through rotor

Fig. 10–14   Induced-air flotation unit.

From: Lash, L. D. and Kominek, E. G. Primary-Waste-Treatment Methods. *Chemical Engineering 82*(No. 21):49–61, 1975.
© McGraw-Hill. Used with permission of the publisher.

of the water. Skimmer paddles push the contaminated froth from the top of the cell into collection launders. The tank is designed with sloping sides at the lower part to recirculate the water for continuous purification. The conical shape of the dispenser hood, with its perforated surface, acts to quiet the liquid surface.

Efficiency of contaminant removal in induced-air devices is often improved with chemical aids injected in the water, upstream in the flotation cell. In addition to breaking oil-water emulsions, they also make the air bubbles more stable and promote formation of froth on the surface of the water. Properly conditioned wastes leave the device nearly 100% oil-free, and the suspended solid load is also significantly reduced.

FILTRATION

Granular-media filtration is a liquid-solids separation method that uses flow through porous media, such as sand, to remove particulates. To be readily filterable, the suspended solids must either be naturally flocculent, or be made so by chemical coagulation. Granular-media filters are used for removal of solids (and oil) from refinery wastewaters, for effluents from activated-sludge treatment plants (see next section) and for effluents from

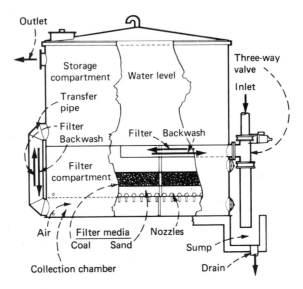

Fig. 10–15   Granular-media filter (shallow bed, gravity type).

From: Lash, L. D. and Kominek, E. G. Primary-Waste-Treatment Methods. *Chemical Engineering* 82(No. 21):49–61, 1975.
©McGraw-Hill. Used with permission of the publisher.

pulp and paper plants. Heavy-metal precipitates may be recovered in granular filters, and process streams are clarified by similar units.

The heart of any granular-media filter is the filter bed. The size and depth of the filter medium are the most important design parameters. As an example of granular-media filtration, the operation of a shallow-bed, gravity, granular-media filter is shown in Figure 10–15. The filter is completely automatic on a controlled cycle. Several filters are customarily used in parallel to avoid interruption of the flow stream when the filter backwashes. During filtration, wastewater is passed through the media, and solids are removed. As the cycle progresses, the pressure drop across the filter increases to a preset maximum, indicating that the bed is full of suspended solids, and a backwash-cycle controller is energized. Alternately, a timed cycle or monitoring of the suspended solids may be used to terminate filtration.

Backwash water flows by gravity through the collection chamber and the filter compartment to remove the suspended solids that were fluidized by the air wash. The backwash discharges by gravity to settling basins, where all of the suspended solids settle out and water is recovered. The addition of polymers to the backwash water accelerates the settling rate of the solids.

BIOLOGICAL TREATMENT

Biological oxidation is the major process for the removal of oxygen demand in sewage wastes; it also finds extensive use for the treatment of industrial wastes containing biodegradable organics. In biological oxidative processes, concentrated masses of microorganisms break down organic matter. These microorganisms are broadly classified as aerobic, facultative, or anaerobic. Aerobic organisms require oxygen for metabolism, anaerobes function in the absence of oxygen, and facultative microbes may function in either an aerobic or anaerobic environment.

The predominant species used in biological systems are known as heterotrophic microorganisms; these require an organic carbon source for both energy and cell synthesis. Autotrophic organisms, in contrast, use an inorganic carbon source, such as carbon dioxide or carbonate. The autotrophs may derive energy for cell synthesis from the oxidation of inorganic compounds of nitrogen or sulfur (chemosynthetic bacteria), or from the sun (photosynthetic bacteria).

Basically, biological oxidative processes involve either of two mechanisms to accumulate and store the microbes: (a) as a flocculated suspension of biological growth known as activated sludge, which is mixed with the wastewaters, or (b) as a biological film fixed to an inert medium, over which the wastewaters pass.

Biological processes may not be applicable to many specific industrial wastewaters. For example, cyclic organics, especially if halogenated, are resistant to biological degradation. Organochlorine pesticides may prove toxic to a biosystem if discharged in high concentrations. Slightly soluble components, such as PCB's or heavy metals, can accumulate in a biological system through adsorption and bioconcentration and may, thereby, reach levels inhibitory to the process.

Microbial populations may be specifically adapted to certain chemicals to successfully achieve oxidation. Examples of materials which are biodegradable under acclimated conditions are cyanide, phenol, formaldehyde, acrylonitrile, and hydroquinone. However, some cyanide compounds, such as the metallic-cyanide complexes, are highly resistant to degradation, even under strongly acclimated conditions.

Activated-Sludge Process   The activated-sludge process is a continuous

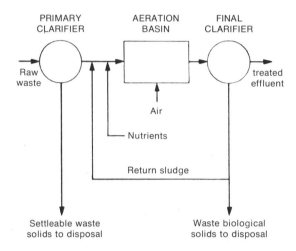

Fig. 10–16   Basic flow sheet for the activated-sludge process.

From: Lesperance, T. W. Biological Treatment. *Chemical Engineering 75* (Oct. 14):89–94, 1968.
© McGraw-Hill. Used with permission of the publisher.

system in which aerobic bacteria are mixed with wastewater and then physically separated by gravity clarification or by air flotation. As shown in Figure 10–16, the concentrated sludge is recycled to the reactor to mix with the incoming waste. Oxygen is provided in a variety of ways, e.g., by diffused aeration, surface aeration, or static mixers, and may be introduced either as air, pure oxygen, or oxygen-enriched air. The waste product from the activated-sludge process is excess sludge, which must ultimately be disposed of, generally after partial reduction by anaerobic digestion.

As the concentration of organics in the remaining wastewaters decreases, the rate of biological removal also decreases, since the remaining organics are progressively more difficult to remove. In the treatment of industrial wastes, a significant fraction of the organics may prove nonbiodegradable. Thus, although the BOD removal may be excellent, the removal of COD may be quite low.

The activated-sludge process is broadly applicable. It is used in the treatment of soluble organic wastes from many industries, including food processing, meat packing, pulp and paper, oil refining, leather tanning, textiles, organic chemical, and petrochemical.

A major modification to the conventional air-aerated activated sludge process has been the use of high-purity oxygen. Some advantages of this compared to the conventional system are lower land requirements, the ability to supply oxygen at high rates (making it amenable to high-strength wastes), and increased self-neutralization of highly alkaline wastes.

Increasingly stringent effluent criteria require the removal of inorganic nutrients, such as nitrogen, prior to stream discharge into a natural receiving water. Activated-sludge systems have been successfully operated for nitrogen removal, via the processes of nitrification and denitrification.

Nitrification is achieved by autotrophic nitrite bacteria, organisms which convert reduced nitrogen (as NH_3) to nitrite (NO_2^-), and nitrate bacteria which further oxidize the nitrite to nitrate (NO_3^-). Sufficient alkalinity must be available to offset the production of nitrous and nitric acids.

The growth rate of the nitrifiers is significantly lower than that of the heterotrophic bacteria used in the breakdown of carbonaceous organic matter. Thus, in the presence of significant amounts of carbonaceous organic material, the nitrifiers are unable to compete successfully with the heterotrophs and cannot accumulate as significant populations. The process may, therefore, have to be a two-stage system, where carbonaceous removal is achieved in the first stage and nitrification is performed in the second. Interstage clarification allows segregation of the two types of bacteria.

Denitrification is a biological process by which nitrite and nitrate are reduced to nitrogen gas. Facultative, heterotrophic bacteria found in activated sludges can produce denitrification. Since these organisms require an organic-carbon source for cell growth, and because this process generally follows secondary treatment, some form of organic compound must be present or added. Approximately 4.5 lb of COD (typically as methanol) per lb of nitrate are required to reduce the nitrate to nitrogen.

Trickling Filters   The predominant fixed-film biological process is the trickling filter. This is a packed bed of some medium (such as stones) covered with a biological slime, over which wastewater is passed. Oxygen and organic matter diffuse into the slime film, where degradation reactions occur. End products such as CO_2, NO_3^-, etc., counter-diffuse back out of the film and appear in the filter effluent.

Trickling filters have found limited application for industrial wastewaters. Removal rates for soluble industrial wastes are typically low, making filters unattractive for high BOD-removal efficiency. The process is, however, capable of accepting highly variable loadings, and, thus, may be used as a roughing filter to provide partial treatment.

Another type of fixed-film biological reactor is the rotating disk process.

Plastic disks are mounted on a shaft and placed in a tank conforming to the general shape of the disks. The disks are slowly rotated while approximately half immersed in the wastewater. Rotation brings the attached bacteria culture into contact with the wastewater for removal of organic matter. Rotation also provides a means of aeration, by exposing a thin film of wastewater on the disk surface to the air.

Other Biological Treatment Processes   Biological oxidation may be achieved in ponds, lagoons, and basins, or in soil after land application of wastewater. The land requirements for oxygen demand removal are greater in these systems than those required for fixed-film processes, and are much greater than those for activated-sludge processes.

In aerated stabilization basins, air is pumped into the water or the water is sprayed through the air to increase the oxygen-demand removal capacity. In ordinary stabilization and oxidation ponds and in land irrigation, natural aeration rates are the limiting factor in determining the capacity for oxygen-demand removal.

COMBINED BIOLOGICAL AND CHEMICAL OXIDATION

The continuous addition of powdered activated carbon to aeration tanks of suspended-bacterial systems can improve the operation and performance of the activated sludge. The activated carbon gives the system the capability of adsorbing nonbiodegradable organic matter present in the wastewaters, thereby providing a degree of tertiary treatment, and resistance to shock-loading (i.e., sudden input of a large amount of a chemical). The carbon is regenerated by pyrolysis.

A wet-air-oxidation-biophysical treatment process is schematically shown in Figure 10–17; this has been used in the treatment of wastewaters containing cyanide, acrylonitrile, and pesticides. The incoming wastes are subjected to wet-air-oxidation under high temperature and pressure, resulting in the breaking down of large molecules and cyclic compounds into intermediates, such as aldehydes and organic acids. These latter chemicals are more amenable to the biological treatment system that follows. The process includes the addition of powdered activated carbon, which serves as a sink for substances that may have passed wet-air oxidation; these components are adsorbed and pass out on the surface of the carbon. The combined biological sludge and carbon are treated by the wet-air-oxidation process, where the sludge is reduced to a small fraction of its original weight, after which the carbon is reactivated and returned to the aeration tank.

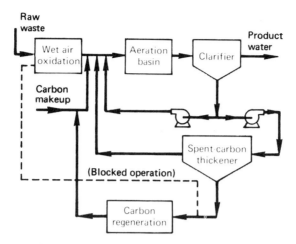

Fig. 10–17 Biophysical treatment by wet-air oxidation.

From: Mulligan, T. J. and Fox, R. D. Treatment of Industrial Wastewaters. *Chemical Engineering 83*(No. 22):49–66, 1976.
© McGraw-Hill. Used with permission of the publisher.

CHEMICAL OXIDATION

The chemical oxidants widely used today are chlorine, ozone, and hydrogen peroxide. Their historical use, particularly for chlorine and ozone, has been for disinfection of water and wastewaters. They are, however, receiving increased consideration for removing from wastewaters organic materials that are resistant to biological or other treatment processes.

Ozone is found to be effective in many applications: for color removal, disinfection, taste and odor removal, iron and manganese removal, and in the oxidation of many complex organics, including lindane, aldrin, surfactants, cyanides, phenols, and organometal complexes. With the latter, the metal ion is released, and can be removed by precipitation.

Chlorine (and its more readily storable form, hypochlorite ion, or bleach) has long been used to purify water, destroy organisms in wastewater and swimming pools, and oxidize chemicals in aqueous solutions. The destruction of cyanide and phenols by chlorine oxidation is well known in waste treatment technology. The continued use of chlorine for these applications is, however, uncertain, because of concern about the toxicity and carcinogenicity of chlorine oxidation products.

ADVANCED WASTEWATER TREATMENT PROCESSES

Most conventional wastewater-treatment facilities use primary and second-treatment stages. Table 10–8 shows typical removal efficiencies for BOD, suspended solids, and dissolved solids by these processes; dissolved solids, which may include many toxic chemicals, are not effectively removed by this treatment.

Various physical and chemical processes are used in tertiary or advanced treatment to bring the water to a higher quality level than that achievable by conventional primary and secondary treatment. The methods include air (ammonia) stripping, electrodialysis, activated carbon adsorption, reverse osmosis, distillation, and ion exchange.

Ammonia can be removed from wastewater by the technique of air stripping, which involves pH control of the water. Ammonium ions in the water exist in equilibrium with ammonia and hydrogen ions. As the pH increases (usually by the addition of lime), the equilibrium shifts to the right, and above a pH of 9, ammonia may be liberated as a gas by agitating the wastewater in the presence of air.

Electrodialysis uses an induced electric current to separate the cationic and anionic components of a solution. It relies on membranes, which are at right angles to the line of electric current flow, that permit ions to pass from a dilute solution on one side to a concentrated solution on the other.

Activated carbon adsorption is used for the removal of soluble organic matter not removed by conventional methods. Water is passed through a bed of granular activated carbon, where organic molecules attach to the carbon surface. When the carbon reaches its adsorptive capacity, it is regenerated by pyrolysis. Very large particles are removed by filtration or coagulation-sedimentation prior to carbon adsorption.

Reverse osmosis forces water at high pressures through cellulose acetate membranes against the natural osmotic pressure. The mechanisms responsible for the action of the membranes in reverse osmosis include sieving, surface tension, and hydrogen bonding. The technique removes dissolved solids.

Distillation is a vapor-liquid transfer operation for removal of dissolved solids, in which vapor is driven from wastewater by heat. Because of its high cost, distillation has limited application in wastewater reclamation.

Ion exchange is a very effective technique for the removal of ions from wastewater. Various natural materials and synthetic ion-exchange resins (see Chapter 8) may be used. Some common applications are for Cl^-, NH_4^+, PO_4^{-3} and metal ion removal.

Table 10.8 Typical Removal Efficiencies of Wastewater Treatment Stages

	Percentage reduction from original wastewater		
Contaminant	Primary treatment	Secondary treatment	Tertiary treatment
BOD	30	90	99.8
Suspended solids	60	90	100
Dissolved inorganic solids	0	5	99.5

Compiled from: Krenkel, P. A. Waste Treatment Methodology, pp. 220–43 in *Industrial Pollution*. N. I. Sax (ed.). New York: Van Nostrand, 1974.
and
Vesilind, P. A. *Environmental Pollution and Control*. Ann Arbor, Mich., Ann Arbor Science Publishers, 1975.

Control of Chemical Contamination from Diffuse Sources

Adequate control of discharges into surface waters from sewage-treatment plants and industrial operations is technically feasible, even if not always achieved. It is more difficult to control the discharge of chemical contaminants associated with diffuse sources, such as the drainage from agricultural land. Such drainage contains a variety of pesticides and nutrients which can adversely affect aquatic organisms and stream productivity.

Some agricultural operations use oxidation lagoons to reduce the oxygen demand of animal wastes. However, the best prospects for reducing the fertilizer and pesticide burdens in runoff water appear to lie in educating farm operators about the problem, with the hope that they will apply these chemicals in quantities and by procedures which will produce the minimum of adverse environmental effects. Some of the problems have been, and will continue to be, resolved by the removal of the more toxic pesticides from the market.

CONTROL OF DRINKING-WATER QUALITY

The provision of bacteriologically pure drinking water is one of the more notable public health success stories in this country. Disease transmission via public drinking-water supplies is, fortunately, a rare event.

There are two ways to accomplish the delivery of high-quality drinking water. One is to have an adequate volume supply of clean natural water. The other is to purify and disinfect available water to the point that it is safe to use. Another key factor is continual monitoring of the purity of the water, not only at central distributing points, but also at various points within the distribution system where it is used, to make sure that clean water is delivered to the consumer.

Selection of Clean Natural-Water Sources

Natural waters, with little or no treatment other than light disinfection with chlorine, are used as drinking-water supplies by a large proportion of the people in this country. Some cities, most notably New York and Boston, maintain protected watershed areas and reservoirs which collect and distribute rainwater. Many other cities and smaller towns use groundwater from deep wells for drinking purposes. Such groundwater may be "hard," but is usually of very good quality in terms of bacterial content. Water from shallow wells in urban areas can readily be contaminated, and must be monitored carefully.

Little attention was paid to the chemical purity of drinking water until the mid-1900's. The extent of the problem arising from the increasing content of organic compounds and trace elements in drinking water has not yet been adequately defined.

Purification of Contaminated Water

Most people drink water which is drawn from rivers and streams and which requires purification to bring it up to current drinking-water standards. The usual treatment procedures involve coagulation, precipitation, sand filtration, and disinfection. These techniques as applied to drinking water are essentially the same as described in the preceding section on wastewater treatment.

IMPACT OF CONTROLS ON ENVIRONMENTAL QUALITY

The intended effect of controls, and generally their main impact, is to reduce or eliminate releases of toxic or otherwise harmful chemicals to the workroom air or general environment. To the extent that controls are successful, they are beneficial. However, the installation and operation of control devices does not ensure that all problems will be totally eliminated or that others will fail to develop.

Controls as Contaminant Sources

Control devices may actually be sources of contaminants. For example, if the collection efficiency for a given contaminant is not high enough, the control device may release the contaminant into the environment. A collector may be the source of a highly concentrated toxic material, or of a large volume of material. In many cases, the collected materials represent significant problems with respect to their handling and ultimate disposal in an environmentally acceptable manner.

There are several other ways in which controls may serve as significant sources of contaminants. Controls may consist of a network of conduits which prevent or control the releases from numerous small sources by combining them into one large source. This may solve some problems but create a new larger one. Problems may also develop when the control procedure involves chemical reactions for the purpose of contaminant degradation or neutralization. The by-products and secondary products of these reactions may themselves be harmful, e.g., organochlorine compounds resulting from the chlorination of drinking water.

Another example of the production of a new problem by a control device is the release of excessive moisture, forming fogs, by some types of cooling towers used to remove heat from cooling waters. They may also release into the ambient air some of the chemicals used to prevent algae growth on the tower walls.

INCOMPLETE CAPTURE

The design efficiency of a contaminant-control device is generally selected on the basis of (a) its ability to meet an effluent standard with an assured margin of safety and (b) the cost of its purchase, installation, and operation. These two considerations become increasingly difficult to reconcile as the desired collection efficiency approaches 100%.

When very high collection efficiencies are needed, it is generally more constructive to consider penetration rather than efficiency; the percent penetration (P) is the complement of the percent efficiency (E), i.e., $P = 100\text{-}E$. There may not seem to be a great difference between efficiencies of 90%, 95%, 98%, 99%, and 99.5%, but their significance becomes clearer when considered as a penetration of 10%, 5%, 2%, 1%, and 0.5% respectively (e.g., see Figure 10–5). Each now differs from the other by a factor of about 2, and the increased cost of achieving each increment may be quite large.

The strengths of collection devices as contaminant sources vary directly as their penetration efficiencies. Thus, for example, a new power-plant electrostatic precipitator having a 99% collection efficiency which replaces one having a 90% efficiency will discharge ten times less fly ash. While this may have a large effect on airborne total suspended particulate and some trace-metal concentrations, it will have relatively little impact on the ash-disposal problem. There would only be a 10% increase in collected fly ash, and less than that for the combined fly-ash and bottom-ash mass.

FORMATION OF SECONDARY CONTAMINANTS AND RELEASE OF TRACE CONTAMINANTS IN EFFLUENTS

In collectors which act as simple traps, where contaminants are removed from the effluent stream by physical processes such as filtration, electrostatic precipitation, scrubbing, adsorption, and condensation, the collected materials are not changed chemically, although in wet collectors they may be dissolved or suspended in water. On the other hand, in collectors utilizing chemical reactions to capture, oxidize, or neutralize the waste, the residual reaction products will differ in composition from the original materials. The major reaction products will usually be innocuous or relatively easy to handle. If this was not the case, the process would not have been selected in the first place. Problems may arise, however, from the presence of trace contaminants in the waste.

The problem of the release of trace contaminants may be exemplified by considering incineration, which is an effective means of reducing the weight and bulk of household refuse and waste paper and, to an increasing degree, may be favored as a means of resource recovery in terms of its heating value. However, mixed refuse (see Table 3 – 13) may contain, for example, PCB-treated papers, halogenated plastics, discarded batteries containing lead, mercury, nickel, and cadmium, and a host of other chemicals; these constituents may be vaporized and injected into the atmosphere along with the waste gases.

A classic example of the formation of a secondary contaminant in a control device involves the adoption of catalytic converters in motor-vehicle exhaust systems. As discussed in Chapter 3, no consideration was initially given to the oxidation of the SO_2 in the exhaust to SO_3, and the hydrolysis of the latter in the exhaust stream to sulfuric acid mist.

Ultimate Disposal of Collected Materials and Process Sludges

The ultimate and safe disposal of liquid and solid wastes and treatment-plant residues (see Figure 10–9) is one of the most important current concerns of environmental protection and public health authorities. There are no easy, foolproof, or inexpensive solutions to most of the current problems.

Control devices such as incinerators, which destroy chemical contaminants, will produce relatively little residue material requiring disposal. On the other hand, most control devices which extract chemicals from air or liquid streams, or react them to form secondary chemicals, create more difficult disposal problems. The wastes may be dry solids, concentrated suspensions of solids in water (slurries and sludges), or concentrated organic or aqueous solutions. These wastes may require further treatment or degradation before they can be disposed of in an environmentally acceptable manner.

INCINERATION

For combustible wastes, incineration has the advantages of relatively low cost and possibilities for heat recovery. The residual ash occupies less than 10% of the volume of the original waste, and is usually a relatively innocuous, soil-like material. However incineration creates smoke and odor nuisances. Even with nominally complete combustion, there will be unburned and noncombustible volatile contaminants in the exhaust, as previously discussed. Exhaust gas cleaners, when installed, reduce the cost advantages of incinerators and may not collect some of the materials of interest, e.g., mercury vapor. Contaminants which escape into the atmosphere will eventually be deposited on land or surface waters, becoming available for uptake into the biosphere.

LAND DISPOSAL

Most solid refuse is disposed of in sanitary landfills. The basic construction of a modern landfill is illustrated in Figure 10–18. Each layer of refuse is compacted and covered with a layer of earth or clean fill. The earth cover absorbs odorous gases of decomposition and acts as a barrier to pests and vermin. Filled land gradually settles as the waste decomposes. Thus, these areas should not be used for residential or commercial purposes without special precautions, not only because of the settlement problem, but because explosive concentrations of methane gas can build up in basement areas. Landfills in urban areas have usually been used for recreational purposes, such as parks and golf courses.

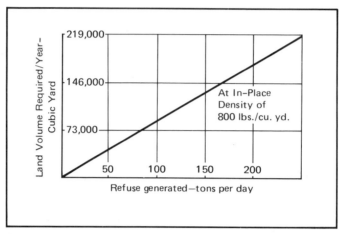

1. Amount of land required for a sanitary landfill operation.

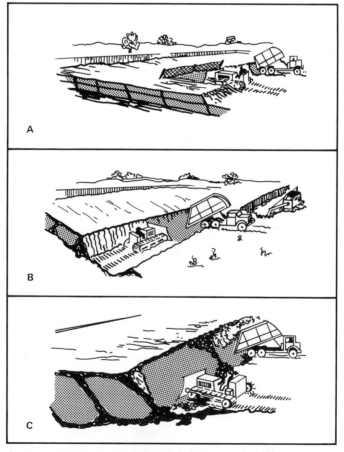

2. In all methods of sanitary landfilling, refuse is formed into compacted cells totally surrounded by cover material. The Area
   Method (A) is used to restore large areas of land to proper grade,
   while the Trench Method (B) can be used for filling ravines and
   ditches. The Ramp Variation (C) can be employed with either
   the area or trench methods. Solid wastes are spread and compacted on a slope.

A major problem in the selection of suitable areas for landfill, and their successful operation, is the isolation from groundwater of the leachate solutions which form in the wastes. It is generally necessary to line the bottom of the landfill with an impervious layer of clay to control the drainage patterns, and to pump out the leachate for subsequent biological treatment.

UNDERGROUND DISPOSAL

Some highly toxic and radioactive solid and liquid wastes are stored in natural underground caverns or in chambers excavated in selected geological strata such as salt domes. The sites are selected for their inaccessibility and isolation from groundwater and for their supposed geologic stability.

Large volumes of liquid wastes are disposed of by deep-well injection into underground strata. Wells are drilled into suitable permeable strata which are isolated from groundwater sources by impermeable strata and the well casing. A schematic diagram of a waste-liquid injection well is shown in Figure 10-19.

Strata suitable for deep-well disposal are located in many parts of the United States, primarily in the western parts of the country, as indicated in Figure 10-20.

Problems occasionally develop due to the instability of the underground rock strata. Liquids pumped into certain strata at high pressures can act as lubricants, causing slippage of contact surfaces; this effect is manifested as earthquakes at the surface. The close correlation between the times of active pumping of wastes by the U. S. Army Rocky Mountain Arsenal near Denver and the occurrence of earthquakes in the Denver region in the mid-1960's led to discontinuance of deep-well disposal there.

DISPOSAL IN THE OCEANS

There are several different kinds of ocean disposal, and it is important not to confuse them. One is the disposal of waste liquid or slurries into surface waters relatively close to shore. Huge volumes of sewage sludge and industrial acid wastes are currently disposed of in this manner (see Table 3-19), with the implicit rationalization that the wastes are degraded

---

Fig. 10-18 Sanitary landfill.

From: Schroering, J. B. Sanitary Landfilling-Applications and Limitations. *Pollution Engineering* v. 2 (No. 11) 32-33, 1969.

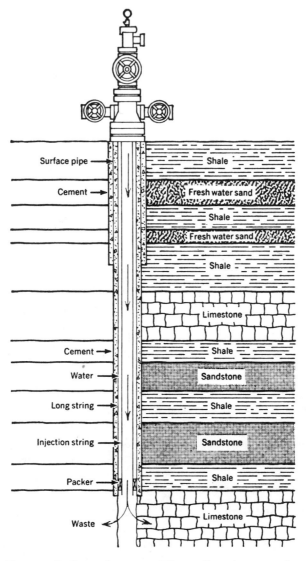

Fig. 10–19 Deep-well disposal system. The well casing must be cemented in place to protect ground water.

From: Talbot, J. S. Deep Wells. *Chemical Engineering* 75(Oct. 14):108– 11, 1968.
© McGraw-Hill. Used with permission of the publisher.

Fig. 10–20 Basins whose geological structures are most suited for deep-well disposal.

From: Sheldrick, M. G. Deep-well Disposal: Are Safeguards Being Ignored? *Chemical Engineering* 76(Apr. 7):74–78, 1969.

and/or neutralized within the ocean and do not accumulate excessively or do permanent damage to marine life. Major dumping grounds are located in the New York Bight, as indicated in Figure 10–21.

Construction and demolition debris which will sink in sea water is also dumped into the New York Bight, but at a different dump site. There is not supposed to be any dumping of wood, primary sewage sludge, or floating refuse, and the appearance of such materials in the waters of the Bight, and occasionally on the adjacent beaches, is generally attributable to surface transport of materials discharged from ships, to raw-sewage outfalls, and to debris from rotting piers.

The extent and significance of the damage that may be caused to the

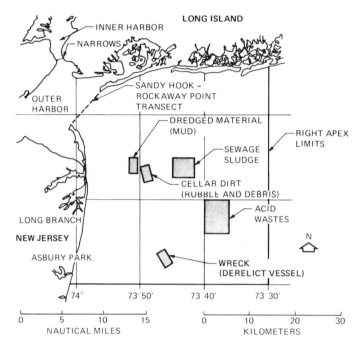

Fig. 10–21 New York Bight showing existing dump sites.
Source: U.S. Environmental Protection Agency.

marine environment by sewage-sludge dumping remains controversial. Dumping will continue to increase in the New York Bight until at least the early 1980's, as secondary-treatment plants currently under construction are completed, thus increasing the volumes of sludge produced. On the other hand, the EPA has ordered an end to sludge dumping in the Bight by 1981, and efforts are under way to develop alternative disposal techniques which are economically realistic.

Near-shoreline marine disposal is also done using pipelines which extend for distances up to several miles out to sea. In this case, the discharge is at the lower, rather than the upper, surface of the water. Such systems have been used for disposal of raw sewage by coastal cities, and may appear to be adequate initially. However, they are usually found wanting as population and sewage volume increase, especially when on-shore winds deposit floating sewage solids on the beaches. Problems may also arise when the pipe develops leaks, since inspection and maintenance are difficult and expensive.

Deep-ocean dumping is theoretically a safe and effective means of disposal for dangerous wastes. As discussed in Chapter 4, vertical diffusivity in ocean water is extremely low, and should provide an effective barrier against transfer to the surface water for periods of hundreds to thousands of years. However, diffusivity is the controlling factor only for dissolved materials; for particulates, their effective density would be controlling. Materials with low density, or particles of high-density materials with attached gas bubbles, could rise rapidly through the water. In addition, if the materials are ingested by marine organisms which migrate vertically or are consumed by predators who do, these chemicals can be readily transported toward the surface.

In order to prevent their dispersion, materials dumped into the ocean depths have generally been enclosed within sealed containers, frequently 55-gallon steel drums which contain concrete liners or covers to ensure their sinking. Unfortunately, the amount of time that steel and concrete can withstand the extreme pressures and corrosive properties of deep-ocean water has sometimes been overestimated.

Appendix

SUPPLEMENTARY BIBLIOGRAPHY

This bibliography lists selected references, most of which have not been previously cited, which the reader may consult for more detailed information on the various topics discussed in the text.

Historical Perspectives

Agricola, G. *De Re Metallica*. Basel, 1556. Translated by H. C. Hoover and L. H. Hoover for the *Mining Magazine*, London, 1912. Reprinted by Dover Press.

Lodge, J. P., Jr. (ed.). *The Smoake of London — Two Prophecies*. Elmsford, N.Y.: Maxwell Reprint Co., 1969.

Paracelsus, T. *Von der Bergsucht und anderen Bergkrankheiten,* translated from the German in *Four Treatises of Theophrastus von Hehenheim, called Paracelsus* H. E. Segerist (ed.). Baltimore: The Johns Hopkins Press, 1941.

Pott, P. Cancer Scroti, pp. 63–68, in *Chirurgical Observations*. London: Hawes, Clarke, and Collins, 1775.

Ramazzini, B. *De Morbis Artificum*. Geneva, 1713. Translated by W. C. Wright. Chicago: U. of Chicago, 1940; New York: Hafner, 1964.

Broad Reviews of Chemicals in the Environment, and Their Effects and Fate

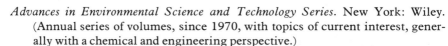
Advances in Environmental Science and Technology Series. New York: Wiley. (Annual series of volumes, since 1970, with topics of current interest, generally with a chemical and engineering perspective.)

Cleaning Our Environment — The Chemical Basis for Action, 1969. American Chemical Society, Special Issues Section, 1155 Sixteenth Street, N.W. Washington, D.C. 20036.

Council on Environmental Quality. *Environmental Quality: The Annual Report of the Council on Environmental Quality.* Washington, D.C.: U.S. Government Printing Office (Published annually since 1970; provides a review of the past year's problems and progress).

Higgins, I. J. and Burns, R. G. *The Chemistry and Microbiology of Pollution.* New York: Academic Press, 1975.

Manahan, S. E. *Environmental Chemistry.* Boston: Willard Grant, 1972.

Man's Impact on the Global Environment: Report of the Study of Critical Environmental Problems (SCEP). Cambridge, Mass., MIT, 1970.

McGraw-Hill Encyclopedia of Environmental Science. New York: McGraw-Hill, 1974.

National Academy of Sciences. *Principles for Evaluating Chemicals in the Environment.* Washington, D.C.: National Academy of Sciences, 1975.

Sax, N. I. (ed.). *Industrial Pollution.* New York: Van Nostrand-Reinhold, 1974.

Second Task Force for Research Planning in Environmental Health Science. *Human Health and the Environment — Some Research Needs.* DHEW Publication No. NIH 77-1277, USGPO, Washington, D.C., 1977.

Singer, S. F. (ed.). *Global Effects of Environmental Pollution.* New York: Springer-Verlag, 1970.

Stoker, H. S. and Seager, S. L. *Environmental Chemistry of Air and Water Pollution.* Glenview, Ill.: Scott, Foresman 1972.

Suffet, I. H. (ed.). *Fate of Pollutants in the Air and Water Environments.* New York: Wiley, 1977.

In-Depth Reviews of Specific Chemical Contaminants

Reviews prepared by: Committee on Medical and Biologic Effects of Environmental Pollutants — Division of Medical Sciences, Assembly of Life Sciences — National Research Council. Published by: National Academy of Sciences, Washington, D.C. 20418.

Arsenic (ISBN 0-309-02604-0) – 1977
Asbestos (ISBN 0-309-01927-3) – 1971
Chlorine and Hydrogen Chloride (ISBN 0-309-02519-2) – 1976
Chromium (ISBN 0-309-02217-7) – 1974
Copper (ISBN 0-309-02536-2) – 1977
Fluorides (ISBN 0-309-01922-2) – 1971
Lead (ISBN 0-309-01941-9) – 1972
Manganese (ISBN 0-309-02143-X) – 1973
Nickel (ISBN 0-309-02314-9) – 1975
Nitrogen Oxides (ISBN 0-309-02615-6) – 1977
Ozone and Other Photochemical Oxidants (ISBN 0-309-02531-1) – 1977
Particulate Polycyclic Organic Matter (ISBN 0-309-02027-1) – 1972
Selenium (ISBN 0-309-02503-6) – 1976
Vanadium (ISBN 0-309-02218-5) – 1974
Vapor-Phase Organic Pollutants (ISBN 0-309-02441-2) – 1976

Specialty Books on Specific Environments

CHARACTERISTICS OF SPECIFIC ENVIRONMENTS

THE BIOSPHERE

Odum, E. P. *Fundamentals of Ecology*, 3rd Ed. Philadelphia: W. B. Saunders, 1971.

Southwick, C. H. *Ecology and the Quality of Our Environment*, 2nd Ed. New York: Van Nostrand, 1976.

The Biosphere. *Scientific American, Vol. 223*, Sept. 1970. (Special issue on the biosphere.)

THE ATMOSPHERE

Junge, C. E. *Air Chemistry and Radioactivity*. New York: Academic Press, 1963.

Lorenz, E. N. *The Nature and Theory of the General Circulation of the Atmosphere*. World Meteorological Organization, 1967.

Rasool, S. I. *Chemistry of the Lower Atmosphere*, New York: Plenum Press, 1973.

Verniani, F. *Structure and Dynamics of the Upper Atmosphere*. New York: Elsevier, 1974.

THE HYDROSPHERE

Hutchinson, G. E. *Treatise on Limnology*. New York: Wiley-Interscience, 1975.

Ross, D. *Introduction to Oceanography*. Englewood Cliffs, N.J.: Prentice-Hall, 1977.

Stumm, W. and Morgan, J. J. *Aquatic Chemistry*. New York: Wiley-Interscience, 1970.

Yen, T. F. (ed.). *Chemistry of Marine Sediments*. Ann Arbor, Mich.: Ann Arbor Science, 1977.

THE LITHOSPHERE

Brady, N. C. *Nature and Properties of Soils*, 8th Ed. New York: Macmillan, 1974.

Foth, H. D. and Turk, L. M. *Fundamentals of Soil Science*, 5th Ed. New York: J. Wiley, 1965.

CONTAMINATION IN SPECIFIC ENVIRONMENTS

AIR CONTAMINATION

Broad Reviews

Air Pollution Engineering Manual, 2nd Ed. *PHS Publication AP-40*, No. EP4·9: 40–2, U.S. Government Printing Office.

Air Pollution Manual — part II — Control Equipment. American Industrial Hygiene Association, Akron, Ohio 44313, 1968.

Perkins, H. C. *Air Pollution*. New York: McGraw-Hill, 1974.

Seinfeld, J. H. *Air Pollution: Physical and Chemical Fundamentals*. New York: McGraw-Hill, 1975.

Stern, A. C. (ed.). *Air Pollution*, 3rd Ed. (5 vols.). New York: Academic Press, 1977.

Williamson, S. J. *Fundamentals of Air Pollution*. Reading, Mass.: Addison-Wesley, 1973.

Some Specific Aspects

Fluorocarbons and the Environment: Report of the Federal Task Force on Inadvertent Modification of the Stratosphere (IMOS). Washington, D.C.: U.S. Government Printing Office, 1975.

Inadvertent Climate Modification: Report of the Study of Man's Impact on Climate (SMIC). Cambridge, Mass.: MIT Press, 1971.

Mansfield, T. A. (ed.). *Effects of Air Pollutants on Plants*. Cambridge: Cambridge University, 1976.

Schneider, S. H. *The Genesis Strategy: Climate and Global Survival*. New York: Plenum Press, 1976.

Stumm, W. (ed.). *Global Chemical Cycles and Their Alterations by Man*. Philadelphia: Heyden and Son, Inc., 1977.

Summer, W. *Odor Pollution of Air: Causes and Control*. Cleveland: CRC Press, 1971.

WATER CONTAMINATION

Surface Waters

Ciaccio, L. L. (ed.). *Water and Water Pollution Handbook* (4 vols.). New York: Marcel Dekker, Inc., 1971.

Eckenfelder, W. *Industrial Water Pollution Control*. New York: McGraw-Hill, 1966.

Fair, G. M., Geyer, J. C. and Okun, D. A. *Water and Wastewater Engineering* (2 vols.). New York: Wiley, 1968.

Klein, L. *River Pollution* (6 vols.). London: Butterworths, 1966.

McCaull, J. and Crossland, J. *Water Pollution*. New York: Harcourt Brace Jovanovich, 1974.

Metcalf and Eddy, Inc. *Wastewater Engineering: Collection, Treatment, Disposal*. New York: McGraw-Hill, 1972.

Miller, S. S. (ed.). *Water Pollution*. Washington, D.C.: American Chemical Society, 1974.

National Academy of Sciences. *Assessing Potential Ocean Pollutants*. Washington, D.C.: National Academy of Sciences, 1975.

Nemerow, N. C. *Theories and Practices of Industrial Waste Treatment*. Reading, Mass.: Addison-Wesley, 1963.

Warren, C. E. *Biology and Water Pollution Control*. Philadelphia: W. B. Saunders, 1971.

Drinking Water

Safe Drinking Water Committee, Advisory Center on Toxicology, Assembly of Life Sciences, National Research Council. *Drinking Water and Health*. Washington, D.C.: National Academy of Sciences, 1977.

LAND AND FOOD CONTAMINATION

Loehr, R. C. *Agricultural Waste Management*. Academic Press: New York, 1974.

National Research Council. *Toxicants Occurring Naturally in Foods*. Washington, D.C.: National Academy of Sciences, 1975.

Newberne, P. M. *Geochemistry and the Environment*, vol. 1: The Relation of Selected Trace Elements to Health and Disease. Washington, D.C.: National Academy of Sciences, 1974.

Newberne, P. M. (ed.). *Trace Substances and Health*. New York: Marcel Dekker, Inc., 1976.

Rodricks, J. V. *Mycotoxins and Other Fungal Related Food Problems*, Advances in Chemistry Series, No. 149. Washington, D.C.: American Chemical Society, 1976.

OCCUPATIONAL ENVIRONMENT

Ashford, N. *Crisis in the Workplace: Occupational Disease and Injury*. A Report to the Ford Foundation. Cambridge, Mass.: MIT Press, 1976.

Cralley, L. V. (ed.). *Industrial Environmental Health*. Industrial Health Foundation, Inc. Pittsburgh, Pa., 1972.

Hunter, D. *The Diseases of Occupations*, 5th Ed. London: English Universities Press, 1975.

Hamilton, A. and Hardy, H. L. *Industrial Toxicology*, 3rd Ed. Acton, Mass.: Publishing Sciences Group, 1974.

Industrial Ventilation, 15th Ed. ACGIH Committee on Industrial Ventilation, P. O. Box 453, Lansing, Mich. 48902, 1978.

Mercer, T. T. *Aerosol Technology in Hazard Evaluation*. New York: Academic Press, 1973.

National Institute for Occupational Safety and Health. *National Occupational Hazard Survey*, vol. 1 HEW Publication No. (NIOSH) 74-127, U.S. Department of Health, Education, and Welfare, Washington, D.C., 1974.

Patty, F. A., (ed.). *Industrial Hygiene and Toxicology*-3rd Ed. New York: Wiley, 1978.

Saffioti, U. and Wagoner, J. K. (eds.) Occupational Carcinogenesis. *Ann. NY Acad. Sci. 271* (May 28):1–516, 1976.

The Industrial Environment—Its Evaluation and Control. No. 1701-00396, U.S. Government Printing Office, Washington, D.C. 20402, 1973.

Environmental Toxicology

GENERAL REVIEWS

Casarett, L. J. and Doull, J. (ed.). *Toxicology: The Basic Science of Poisons*. New York: Macmillan, 1975.

De Bruin, A. *Biochemical Toxicology of Environmental Agents*. New York: Elsevier/North-Holland, 1976.

Lee, D. H. K., Falk, H. L., and Murphy, S. D. (eds.). *Handbook of Physiology*, Sect. 9: Reactions to Environmental Agents. Bethesda, Md.: American Physiological Society, 1977.

Lee, D. H. K. and Kotin, P. *Multiple Factors in the Causation of Environmentally Induced Disease*. New York: Academic Press, 1972.

Loomis, T. A. *Essentials of Toxicology*, 2nd Ed. Philadelphia: Lea and Febiger, 1974.

Patty, F. A. (ed.). *Industrial Hygiene and Toxicology*, vol. II: Toxicology, 2nd Ed. New York: Interscience, 1962.

Waldbott, G. L. *Health Effects of Environmental Pollutants*, 2nd Ed. St. Louis: C. V. Mosby Co., 1978.

Williams, R. T. *Detoxication Mechanisms*, 2nd Ed. London: Chapman and Hall, 1959.

SPECIFIC ASPECTS OF ENVIRONMENTAL TOXICOLOGY

Becker, F. F. (ed.). *Cancer: A Comprehensive Treatise*, vol. 1: Etiology: Chemical and Physical Carcinogenesis. New York: Plenum Press, 1975.

Essays in Toxicology Series, vols. 1–6. New York: Academic Press. (Contains articles on current topics of interest.)

Lee, D. H. K. (ed.). *Metallic Contaminants and Human Health*. New York: Academic Press, 1972.

Lee, D. H. K. (ed.). *Environmental Factors in Respiratory Disease*. New York: Academic Press, 1972.

MacMahon, B. and Pugh, T. F. *Epidemiology: Principles and Methods*. Boston: Little, Brown and Co., 1970.

Searle, C. E. (ed.). *Chemical Carcinogens*, ACS Monograph No. 173. Washington, D.C.: American Chemical Society, 1976.

Sutton, H. E. and Harris, M. I. (ed.). *Mutagenic Effects of Environmental Contaminants*. New York: Academic Press, 1972.

Wilson, J. G. *Environment and Birth Defects*. New York: Academic Press, 1973.

Broad Reviews of Contaminant Evaluation Techniques

Ewing, G. W. *Instrumental Methods of Chemical Analysis*, 4th Ed. New York: McGraw-Hill, 1975.

Pickering, W. F. *Pollution Evaluation: The Quantitative Aspects*. New York: Marcel Dekker, 1977.

Willard, H. H., Merritt, Jr., L. L. and Dean, J. A. *Instrumental Methods of Analysis*, 5th Ed. New York: Van Nostrand, 1974.

Index

(F) and (T) indicate references to figures and tables, respectively.

Accumulation mode, 20
Acid-base reactions, 165
Acid mine drainage, 260
Acid rain, 246–47, 247(F)
Activated carbon, 337, 355, 364, 401, 428, 430
Activated sludge process, 425–27, 426(F)
Active transport, 191, 192, 198
Adsorption, 143, 313, 335, 337, 355, 401, 402(T)
Aerodynamic diameter, 19
Aerosol, 16. *See also* Particles
  properties of, 17–21
  types of, 19–21
Aflatoxin, 108, 270, 309
Agricola, 6, 7(F), 8(F)
Agriculture, 91–93
Air cleaning techniques, 397–406, 398(F), 399(F). *See also specific techniques*
Air flotation, 421–23, 422(F), 423(F)
Air pollution episodes, 12, 218–19(T), 219–20, 232–33
Air quality act (1967), 295, 296
Air quality control regions, 295
Air quality criteria and control technology documents, 295, 296, 297
Air quality standards, community, 294–95, 296(T)
  federal, 298(T)
  primary, 284
  secondary, 284
Air-sampling train, 332–35, 334(F)
Air stripping, 430
Aircraft, 73–74, 74(T), 75(F)
Aitken nuclei. *See* Condensation nuclei
Albedo, 35, 244
Aldrin, 146, 309
Allergen, 210, 211(T)

Allergy, 210–11
Alpha particles, 282
Aluminum production, 82–86, 85(F), 86(F)
Alveoli, 194, 196, 209
American Conference of Governmental Industrial Hygienists (ACGIH), 286
American Industrial Hygiene Association (AIHA), 287
American National Standards Institute, Inc. (ANSI), 286
American Standards Association (ASA), 286
Ammonia, 60, 156, 157, 160, 241
Ammonification, 173
Analytic instrumentation 366–69(T), 368. *See also specific instruments*
Anaphylactic shock, 211
Animals, effects of air pollution on, 232–33, 233(T)
  use in toxicological studies, 276–80, 277(T)
Antagonism, 216
Antibody, 210–11
Aquifer, 45, 46(F), 300
Argon, 33
Arsenic, 233
Artesian well, 46, 46(F)
Asbestos, 53, 213, 216, 226, 293, 387
Asbestosis, 209
Asphyxiant, 210
Asthma, 211, 222
Atmosphere, 32
  chemical composition of, 33–35, 59(T)
  contaminant dispersion in, 117–35
  earth-atmosphere energy balance, 35–37, 35(F), 36(F)
  general circulation of, 37–38, 38(F)
  structure of, 33

Automobile
    exhaust emission controls, 69–73,
        71(F), 72(F), 297–98
    exhaust emission standards, 299(T)
    exhaust products, 69–73

Bag filters, 403–5, 404(F), 406, 407
Bar screens, 417
Bauer, George. *See* Agricola
Beilstein reaction, 372
Bernoulli's theorem, 321, 329
Berylliosis, 53
Beta particles, 282
BHT, 203
Bile, 199
Biochemical oxygen demand (BOD), 22,
    43, 95, 106(T), 166, 256
Bioconcentration, 51, 146–49, 258
Biodegradability, 92, 175–76
Biogeochemical cycles, 51, 171–74,
    260–63, 264
Biologic limit values (BLV), 290, 362
Biological concentration. *See*
    Bioconcentration
Biological discrimination, 146
Biological magnification. *See*
    Bioconcentration
Biosphere, 32, 51, 136(F), 144–49
Biotransformation, 193, 202–4, 206, 213,
    277
Bronchitis, 220–21, 221(F), 226
Bronchoconstriction, 197, 209
Bubblers, 335, 373
"Bucket and stopwatch" sampling, 324

Cadmium, 67, 227
Calibration, 318
Cancer, 53, 74, 183, 183(T), 212–13,
    213(T), 216, 220, 222, 224, 224(T),
    226, 245
Carbon, 171–72, 172(F), 260–61. *See
    also specific carbon compounds*
Carbon dioxide, 33, 36–37, 62–63, 69,
    158–59, 171, 235, 237, 240–43,
    260–61
Carbon-14, 64
Carbon monoxide, 60, 69, 70, 158–59,
    179(F), 180, 296(T), 369, 407
Carbon tetrachloride, 200
Carcinogen. *See* Cancer
Cardiovascular disease, 224, 226, 227
Cascade impactor. *See* Impactor
Catalytic converters, 70, 298, 401, 407,
    434
Centrifuge, 337, 340

Chelation, 170
Chemiluminescent analyzer, 373, 372(F)
Chlorinated hydrocarbons, 58
    in drinking water, 113–14, 114(T)
Chlorination, 433
Chlorine, 429, 432
Chloroform, 107–8
Cigarette smoking, 220–21, 221(F), 222,
    216
Cilia, 196
Clarifier, 420, 420(F)
Clay, 49, 50(F), 164
Clean Air Act (1968), 70
Clean Air Act Amendments (1970), 297,
    407–8
Clean Air Act Amendments (1977),
    298–300
Climate, 236–46
Clouds, 239
Coagulation, 154, 164, 419
Coal, 66–67, 68(T), 299–300
Coarse particle mode, 20
Coastal waters, 46–47
Co-carcinogen, 216
"Cold trap," 337
Colloid, 23, 164–66, 170–71
Colorimetry, 373
Combustion, 401
Combustible gas meter, 371–72, 371(F)
Comminution, 19
Concentration factor, 146–49, 148(T)
Condensation, 335, 401
Condensation nuclei, 20, 154
Contaminant residues
    in the atmosphere, 176–80, 181
    in the biosphere, 184–86
    in the hydrosphere, 181, 182–84
    in remote areas, 186–87
Contaminants
    secondary, 107–13, 113–15
Contamination, 3–4, 16. *See also
    specific sources*, e.g., Automobile, etc.
Contamination criteria, 270, 271–84
Continuous analyzers, 317–18
Controls. *See also specific control devices*
    administrative, 388–89
    effluent, 397–411, 416–31
    emission, 285
    federal legislation of source, 297–98,
        386–87, 391–92; mobile source,
        406–7, 407(F), 408(F)
    process, 386–87, 389–91, 390(T)
    regulatory, 386, 387–88
    source, 389, 391–2
    stationary combustion, 407–11
Cooling towers, 68, 251, 433
Coordination compounds, 165, 170
Coprecipitation, 166

Core-sampling, 359
Corrugated-plate interceptor, 421, 421(F)
Cosmic rays, 64–65, 65(T)
Coulometry, 378
Criteria documents, 292(T), 293, 397
Cyclamates, 270
Cyclone, 337, 339(F), 340, 406

DDT, 143, 149, 175, 186, 199, 200–201,
    203, 205, 266, 284, 309
Decay, 63
Decomposition, 63, 253, 254
Deep-well injection, 437, 438(F), 439(F)
Delaney Clause, 5, 15, 309
Denitrification, 173, 261–62, 427
Deposition
  of particles in respiratory tract, 194–95
Deposition velocity, 132
Dermatitis, 211
Dermis, 200
Detector tubes, 374, 375(F), 377
Detention basin, 392
Diesel engine, 73, 74(T)
Diethylstilbestrol (DES), 30
Diffusion
  Brownian, 194
  molecular, 117
  turbulent, 117–18
Dilution, 387, 392, 397, 411
Dissolved oxygen, 22, 44, 252, 256
Distillation, 430
Dose, 56
Dose-response relationship, 54–56, 55(F)
Dredges, 359, 360
Drinking water, 42, 182, 183
  standards, 300, 301(T)
Dry-collection techniques, 337. See also
    specific collection devices
Dry deposition, 154
Dry gas meter, 319, 320(F), 321
Dust, 19, 209, 210(T)
Dust fall, 21, 295, 314

Earth-atmosphere energy balance, 239
Earth-surface temperature, 242,
    243(F), 244
Ecological pyramid. See Trophic
    structure
Ecosystem, 264–65
Eddies. See Diffusion, turbulent
EDTA, 166, 200
Effective stack height, 392
Effluent guidelines and standards,
    306–7, 307(T), 308
Effluent-quality regulations, 300, 306–9
Ejector, 333

Electrodialysis, 430
Electrostatic precipitation, 313, 337,
    343–45, 345(F), 400, 405–6, 407
Ellenbog, Ulrich, 6
Emergency exposure levels (EEL),
    287–88, 290
Emission standards, 297
Emission units, 285
Emphysema, 221–22, 223(F), 226
English Factory Acts, 11
Enterohepatic circulation, 189
Environmental impact statement, 306
Environmental Protection Agency (U.S.),
    70, 284, 294, 297, 299, 300, 306, 309
Environmental quality, 232
Epidemiology, 53–54
  of air pollution, 217–26
  as basis for establishment of
    contaminant criteria and exposure
    limits, 271–72, 272–76
  of water pollution, 226–27
Epidermis, 200
Epilimnion, 43–44, 43(F)
Ergonomics, 389
Erosion, 63–64
Estuary, 46–47, 138, 139(F), 164, 256
Eutrophication, 255–57
Evelyn, John, 12
Evolution, 265–66
Excretion, 193
Exhaust hood, 393, 394(F), 395(F)
Exhaust ventilation, 393
Exposure limits, 271, 271–84, 286–94
Eyes, 201

Fallout, 148. See also Nuclear weapons
    tests
Federal legislation, 14–15, 382–85(T).
    See also specific laws
Fertilizers, 92
Fibrosis, 209
Filter
  fiber, 340, 343, 355
  granular media, 340, 343, 423–25,
    424(F)
  membrane, 340, 355
  Nuclepore, 340, 341, 341(F), 343
Filtration, 337, 340–43, 342(F), 423–25,
    424(F). See also specific types of
    filters
Fire, 64, 91
Fish-kills, 254, 258
Fission, 96–103
Flocculation, 164, 419
Flow nozzle, 321, 323, 325(F)
Flow velocity meter, 329
Flow meters. See specific types

Flue-gas desulfurization, 409–11, 409(T)
Flume, 7, 321, 321(F), 323
Fluoride, 85, 233
Fluorocarbons, 241, 246, 382
Fly-ash, 67
Fog, 19
Food
    additives, 15, 31–32
    contaminants in 24–27
    drug residues in, 27–30, 29(T)
    packing material residues in, 30
    pesticide residues in, 27, 28(T)
    preservation, 13–14
Food chains, 144–49, 145(F)
Food and Drug Administration (U.S.), 14, 27, 31, 309, 361
Food, Drug and Cosmetic Act (1938), 14
Food web, 144
Fossil fuel, 66–68, 68(T)
Fulvic acid, 165
Fume, 19
Fumes, metallic, 214
Fumigation, 123, 130(F)

Galen, 6
Gamma rays, 282
Gas chromatographic flame photometer, 374, 375(F)
Gas washer, 335, 335(F), 336(T), 373, 397
Gastrointestinal tract, 190, 197–200, 198(F), 199(F), 203, 205, 206, 210
Gravel, 49
Gravitational settlement. See Sedimentation
Greenhouse effect, 37, 240, 244
Groundwater, 44–46, 45(F), 46(F), 169, 432

Halide meter, 372–73
Hamilton, Alice, 11
Hard water, 257
Hazard, 24
Haze, 20
Health, 52
Heat. See Thermal pollution
Hippocrates, 5
Histamine, 209
Humic acid, 165, 171
Humus, 49
Hydrocarbons, 70, 162, 407
Hydrocyclone, 417(F), 418
Hydrogen peroxide, 429
Hydrogen sulfide, 60, 159–60, 174, 208
Hydrologic cycle, 38, 40, 40(F), 46(F)

Hydrosol, 21
Hydrosphere, 32
    contaminant dispersion in, 135–42
    general characteristics, 38–47
Hyperplasia, 208
Hypolimnion, 43(F), 44, 137–38, 256

Immune system, 217
Impaction, 153–54, 194, 313
Impactor, 337–40, 338(F)
Impinger, 335, 337
Incineration, 93, 95, 434, 435
Indicator organisms, 360
Industry. See also specific industries
    contaminants from, 76–91
Inertial separators, 406
Infrared analyzer, 369–70, 370(F)
Integral-volume flowmeter, 319–21
Interception, 153–54, 194
Intermittent discharge, as control, 411
International Commission on Radiological Protection (ICRP), 283
Intestines, 197–98, 198(F)
Inversion, 122–23, 125(F), 130(F), 392
Ion exchange, 49–50, 164–65, 313, 355–56, 364, 430
Ion-flow meter, 329
Ion-selective electrode, 377
Irritants, 208–9, 208(T), 217
Isokinetic sampling, 349, 351(F)

Jet stream, 37, 135

Kepone, 386
Kidneys, 203, 205–6

Lagoons, 428, 431
Lake Erie, 256–57
Lake Washington, 257
Lakes, 42–44, 137–38, 167, 256
Landfill, 94–95, 435, 436(F), 437
Lanza, Anthony J., 11
Lapse rate, 33, 121–26, 312
Lead, 6, 67, 69, 70, 134, 176, 187, 187(F), 191(F), 192, 194, 200, 233, 360
Lehmann, K. L., 10
Light-scatter, 21
Light-scatter particle analyzer, 374(F)
Lithosphere, 32, 47. See also Soil
Liver, 199, 202
Looping, 123–24
Lungs. See Respiratory tract
Lymphatic system, 196, 209

Macrophages, alveolar, 196, 197
Magnetic flowmeter, 332, 330(F)
Materials, effects of air contaminants on, 236, 237(T)
Maximum acceptable concentrations (MAC), of air contaminants, 286, 287, 288, 290, 291
Maximum permissible concentrations (MPC), for radionuclides, 283
Melanism, 266
Membranes, biological, 191–192
Mercury, 31, 67, 114–15, 149, 214, 227, 262, 386
Metal fume fever, 214
Metalimnion, 43(F), 44
Metals, 165, 170, 176, 176(T), 182, 182(T), 185(T), 186. See also specific metals
Metaplasia, 209
Meteorology
    role in removal of air pollutants, 133(T)
Methane, 63, 162
Methemoglobinemia, 204, 227, 261
Microsome, 202
Microstrainers, 417
Minimata Bay, 227
Mirex, 386
Mist, 19
Mixture metering, 329
Morbidity, 52, 54, 273
Mortality, 52, 53–54, 273
Mucociliary clearance, 209
Mucus, 196
Mutagenicity, 211
Mutagens, 211(T)
Mycotoxin, 26(T)

National Committee on Radiation Protection (NCRP), 283
National Environmental Policy Act, 306
National Institute for Occupational Safety and Health (NIOSH), 274, 290, 291, 293
National Pollutant Discharge Elimination System (NPDES), 306
Natural gas, 66, 68(T)
Natural selection, 265
Neutralization reactions, 418
Neutron activation, 96
New York Bight, 440, 440(F)
Nitrate, 92, 95, 143, 156, 157, 172, 173, 204, 227, 255, 261
Nitric acid, 156–57, 173, 246
Nitric oxide. See Nitrogen oxides
Nitrification, 173, 427

Nitrite, 173, 227
Nitrogen, 33, 63, 106(T), 172–73, 173(F), 261–62. See also specific nitrogen compounds
Nitrogen dioxide. See Nitrogen oxides
Nitrogen dioxide sampler, 336(F), 337
Nitrogen fixation, 172, 261
Nitrogen oxides, 60, 66, 69, 70, 74, 109–12, 156–57, 161, 162, 163, 173, 180, 180(T), 197, 223, 226, 245, 246
Nitrogen pentoxide, 157
Nitrogen tetroxide, 156
Nitrosamine, 204, 227, 261
Nitrous acid, 156–57
Nitrous oxide, 156, 173, 241, 245, 246
NTA, 166
Nuclear energy, 68
Nuclear fuel cycle, 98–103, 100–103(F), 102(T)
Nuclear reactors, 99–101
Nuclear waste disposal, 103
Nuclear weapons tests, 96–98, 134–35, 140, 141(F)

Occupational environment, 51–52
Occupational exposure limits, 275
Occupational health standards,
    federal, 290–93
    in Soviet Union, 293–94
Occupational Safety and Health Act (1970), 11, 274, 291
Occupational Safety and Health Administration (U.S.), 290, 291
Oceans, 38, 47, 48(F), 142, 169–70, 259
    contaminant dispersion in, 138–42, 140(F), 141(F), 142(F)
    dumping in, 95, 107(T), 437–41
Odors, 249–50, 249(T)
Oil. See Petroleum
Oil/water separator, 421
Orifice meter, 321, 323, 324(F)
Osmosis, 191
Otto cycle engine, 70–71, 73(F), 74(T)
Oxidation, 165, 429
Oxidation pond, 428
Oxygen, 33
Oxygen demand, 22(T)
Oxygen sag curve, 167, 169(F)
Ozone, 58, 74, 107, 108–12, 156, 157, 160, 161, 162, 163, 181(F), 234, 244, 245, 246, 360, 373, 429

Paracelsus, 6–7
Particle analyzers, 373

Particles
  in atmosphere, 60(T), 61–62, 61(T),
    153–55
  role in climate, 35, 237, 238–40, 243
Passive transport, 191–92, 198, 200
PBB, 30
PCB, 26, 27(T), 143, 149, 175, 309, 386
Performance standards, 297
Peroxyacetylnitrate (PAN), 110
Personal protective devices, 388–89, 393.
    See also specific types, 396–97
Pesticides, 29(T), 92–93, 134, 149, 170,
    171, 175(T), 177(T), 178, 184(T),
    200, 361
Petroleum, 23, 66, 68(T), 74, 108(T), 169,
    185(T), 259
Petroleum refining, 76–82, 80(F), 81(F),
    82(T), 83(T), 84(T)
Phagocytosis, 192
Phenol, 200
Phosphate, 92, 95, 143, 255
Phosphorus, 106(T)
Photochemical reactions
  in atmosphere, 109–12, 111(F), 112(F),
    156, 159, 161, 162
  in water, 166
Pinocytosis, 192, 198
Pitot tube, 329, 330(F)
Pliny the Elder, 5–6
Plutonium, 96, 103
Pneumoconiosis, 209, 210(T), 347
Pneumonitis, 211
Polarography, 378
Pollution (definition), 3
Polycyclic aromatic hydrocarbons, 67,
    162, 203
Positive-displacement flowmeter, 321
Potassium-40, 64
Potentiometry, 378
Pott, Sir Percival, 9
Precipitant scavenging, 155. *See also* Wet
    deposition
Precipitation (meteorology), 131–33,
    132(F), 137
Precipitation reactions, 418–19
Preconcentration techniques, 364
Pretreatment standards, 306, 308
Procarcinogen, 212
Public Health Service, 4–5, 11, 294
Pulp and paper manufacture, 87–91,
    88(F), 89(T), 90(F)
Putrefaction, 63
PVC, 175

Quality assurance, of chemical analyses,
    378–79
Quality factor, 282

Rad, 281, 282
Radiation
  natural sources, 64–66
  population dose, 96
Radionuclides
  standards for, 281–83
  in transport of food chains, 146–49
Radium, 65, 141, 281
Radon, 65, 98
Radon daughters, 65–66
Ragweed pollen, 61
Rainout, 154
Ramazzini, Bernardino, 9
Recharge basins (wells), 46, 46(F)
Refuse, 93, 94(T)
Refuse Act (1899), 306
Rem, 282
Respirator, 396–97
Respiratory tract, 190, 194–97, 203, 205,
    206, 208, 209, 210–11, 215, 222–23
Reverse osmosis, 436
Ringelmann chart, 294–95, 296(F)
Rivers, 42(T), 137, 253, 254–55, 257
Rotameter, 321, 322(F), 323, 324, 327(F)
Rotating disk process, 427–28
Rotating vane flowmeter, 329

Saccharin, 309
Safe Drinking Water Act (1974), 300
Salinity, 257
Salt, road, 95–96
Samplers
  for aerosols, 337–45
  for gases, 335–36
  size selective, 347–49, 348(F)
Samples
  ambient, 314–15, 346–47
  biota, 358–60
  biological, 361–63, 363(T)
    continuous, 352, 353(F)
    effluent, 356, 357(F), 358
  extractive, 313, 364, 315–16, 318, 319,
    355–56
  food, 361
  grab, 316, 351, 362
  handling and preservation of, 314, 358,
    359(F)
  integrated, 316
  intermittent, 351–52, 352(F)
  Market-basket, 361
  for occupational hygiene, 346–47
  personal, 314, 315, 346–47, 347(T)
  probe, 332
  process, 314, 315
  proportional, 334(F), 352–55
  sediment, 359
  source, 314, 349–50

stream, 356, 357(F)
types of, 313–14
Sampling sites, 313
Sand, 49, 50(F)
Screening, 416–18
Scrubbers, 313, 318, 397, 400, 402–3, 403(T)
Sea salt spray, 64
Sedimentation, 131–32, 137, 153–54, 163, 164, 194, 419–21
Sediments, 24, 145, 146, 164–65, 169–70, 314
Semicontinuous analyzers, 318–19
Separation techniques, 364, 365(T)
Settling tanks, 419–21
Sewage, 93, 94(T), 95, 166–69, 412
Sewage fungus, 254
Sewage sludge. *See* Sludge
Short-term exposure limits (STEL), 287
Silicosis, 209
Silt, 49, 50(F)
Sink, 32, 152
Skin, 190, 200–201, 201(F), 203, 205, 209
Sludge, 95, 419, 425, 426, 435, 437–38, 440
Slurry, 435
Smog, 20, 108–12, 226. *See also* Photochemical reactions
Smoke, 19
Snow, John, 12
Soil, 47, 49–50, 50(F), 142–43, 169, 170–71, 175(T)
Soil tillage, 91–92
Solar flux, 37
Source inventory, 285
Sources
    area, 120, 128
    line, 120, 128
    point, 120, 128, 307
Spectrophotometry, 377–78
Stabilization basins, 428
Standard methods of analysis, 379(T)
Storage, of chemicals in body, 204–5, 205(F)
Stratified charge engine, 73, 73(F), 407
Stratosphere, 33, 37, 15, 244–46
Strontium, 143
Sulfate, 113, 134, 160, 174
Sulfite, 160
Sulfur 63, 174, 174(F). *See also specific sulfur compounds*
Sulfur dioxide, 60, 69, 70, 113, 157, 159, 160, 161, 162, 178(F), 180, 197, 206, 220, 234, 360, 373, 407
Sulfur trioxide, 113, 159, 160, 161
Sulfuric acid, 70, 113, 160, 161, 208, 246
Supersonic transport, 246

Surface runoff, 42, 137
Surface water, 38, 40, 40–44. *See also* Rivers; Lakes; Oceans
    standards, 300, 301(T), 302–3(T), 304–5(T), 306
Sutton, O. G., 128
Synergism, 125–16

Tailings (radioactive), 98–99
Teratogen, 211–212, 212(T)
Terminal settling velocity, 19
Terpenes, 63, 312
Thackrah, Charles Turner, 9
Thermal meter, 329
Thermal pollution
    of atmosphere, 237–38, 243
    of water, 68, 251–53
Thermal precipitation, 313, 337, 343, 344(F)
Thermal stratification, in lakes, 137–38
Thermocline, 44
Thermophoresis, 343
Thompson, W. Gilman, 11
Thorium, 64, 141
Thoron gas. *See* Radon
Threshold limit values (TLV), 286, 287, 288, 289, 290, 291, 293
Tolerance, 216
Tolerance limits, 27, 270, 271, 309
Topography, effects on plume dispersion, 129–31, 130(T), 130(F)
Total suspended particulates, 177(F), 178
Toxicity, 24
Toxicologic tests, 272, 276–80, 278(T), 279(T), 280(T)
Transport of air contaminants, long range, 134
Traverse, 349, 350(F), 356
Trickling filters, 427
Trihalomethanes, 114
Tritium, 64
Trophic structure, 144, 145(F), 171, 264
Tropopause, 33, 37, 39(F)
Troposphere, 33, 36, 37, 39(F)
Turbidity, 253
Turbine flowmeter, 329
Turbulence
    atmospheric, 118–20, 119(F)
    in water, 163–64
Two-mass sampler, 376(F), 377

Ultraviolet analyzer, 370
Ultraviolet radiation, 244, 246
Uranium, 64, 96, 98–103, 141
Urban heat island, 238
Urticaria, 211

Variable-area meter, 323
Variable-head meter, 321
Vegetation, effects of air contaminants, 233–35, 235(T)
Venturi meter, 321, 323, 325(F)
Venturi scrubbers, 402, 406
Villi, 197–98, 199(F)
Vinyl chloride, 53
Visibility, 248–49
Volcanoes, 66
Volumetric flowrate meters, 321–37

Walsh-Healey Act, 291
Washout, 154
Washout factor, 133
Waste heat. See Thermal pollution
Wastewater treatment, 13
   advanced. See Wastewater treatment, tertiary
   biological. See Wastewater treatment, secondary

Wastewater treatment, primary, 412–14, 430, 431(T)
Wastewater treatment, secondary, 412–14, 425–28, 430, 431(T)
Wastewater treatment, tertiary, 413(T), 414–15(T), 415, 430, 431(T)
Water Pollution Control Act (1972), 306–9
Water quality criteria, 300
Water quality parameters, measurement instruments, 378(T)
Water table, 45–46, 45(F)
Water vapor, 33, 35, 37
Weirs, 321, 323, 326(F)
Wet-air oxidation, wastewater treatment, 428, 429(F)
Wet deposition, 154
Wet-test meter, 319, 321, 320(F)
Wiley, Harvey W., 14
Workman's Compensation, 11

Zeolite, 356